AF786098

Current Topics in Microbiology and Immunology

Volume 367

Series Editors

Klaus Aktories
Medizinische Fakultät, Institut für Experimentelle und Klinische Pharmakologie und Toxikologie, Abt. I Albert-Ludwigs-Universität Freiburg, Albertstr. 25, 79104 Freiburg, Germany

Richard W. Compans
Department of Microbiology and Immunology, Emory University, 1518 Clifton Road, CNR 5005, Atlanta, GA 30322, USA

Max D. Cooper
Department of Pathology and Laboratory Medicine, Georgia Research Alliance, Emory University, 1462 Clifton Road, Atlanta, GA 30322, USA

Jorge E. Galan
Boyer Ctr. for Molecular Medicine, School of Medicine, Yale University, 295 Congress Avenue, room 343, New Haven, CT 06536-0812, USA

Yuri Y. Gleba
ICON Genetics AG, Biozentrum Halle, Weinbergweg 22, 06120 Halle, Germany

Tasuku Honjo
Department of Medical Chemistry, Faculty of Medicine, Kyoto University, Sakyo-ku, Yoshida, Kyoto 606-8501, Japan

Yoshihiro Kawaoka
School of Veterinary Medicine, University of Wisconsin-Madison, 2015 Linden Drive, Madison, WI 53706, USA

Bernard Malissen
Centre d'Immunologie de Marseille-Luminy, Parc Scientifique de Luminy, Case 906, 13288 Marseille Cedex 9, France

Fritz Melchers
Max Planck Institute for Infection Biology, Charitéplatz 1, 10117 Berlin, Germany

Michael B. A. Oldstone
Department of Immunology and Microbial Science, The Scripps Research Institute, 10550 North Torrey Pines Road, La Jolla, CA 92037, USA

Rino Rappuoli
Novartis Vaccines, Via Fiorentina 1, Siena 53100, Italy

Peter K. Vogt
Department of Molecular and Experimental Medicine, The Scripps Research Institute, 10550 North Torrey Pines Road, BCC-239, La Jolla, CA 92037, USA

Honorary Editor: Hilary Koprowski
Biotechnology Foundation, Inc., 119 Sibley Avenue, Ardmore, PA 19003, USA

For further volumes:
http://www.springer.com/series/82

Ellen Heber-Katz · David L. Stocum
Editors

New Perspectives in Regeneration

Responsible series editor: Hilary Kropowski

Springer

Editors
Prof. Dr. Ellen Heber-Katz
Wistar Institute
Philadelphia, PA
USA

Prof. Dr. David L. Stocum
Indiana University, Purdue University
Indianapolis, IN
USA

ISSN 0070-217X
ISBN 978-3-642-35809-8 ISBN 978-3-642-35810-4 (eBook)
DOI 10.1007/978-3-642-35810-4
Springer Heidelberg New York Dordrecht London

Library of Congress Control Number: 2013930355

© Springer-Verlag Berlin Heidelberg 2013
This work is subject to copyright. All rights are reserved by the Publisher, whether the whole or part of the material is concerned, specifically the rights of translation, reprinting, reuse of illustrations, recitation, broadcasting, reproduction on microfilms or in any other physical way, and transmission or information storage and retrieval, electronic adaptation, computer software, or by similar or dissimilar methodology now known or hereafter developed. Exempted from this legal reservation are brief excerpts in connection with reviews or scholarly analysis or material supplied specifically for the purpose of being entered and executed on a computer system, for exclusive use by the purchaser of the work. Duplication of this publication or parts thereof is permitted only under the provisions of the Copyright Law of the Publisher's location, in its current version, and permission for use must always be obtained from Springer. Permissions for use may be obtained through RightsLink at the Copyright Clearance Center. Violations are liable to prosecution under the respective Copyright Law.
The use of general descriptive names, registered names, trademarks, service marks, etc. in this publication does not imply, even in the absence of a specific statement, that such names are exempt from the relevant protective laws and regulations and therefore free for general use.
While the advice and information in this book are believed to be true and accurate at the date of publication, neither the authors nor the editors nor the publisher can accept any legal responsibility for any errors or omissions that may be made. The publisher makes no warranty, express or implied, with respect to the material contained herein.

Printed on acid-free paper

Springer is part of Springer Science+Business Media (www.springer.com)

Preface

Regeneration, the homeostatic ability to maintain tissue structure in the face of normal cell turnover or loss of tissue damaged by trauma or disease, is a developmental process that continues throughout life and is essential for life. Investigation of the cellular mechanisms of natural regeneration, including the molecular properties of regeneration-competent cells permissive for regeneration, and the interactions of these cells with surrounding instructive or support cells has become one of the hottest topics in the field of regeneration and developmental biology. It is now understood that widely different regenerative phenomena, from epithelial cell turnover to limb regeneration in amphibians share many common features. The value of comparative molecular analysis of regeneration-competent versus regeneration-deficient tissues, via bioinformatics and systems biology approaches is being recognized as a way to provide insights into why a tissue or appendage regenerates in some species or mutants, but not in others.

Fifty years ago, John Gurdon showed that the nuclei of frog tadpole intestinal cells could be reprogrammed by egg cytoplasm to a zygotic state that would support the development to adulthood of a clone of the donor. Sixteen years ago, Ian Wilmut and Keith Campbell showed the same thing for mammalian nuclei in cloning Dolly, the sheep. And just 6 years ago, Shinya Yamanaka's laboratory identified four transcription factors involved in the pluripotency circuits of embryonic stem cells (ESCs) that, when transfected into skin fibroblasts, reprogrammed them to ESC-like cells called induced pluripotent stem cells (iPSCs) that could be directed to differentiate into a wide variety of adult cell types. Gurdon and Yamanaka were awarded the 2012 Nobel Prize in Medicine for their work, which has opened the door to potential regenerative medical therapies for many types of injuries and degenerative diseases, as well insights into disease etiology and the ability as to screen for therapeutic molecules and compounds. Insights into both the mechanisms of nuclear reprogramming and how to use dedifferentiated cells are growing rapidly.

This book entitled "New Perspectives in Regeneration" is a compendium of current findings in vertebrate and mammalian wound healing and regeneration. In this volume we present 11 reviews that cover a wide range of regenerative topics, from wound repair and its relation to regeneration, through the regeneration of lenticular, neural, and musculoskeletal tissues and limb regeneration, to the epigenetics of regeneration and the role of the cell cycle. Nuclear reprogramming and cellular plasticity are recurring themes throughout the volume.

We begin in the first part with two papers on wound repair. The first by Wietecha, Cerny, and DiPietro describes in-depth the important process of angiogenesis. One of the most novel parts of this chapter is the focus on how vessels are degraded and the molecules involved. This is key during the remodeling phase of the healing response and of course is what is important to block tumor formation. The second is a general and thorough review by Kawasumi, Sagawa, Hayashi, Yokoyama, and Tamura comparing wound repair and regeneration in amphibians and mammals, with a focus on limb regeneration.

In the first paper of the second part, Monaghan and Maden examine the role and importance of biological plasticity in vertebrate appendage regeneration to provide a framework for asking questions about how to look at models of regeneration. This is especially important in the lens regeneration model presented in the next chapter by Henry, Thomas, Hamilton, Moore, and Perry. Only certain larval and adult fish, salamanders, and frogs are known to regenerate their lens and these include newts, salamanders, fish, and Xenopus. A thorough review of the molecular pathways involved in this process in Xenopus is presented.

Part III begins with a paper by Viswanathan and Joshi that explores the role of stem cells in all aspects of regenerative medicine, including cell-based therapies, tissue engineering, and the activation or recruitment of stem cells to organs of interest. Next, reviews by Cameron, Milner, Lee, Cheng, Fang, Jasiuk, and by Milner and Cameron, address musculoskeletal regeneration and compare and contrast amphibian and mammalian systems. Both these papers review the use of stem cells and dedifferentiation and address current important issues of tissue engineering.

In the fourth part are two reviews of CNS regeneration. The first by Steward, Sridhar, and Meyer, compares the successful regeneration of mammalian PNS to the much poorer regeneration of the CNS, and reviews the reprogramming strategies used to convert non-neural cells to neural cells. The second, by Zupanc and Sirbulescu, explores the teleost fish as a model to study the regeneration of CNS tissue. Teleost fish are particularly interesting because their tails are often lost to predator fish and they regenerate not only their tails but also their spinal cord. Thus, learning what they use to regenerate is intriguing and revealing.

Finally, in the last part, we present two papers exploring subcellular events that support a regenerative response. The first paper by Maki and Kimura describes a study exploring the epigenetic changes occurring during regeneration. Here, besides a review of general epigenetic changes previously reported is a discussion of a highly novel finding, the use of an oocyte-type linker histone protein, B4, that is specifically used in the embryo and then reappears in regenerating newt lens tissue during trans-differentiation. The second paper by Heber-Katz, Zhang, Bedelbaeva, Song, Chen, and Stocum is a comparison between amphibian and mouse regenerative tissue and examines cell-cycle regulation in the axolotl limb and MRL mouse ear tissue and finds striking similarities in terms of G2 arrest and the expression of proteins specific to this process, including Evi-5.

We hope you enjoy it!

Ellen Heber-Katz
David L. Stocum

Contents

Part I
Wound Repair and Regeneration

Mechanisms of Vessel Regression: Toward an Understanding of the Resolution of Angiogenesis

Mateusz S. Wietecha, Wendy L. Cerny and Luisa A. DiPietro

Abstract Physiological angiogenesis refers to a naturally occurring process of blood vessel growth and regression, and it occurs as an integral component of tissue repair and regeneration. During wound healing, sprouting and branching results in an extensive yet immature and leaky neovascular network that ultimately resolves by systematic pruning of extraneous vessels to yield a stable, well-perfused vascular network ideally suited to maintain tissue homeostasis. While the molecular mechanisms of blood vessel growth have been explored in numerous cell and animal models in remarkable detail, the endogenous factors that prevent further angiogenesis and control vessel regression have not received much attention and are largely unknown. In this review, we introduce the relevant literature from various disciplines to fill the gaps in the current limited understanding of the major molecular and biomechanical inducers of vascular regression. The processes are described in the context of endothelial cell biology during wound healing: hypoxia-driven activation and sprouting followed by apoptosis or maturation of cells comprising the vasculature. We discuss and integrate the likely roles of a variety of endogenous factors, including oxygen availability, vessel perfusion and shear stress, intracellular negative feedback mechanisms (Spry2, vasohibin), soluble cytokines (CXCL10), matrix-binding proteins (TSP, PEDF), protein cleavage products (angiostatin, vasostatin), matrix-derived anti-angiogenic peptides (endostatin, arresten, canstatin, tumstatin), and the biomechanical properties of remodeling the extra-cellular matrix itself. These factors aid in the spatio-temporal control of blood

M. S. Wietecha · W. L. Cerny
Center for Wound Healing and Tissue Regeneration, College of Dentistry,
University of Illinois at Chicago, 60612 Chicago, IL, USA

L. A. DiPietro (✉)
College of Dentistry, Center for Wound Healing and Tissue Regeneration (MC 859),
University of Illinois at Chicago, 801 S. Paulina—Room 401B,
Chicago, IL 60612-7211, USA
e-mail: ldipiet@uic.edu

Current Topics in Microbiology and Immunology (2013) 367: 3–32
DOI: 10.1007/82_2012_287

© Springer-Verlag Berlin Heidelberg 2012
Published Online: 7 December 2012

vessel pruning by inducing specific anti-angiogenic signaling pathways in activated endothelial cells, pathways which compete with pro-angiogenic and maturation signals in the resolving wound. Gaining more insight into these mechanisms is bound to shed light on unresolved questions regarding scar formation, tissue regeneration, and increase our understanding of the many diseases with angiogenic phenotypes, especially cancer.

Contents

1 Introduction

1.1 Blood Vessel Biology: A Matter of Life and Death

Blood vessels are vital to the development and survival of complex multi-cellular life. The marvelously efficient method devised by nature to deliver oxygen to cells, tissues, and organs, most of which are located in otherwise anoxic environments, is what makes possible the vertebrates' evolutionarily advanced forms and functions. Indeed, because of the physical constraints of gas diffusion, organisms without a functioning vascular system would not be able to grow beyond a dozen cells in any direction (Phelps and Garcia 2010); in humans, practically no cell is ever more than four cells away from a blood vessel (Fukumura and Jain 2008). The fractal-like branching of the cardiovascular system, from the largest vessel (the aorta) to the smallest (capillary) and back to the largest (the vena cavas), ensures that every one of the roughly 50 trillion cells in the human body is adequately supplied with oxygen and nutrients, and can dispose of its waste products. Given this seminal importance to life, it should not be surprising that mammals have evolved complex mechanisms for the intricate control of blood vessel growth and maintenance. During development of the organism, blood vessels form de novo in a process called vasculogenesis, whereas further sprouting of vessels from pre-existing tubules is called angiogenesis (Ribatti et al. 2009). Besides development, angiogenesis occurs physiologically in adults during the female menstrual cycle in the ovary (Modlich et al. 1996) and the endometrium (Rogers 1996), in skeletal muscle remodeling during exercise (Bloor 2005; Olfert and Birot 2011), and during tissue repair and regeneration (Eming et al. 2007). In all of these cases of naturally occurring, or physiological, angiogenesis, sprouting of tubules is followed by a controlled phase of blood vessel maturation and regression which results in a spatially distributed and completely perfused network that optimally meets a tissue's metabolic demands as required to preserve homeostasis (Egginton and Gaffney 2010; Owen et al. 2009).

Angiogenesis is an integrated function of a variety of cell types and their microenvironment (Schultz et al. 2010) and involves the following stages: activation, sprouting, regression, and maturation (Fig. 1). Endothelial cells (ECs) form the inner wall of every blood vessel and are the primary cell type involved in the angiogenic process. During angiogenesis, ECs are induced to proliferate, migrate out of an existing vessel, differentiate, and assemble to form branches of tubules capable of carrying blood and its constituents (Geudens and Gerhardt 2011; Potente et al. 2011; Wacker and Gerhardt 2011). EC–EC connections are mediated by intercellular junctional protein complexes called vascular-endothelial cadherins (VE-Cadherins) which regulate tubule leakage or vasopermeability (Dejana et al. 2008). Destabilization of VE-Cadherins promotes blood vessel sprouting, whereas strengthening the EC–EC connections is indispensable during vessel maturation and maintenance. While ECs by themselves initially form tubules, these fragile structures need to be stabilized by supporting mural cells. The pericyte is the most intimate partner of the EC and, by binding directly to multiple ECs through a shared basement membrane, is capable of direct cell–cell communication, promoting vessel barrier integrity and vascular network stability (von Tell et al. 2006). Another type of mural cell is the vascular smooth muscle cell, which envelopes the basement membrane of the EC tubule and mediates vessel tone for fine control of blood flow to meet local tissue metabolic demands (Hungerford and Little 1999). The entire blood vessel network is surrounded by the extracellular matrix (ECM), which acts as a bioactive heterogeneous scaffold transmitting molecular signaling and mechanical cues to the blood vessels through cell membrane integrins and other receptors (Bou-Gharios et al. 2004; Califano and Reinhart-King 2010; Davis and Senger 2005; Eble and Niland 2009). The ECM composing the microenvironment of blood vessels is highly dynamic, and it is constantly undergoing remodeling through the actions of fibroblasts, cells located within the ECM which are responsible for its maintenance, regulating its composition and biomechanical properties, thus directly influencing blood vessel biology (Hurley et al. 2010; Kniazeva and Putnam 2009; Pollina et al. 2008). During vessel sprouting, the ECM microenvironment is an important source of pro-angiogenic factors in addition to its role as a permissive scaffold on which the network can form. During vessel maturation and regression, the ECM also undergoes remodeling into a more mature phenotype, providing crucial anti-angiogenic cues, both molecular (Cheresh and Stupack 2008) and biomechanical (Califano and Reinhart-King 2010), which guide the vessels into a spatially distributed and functional network that is optimally suited to maintain tissue homeostasis (Egginton and Gaffney 2010). Thus, angiogenesis refers not only to the growth but also to the maturation and regression of the blood vessel network, a superbly intricate process involving coordination between multiple cell types and their dynamic microenvironment (Fig. 1).

Given the complexity of angiogenesis and its importance to life, it should not be surprising that many prominent human diseases feature angiogenic phenotypes (Bhadada et al. 2011), whose underlying pathology involves a dysfunction of the myriad of mechanisms that comprise the angiogenic process. Excessive blood vessel growth is a hallmark of all malignancies, and it is an important phenotype in arthritis, psoriasis, and macular degeneration effecting the onset of blindness. Insufficient angiogenesis occurs in chronic wounds of diabetic patients, and non-healing often leads to amputations of affected limbs (Schramm et al. 2006). The increasing public health burdens of obesity and diabetes mellitus go hand-in-hand with multiple cardiovascular diseases, including myocardial and peripheral ischemia, which present with deficient angiogenesis leading to infarcts and necrosis of affected tissues (Bhadada et al. 2011). In many explored cases of pathology, there is an imbalance between vessel growth and regression, which perturbs tissue homeostasis. The paramount importance of angiogenesis in health and disease accentuates the necessity for rigorous research in this area—it is quite literally a matter of life and death. Scientific progress here, with a goal to fully understand the underlying physiological and pathological processes, will make it possible to successfully intervene therapeutically and to promote systemic health in millions of current and future patients.

1.2 Angiogenesis: A One-Sided Understanding

It is widely accepted that angiogenesis is regulated in a spatial and temporal manner via a dynamic balance between pro- and anti-angiogenic factors (Chen et al. 2007; Distler et al. 2003; Yuen et al. 2010). The particular inducers and inhibitors of angiogenesis belong to many classes of molecules, depending on the specific cellular function which they regulate, and range from large soluble proteins and cleaved ECM-derived peptides to intracellular molecules that modulate signaling pathways. Because angiogenesis is an integrated function of multiple cell types as well as their microenvironment (Schultz et al. 2010), the overall process

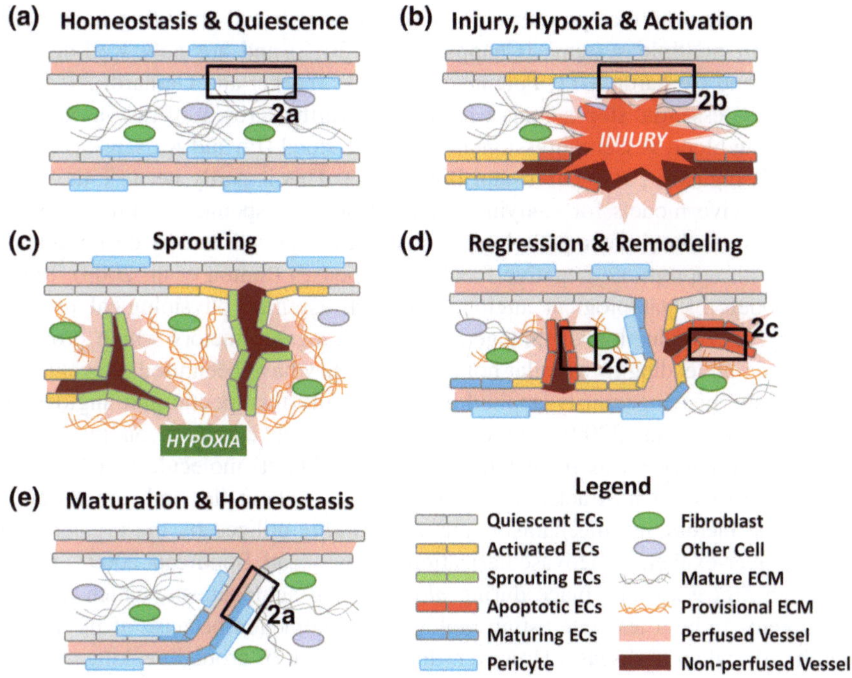

Fig. 1 Overview of the angiogenic process during wound repair, showing relationships between endothelial cells (EC), mural cells (pericytes and fibroblasts), vessel perfusion, and extracellular matrix (ECM) remodeling. *Black boxes* correspond to the detailed mechanistic images in Fig. 2. **a** The default state of the vasculature when tissue is at homeostasis is that of quiescence, maintained in part by adequate perfusion and ample pericyte coverage. See Fig. 2a for detailed representation of the mechanisms of EC survival and quiescence. **b** Injury perturbs the homeostasis of affected tissue, leading to inadequate oxygen supply and hypoxia-driven EC activation in nearby vessels. See Fig. 2b for detailed representation of the mechanisms of EC priming and activation. **c** Pro-angiogenic factors and an angio-permissive provisional ECM promote robust sprouting of neovessels toward the hypoxic gradient of the wound, resulting in a chaotic network of immature and leaky tubules characteristic of granulation tissue formation. **d** As the wound enters the remodeling phase, vessels that have reconnected to a viable network and are perfused are selected for maturation via pericyte coverage, while those inessential tubules that are not perfused are eliminated via systematic pruning by various anti-angiogenic factors. See Fig. 2c for detailed representation of the mechanisms leading up to EC apoptosis during vessel regression. **e** Over time, all extraneous tubules are eliminated, and the vessel network and the ECM are remodeled and return to an architecture and functionality resembling the original, pre-wound homeostatic state

as it occurs in vivo is extremely complex. Nonetheless, over the last couple of decades, using a multitude of clever in vitro and animal models (Staton et al. 2009), investigators have made tremendous progress in unraveling the molecular mechanisms behind the many aspects of angiogenesis (Ribatti et al. 2009).

The mechanisms that promote blood vessel growth and sprouting have been explored in multiple systems (Chappell et al. 2012) and are now well-defined (Geudens and Gerhardt 2011; Potente et al. 2011; Wacker and Gerhardt 2011). In contrast, the essential stop-signals which inhibit further growth of the vasculature and mediate blood vessel regression are not understood (Im and Kazlauskas 2006; Korn and Augustin 2012; Olfert and Birot 2011). This is partly because the most popular in vivo models for studying angiogenesis, like sponge or tumor implantation, are ones that induce pathological blood vessel growth and by design do not present with appropriate negative feedback responses (Staton et al. 2009). Since these models of pathological angiogenesis lack essential physiological responses—e.g., the production of endogenous anti-angiogenic factors which promote vessel pruning leading to vascular homeostasis—they have been used to evaluate a whole range of exogenously applied and potentially therapeutic anti-angiogenic agents (Staton et al. 2009). Ironically, with so much emphasis on models of pathological angiogenesis in search of the 'magic bullet' molecule which would stop the process in its tracks, very little is known about the endogenous anti-angiogenic factors and mechanisms that are *actually* utilized by the organism to counter over-exuberant neovascularization (Cuevas and Boudreau 2009). These resolving processes take place during *physiological* angiogenesis, which historically seems to be a less interesting biological phenomenon because of its link to health rather than to disease. However, due to the lack of groundbreaking clinical successes in combating vascular pathologies thus far, it is becoming apparent that we are missing an important piece of the angiogenesis puzzle. A question could be raised: how are we to therapeutically direct such a complex biological process as angiogenesis without truly understanding how it naturally resolves? We argue here that knowing only one side of the equation is not enough.

The purpose of this review is to shed some light on the relatively unknown but crucial aspect of the angiogenic process: blood vessel regression following robust vessel sprouting. The question we wish to answer is: how does the repaired blood vessel network return to vascular homeostasis? To get at a more global view, we will explore vessel regression from multiple perspectives and attempt to fit the pieces back into a framework of basic understanding. This framework will be modeled after a biological process which encompasses the entire angiogenic response, from induction to resolution: wound healing in the skin (Eming et al. 2007; Wong et al. 2011). The relevant literature from assorted disciplines, featuring studies utilizing various in vitro and animal models, will be reviewed and synthesized in the context of repair and regeneration, assuming for the purpose of this review that physiological anti-angiogenic processes share common mechanisms in different tissues. With the global framework in mind, the authors will identify gaps in our current understanding and suggest directions for further research in this area, particularly for the use of cutaneous wound repair as a robust and versatile model for investigating endogenous mechanisms of blood vessel regression (Wong et al. 2011).

2 A Distinct Pattern of Angiogenesis and Regression During Wound Healing

Wound repair follows a well-characterized sequence of events or phases: hemostasis, inflammation, proliferation, remodeling (Gurtner et al. 2008; Shaw and Martin 2009). In the many areas and microenvironments of the wound, each phase—through complex coordination between the cells and ECM of the healing wound—mechanistically brings about the next phase in the sequence (Barrientos et al. 2008; Schultz et al. 2010). The phases of wound repair are best described as distinct yet overlapping, and they may be illustrated as gradients, both spatially and temporally (Gurtner et al. 2008; Schultz et al. 2010).

The proliferative phase of healing is characterized by the formation of granulation tissue, a rather chaotic collection of cells (macrophages, fibroblasts, ECs) and loose connective tissue that quickly replaces the fibrin clot in the wound bed. One of the primary components of granulation tissue is a dense network of capillary loops which comes about as a result of a vigorous angiogenic response starting at the wound margins and spreading into the wound bed (Eming et al. 2007) (Fig. 1a–c). In a little over a week after injury, the density of blood vessels in the wound bed is over three times higher than that of the original, uninjured tissue (Swift et al. 1999; Szpaderska et al. 2005). The vast, rapidly forming neovascular network is supported by a fibroblast-secreted temporary fibrous scaffold that is rich in Collagen Type III and other ECM components that promote EC invasion through the matrix during sprouting (Schultz et al. 2010).

What stimulates and directs blood vessel growth in wounds has been explored in much detail (Eming et al. 2007). The critical initial stimulus is low oxygen tension or hypoxia, experienced by the tissue as a result of damage to the existing blood vessel network (Rodriguez et al. 2008). The stress of hypoxia stimulates substantial transcriptional changes in the affected cells of the wound (Semenza 2010). Nearly all types of cells in and around the wound are induced to produce massive amounts of various pro-angiogenic factors, which act to promote EC proliferation, migration, and differentiation into a branched tubular network that sprouts toward the hypoxic gradient (Bao et al. 2009) (Fig. 1b, c).

The majority of the newly sprouted vessels are mere EC tubules whose network architecture is tortuous and often blind-ended (Fukumura and Jain 2008). Functionally, these neovessels are immature, lacking tight EC–EC contacts and possessing scant coverage by pericytes, and have been found to be barely perfused and leaky, contributing to granulation tissue's moist appearance (Bluff et al. 2006; Shaterian et al. 2009) (Fig. 1c). Importantly, the disorganized and non-functional nature of the neovessel network during this phase of wound healing has been compared to the vasculature of a solid tumor (Cuevas and Boudreau 2009; Schafer and Werner 2008).

Once wound closure and maximum granulation tissue formation are achieved, the wound enters the final phase of healing: remodeling (Fig. 1d). An important component of the remodeling phase is the systematic breakdown of the provisional, angio-permissive Collagen Type III-rich ECM and its gradual replacement with one that is composed primarily of Collagen Type I, initiating a lengthy process of modification of the wound ECM to an architecture whose composition and bio-mechanics resemble the pre-wounded, angio-restrictive state (Gurtner et al. 2008). With the completion of wound closure, the barrier function of the epithelium is achieved, leading to decreased microbial invasion and thus a much reduced oxygen demand by the inflammatory cells (Rodriguez et al. 2008). As granulation tissue fills up the space of the wound bed, the cells involved experience contact inhibition and begin to wind down their metabolically demanding proliferative and migratory behaviors. With decreased oxygen demand in the wound, there is a decline in tissue hypoxia, which results in a diminished production and activity of soluble pro-angiogenic mediators. Thus, during the remodeling phase, the wound switches to an anti-angiogenic state, and this switch is characterized by a number of critical changes to the vessel microenvironment.

After a peak in vessel density is reached in the wound bed, the anti-angiogenic phenotype prevents further vessel sprouting and mediates the regression of blood vessels back to baseline levels (Gurtner et al. 2008; Swift et al. 1999). While a minority of newly sprouted vessels that have successfully integrated into the existing perfused network undergo maturation (Chen et al. 2007), the majority that are not perfused and functional are predisposed to pruning (Ando and Yamamoto 2009) (Fig. 1d). The most accepted theory for how pruning of extraneous vessels occurs is that the ECs comprising these tubules are induced to undergo programmed cell death, or apoptosis (Dimmeler and Zeiher 2000; Sakamaki 2004). Though the exact mechanisms leading to and mediating EC apoptosis during wound healing are largely unknown, they likely come about as a result of a convergence of a number of biological processes. Besides the oxygen-associated changes to the vessel microenvironment, certain active anti-angiogenic signals are thought to fine-tune the proper pruning of the blood vessel network during the remodeling phase. The known regression signals include soluble and ECM-derived anti-angiogenic mediators which lead to specific intracellular signaling pathways that result in the cellular and microenvironmental changes associated with vessel regression.

3 To Be or Not to Be: The Endothelial Cell as Battleground for Competing Signals

3.1 *Vessel Maintenance: Mechanisms of Homeostasis*

In the adult, the default state for blood vessels is that of quiescence (Fig. 1a). Quiescent ECs are characterized by a remarkably low proliferative potential, stable VE-Cadherin complexes, and ample coverage by mural cells, especially by pericytes (Dejana et al. 2008; von Tell et al. 2006). Importantly, stabilized vessel networks have been shown to be resistant to both pro- and anti-angiogenic stimuli (Benjamin et al. 1998). This protects tissues from random sprouting of superfluous vasculature, while at the same time protecting the existing blood vessel network from inadvertent pruning. This state of quiescence, critical for homeostasis, is maintained by multiple mechanisms (reviewed in (Murakami 2012)) (Fig. 2a). Major disturbances to the homeostatic state, such as injury and/or hypoxia, can prime ECs comprising the affected tubules for either angiogenesis or pruning. Thus, the mechanisms that have been identified for blood vessel network stability are likely lost or lacking in areas of active angiogenesis and regression (Murakami 2012). During wound healing, one of the central endpoints of physiological vessel growth and regression is a return to a vascular state of quiescence, as this state contributes largely to tissue homeostasis (Fig. 1e).

Vascular Endothelial Growth Factor (VEGF) is thought to be the most potent pro-angiogenic mediator during wound repair (Eming and Krieg 2006), but it also has significant roles in vessel maintenance (Murakami 2012) (Fig. 2a). When tissue is at homeostasis, basal levels of VEGF support the survival of the stable vascular network in an autocrine fashion by maintaining critical anti-apoptotic Akt signaling in ECs (Potente et al. 2011). Basal levels of the primary pro-angiogenic VEGF receptor (VEGFR2) are also required for survival of quiescent vessels, as the shear stress in these perfused tubules activates VEGF-independent anti-apoptotic signaling (Ando and Yamamoto 2009). Quiescent vasculature is protected from VEGF overstimulation via modification of signaling events by VE-Cadherin, including its deactivation of overexpressed VEGFR2 (Murakami 2012). Simultaneously, a competing VEGF receptor, the soluble VEGFR1, is expressed in quiescence and functions as a VEGF trap, further limiting stimulation of ECs by VEGF (Potente et al. 2011). In addition, vessels surrounded by pericytes are remarkably immune to VEGF overstimulation and EC destabilization (Benjamin et al. 1998). These and other mechanisms assure the long-term survival of a quiescent, properly perfused blood vessel network that optimally supplies the surrounding tissues (Fig. 2a).

3.2 Loss of Quiescence: Endothelial Cell Priming, Activation, and Sprouting

Loss of EC quiescence following injury is largely mediated by hypoxia and the ischemic state of the affected cells (Semenza 2010) (Figs. 1b, 2b). The oxygen-sensing Prolyl Hydroxylase Domain (PHD) family of proteins are important intracellular mediators of vessel quiescence through their suppression of pro-angiogenic master transcriptional regulator Hypoxia Inducible Factor 1 Alpha (HIF-1α) (Hickey and Simon 2006). Whereas under conditions of homeostasis PHDs continually target HIF-1α for proteasomal degradation, in hypoxia PHDs become inactivated, allowing HIF-1α to accumulate and overturn the transcriptional profile of the EC

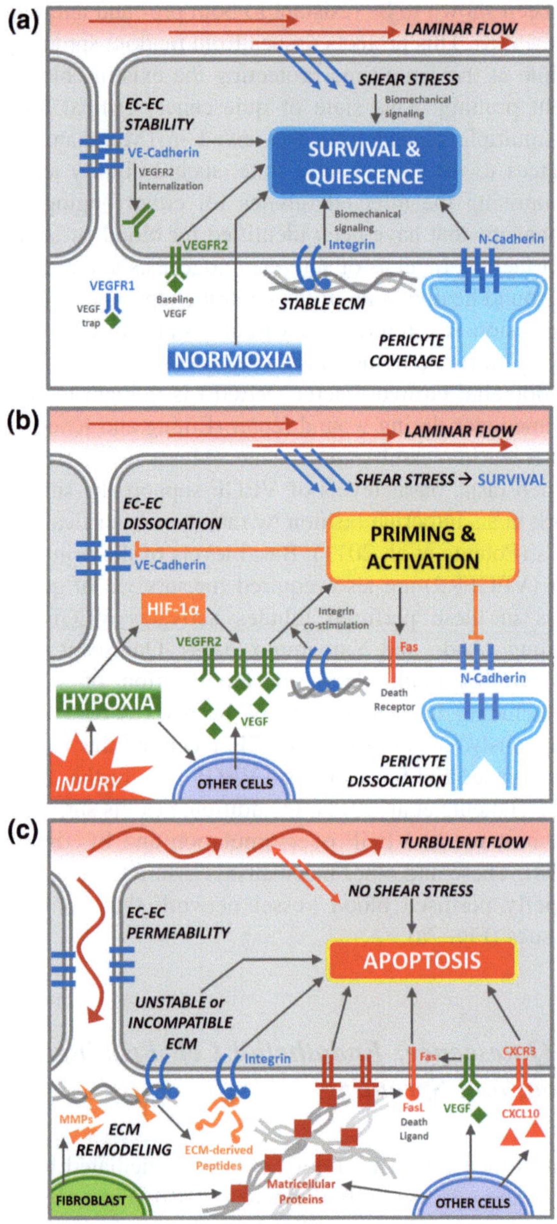

from quiescent to what might be called 'primed'. Priming makes ECs much more sensitive to outside stimuli, especially pro-angiogenic ones.

Multiple cell types including keratinocytes, fibroblasts, and macrophages in the early hypoxic wound produce large amounts of VEGF, while ECs comprising

◄ **Fig. 2** Detailed representations of the essential mechanisms involved in the regulation of endothelial cell (EC) phenotypes. **a** Factors contributing to EC survival and quiescence in homeostasis include normoxia, stable EC integrin contacts with a mature extracellular matrix (ECM), stabilization by adherent pericytes, laminar flow causing shear stress on the luminal surface, tight EC–EC connections via VE-Cadherin junctional complexes, and maintaining baseline autocrine VEGF signaling while preventing pro-angiogenic overstimulation by trapping extracellular VEGF with soluble VEGFR1 and internalizing extraneous membrane VEGFR2. **b** Hypoxia leads to the priming and activation of the EC, resulting in the activation of master transcription factor HIF-1α, ample production of pro-angiogenic VEGFR2 that binds to abundant VEGF secreted by nearby hypoxic cells, co-activation by integrin binding to ECM, and leading to the loss of EC quiescence by dissociation of EC–EC and EC-pericyte junctional complexes. The activated EC is not only more sensitive to pro-angiogenic stimuli but also to apoptotic signals by concurrent production of the death receptor, Fas. **c** Factors contributing to EC apoptosis during vessel regression consist of several anti-angiogenic stimuli in the immediate wound microenvironment, including: soluble and matricellular molecules that can trigger the production of the death ligand, FasL, inducing apoptosis upon binding to Fas receptor; fibroblast-secreted matrix metalloproteinases (MMPs) that cleave ECM molecules as part of matrix remodeling, resulting in unstable ECM contacts with EC integrins, in the collapse of a stable scaffold, as well as in the release of anti-angiogenic ECM-derived peptides; turbulent or stagnant flow on the luminal surface causing loss of survival signaling

vessels in the wound margins react to hypoxia by overexpressing VEGFR2 on their membranes (Hickey and Simon 2006). Binding of VEGF to VEGFR2, a receptor tyrosine kinase, initiates mitogen-activated protein kinase (MAPK) signaling pathways, further overturning the cellular machinery and effectively 'activating' the ECs (Fig. 2b). Activated ECs lose their protective mural cells and loosen up their stabilizing VE-Cadherin connections, contributing to their increased sensitivity to pro-angiogenic factors.

Additional stimulation by VEGF and co-activation by sufficient matrix contacts to EC integrins leads to EC proliferation, migration, and differentiation into a branched tubular network that sprouts toward the VEGF gradient in the wound bed (reviewed in (Bao et al. 2009; Eming and Krieg 2006; Potente et al. 2011; Wacker and Gerhardt 2011)) (Fig. 1c). The leakiness of the neovessels in granulation tissue is mediated in large part by VEGF, which, through nitric oxide synthase and cyclooxygenase signaling (Bao et al. 2009), further dissociates VE-Cadherin junctional complexes (Dejana et al. 2008). The net result of hypoxia-induced pro-angiogenic activity is the remarkably vessel-rich, moist granulation tissue that quickly fills the wound bed.

Since tumors are known to stall at this stage of the angiogenic process (Schafer and Werner 2008), the vast majority of research in the field of angiogenesis also stops at this point (Cuevas and Boudreau 2009). Models of physiological vascular regression, such as healing wounds, provide knowledge about the mechanisms by which the tortuous and leaky neovessel network undergoes pruning back to baseline, leading to vascular and tissue homeostasis (Fig. 1d).

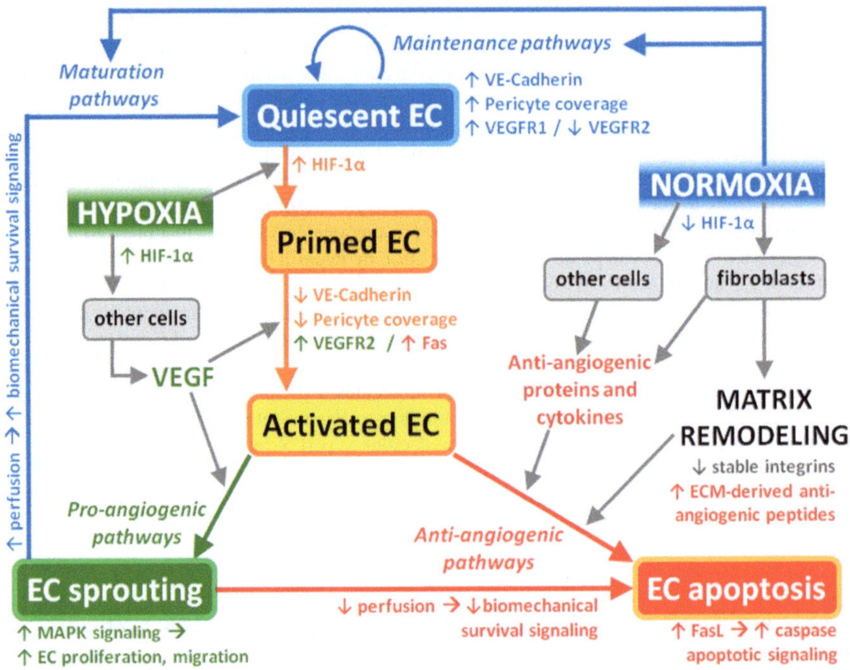

Fig. 3 Diagram summarizing the major converging pathways that direct endothelial cell (EC) phenotypes of quiescence, priming/activation, sprouting, maturation, and apoptosis. The default EC state is that of quiescence, which is maintained by several mechanisms, including pericyte coverage and normoxia. Hypoxia triggers changes in the transcriptional profile of the EC, priming it for stimulation by hypoxia-induced extracellular VEGF. VEGF activates the EC, which enters into a state of plasticity, becoming sensitive to both pro-angiogenic (via VEGFR2) and anti-angiogenic (via Fas receptor) stimuli. The dynamic balance between these two sets of stimuli in the immediate microenvironment of the activated EC determines its fate: sprouting, maturation, or apoptosis. Sprouting results from hypoxia-driven pro-angiogenic signaling pathways like the mitogen-activated protein kinase (MAPK) pathway, which induces EC proliferation and migration. Apoptosis results from anti-angiogenic signaling pathways like those induced by soluble and ECM-derived molecules and by the remodeling of the extracellular matrix (ECM) itself leading to reduced integrin stability. The major determining factor for whether an EC undergoes apoptosis or maturation seems to be vessel perfusion, and maturation pathways aim to return the EC to a quiescent state

3.3 Induced Plasticity and Dynamic Balance of Competing Signals

Paradoxically, the pro-angiogenic mediator VEGF may be one of the factors responsible for the initiation of vessel regression in the post-proliferative phase of healing. Studies have found that activation of ECs by VEGF simultaneously marks these cells for death by inducing the expression of the death receptor Fas, also known as CD95, which, if bound by its ligand, initiates apoptotic signaling pathways (Stoneman and Bennett 2009; Volpert et al. 2002) (Fig. 2c). As a consequence, activated ECs enter a state of plasticity, making them not only more sensitive to pro-angiogenic stimuli but also less resistant to death by apoptosis-promoting signals. Once the ECs are primed by hypoxia and activated by VEGF during the proliferative phase, competing signaling pathways steer them toward multiple potential cellular behaviors: sprouting, maturation, or apoptosis (Fig. 3). The resulting phenotype may be a consequence of the dynamic balance between pro- and anti-angiogenic stimuli acting upon the ECs from their immediate microenvironment.

The microenvironment in the wound is highly dynamic and changes drastically from the proliferative to remodeling phase; whereas the balance clearly favors pro-angiogenesis early on, the scale tips toward anti-angiogenesis in the later wound. What happens at the junction between these two extremes, when pro-angiogenic factors are still present and anti-angiogenic mediators have not yet become dominant, is an area of active investigation. How do the ECs of the wound neovessel network integrate multiple competing stimuli and eventually take the plunge toward regression?

3.4 The Anti-angiogenic Switch: Tipping the Balance toward Regression

A type of mechanism that likely aids during this critical period of the wound angiogenic profile—a period that might be called the 'anti-angiogenic switch'—is a well-conserved one in biology: negative feedback. Indeed, there are probably several redundant intracellular negative feedback mechanisms protecting ECs from VEGF overstimulation during the post-proliferative and remodeling phases of healing which help to guide them into regression. One such mechanism is mediated by a family of intracellular Sprouty proteins, which are known to inhibit pro-angiogenic MAPK signaling pathways like those initiated by VEGF in ECs (Cabrita and Christofori 2008). One Sprouty homolog, Spry2, is produced in dermal wounds during the period of the 'anti-angiogenic switch'. Exogenous application of cell-permeable Spry2 to the healing wound causes a reduction in MAPK signaling and significantly inhibits blood vessel growth during the proliferative phase (Wietecha et al. 2012). In a similar fashion, vasohibin may modulate VEGF signaling in a negative feedback, although this mechanism is yet to be explored in the context of physiological angiogenesis (Sato and Sonoda 2007). These and other negative feedback mechanisms may play an important role as buffers to pro-angiogenic stimuli, making ECs immediately less sensitive to further pro-angiogenic stimulation, thus downregulating the signaling pathways leading to mitogenic cell behavior, while simultaneously giving anti-angiogenic stimuli and their corresponding signaling pathways an advantage during the critical transitional period into the remodeling phase of healing.

3.5 Regression: Streamlining the Blood Vessel Network

The leaky neovessel network of the granulation tissue is largely non-perfused (Bluff et al. 2006) (Fig. 1c). Insufficient perfusion likely leads to reduced shear stress and inadequate biomechanically induced EC survival signaling, and this may be an essential mechanism mediating physiological vessel regression (Ando and Yamamoto 2009; Califano and Reinhart-King 2010) (Fig. 2c). Furthermore, it is well-established that pro-angiogenic factors such as VEGF are reduced during the vessel regression phase of wound repair (Nissen et al. 1998; Swift et al. 1999), although it is unknown whether the autocrine VEGF survival signaling is reduced along with the potent extracellular component of the VEGF stimulus. Several studies have shown that VEGF is required at baseline levels for the continued stability of some established vessel networks and that total VEGF inhibition causes EC apoptosis and regression (Baffert et al. 2006; Murakami 2012). These data certainly point to a potentially pivotal role for the loss of molecular and especially biomechanical survival signals in the process of physiological vessel involution, although these ideas are yet to be investigated in an appropriate in vivo model, such as wound healing.

In contrast, there is good evidence that wound resolution involves the generation of an actively anti-angiogenic environment. Exogenous application of VEGF and other pro-angiogenic factors into a resolving wound does not prevent blood vessel regression from naturally occurring (Gosain et al. 2006). Indeed, while exogenous pro-angiogenic factors cause an even more exuberant angiogenic response during the proliferative phase of healing, the extra blood vessels subsequently undergo remarkably rapid pruning, and vessel density levels of treated wounds match those of the controls within a few days following the peak of granulation tissue formation (Gosain et al. 2006). It appears that the post-proliferative wound not only becomes more immune to pro-angiogenic stimuli by the negative feedback mechanisms just discussed, but also generates active anti-angiogenic signals. During periods of active vessel regression, as occurs during the remodeling phase of wound repair, there exist potent endogenous anti-angiogenic factors that ensure proper pruning of the blood vessel network regardless of the presence of competing pro-angiogenic signals (Fig. 2c).

3.6 Endogenous Mediators of Vessel Regression

Spurred by the promise of anti-angiogenesis in cancer therapy (Folkman 1974), dozens of endogenous inhibitors of angiogenesis have been identified over the last three decades (Nyberg et al. 2005). Even though these molecules are derived from naturally occurring circulating or matrix components, the majority of characterization has been performed either exclusively in vitro or in models of pathological angiogenesis, e.g., in xenografted tumors. Thus, while the list of endogenous anti-angiogenic factors is extensive, very little is known about their actual functions in the control of physiological angiogenesis and vessel regression. How they may fit mechanistically into the larger process of remodeling during wound repair can only be speculated based on their origins (circulating, soluble or matrix-derived) and what limited information can be gleaned from available studies (Nyberg et al. 2005). We will focus our discussion on those anti-angiogenic factors that have been explored in multiple systems and are thus most likely to play significant roles in physiological vessel regression.

Cytokines, or small soluble molecules secreted by various cell types, are probable mediators of vessel regression, as many well-known cytokines that are ubiquitous in the wound environment have been found to have anti-angiogenic properties (Naldini et al. 2003). One of the best studied cytokines in the context of vessel involution is the C-X-C motif chemokine 10 (CXCL10), also known as Interferon Gamma-induced Protein 10 (IP-10) (Fig. 2c). CXCL10 binds CXCR3, a receptor commonly expressed on ECs (Lasagni et al. 2003), and knock-out studies for CXCR3 revealed a significant delay in physiological vessel regression in murine dermal wound models (Bodnar et al. 2009; Yates et al. 2007). Interestingly, CXCR3-null wounds also exhibited a marked delay in ECM remodeling to a mature phenotype, further suggesting a strong association between these two processes. Mechanistic studies in vitro and in the matrigel plug model of pathological angiogenesis revealed that CXCL10 caused a dissociation of newly formed vessels via disruption of critical integrins, secondarily leading to EC apoptosis (Bodnar et al. 2009). The anti-angiogenic effects were found to be strong even in the presence of pro-angiogenic factors, giving yet more credence to the view that a potent anti-angiogenic phenotype dominates in the resolving wound microenvironment. Finally, since CXCL10 is known to be produced in the post-proliferative and beginning of the remodeling phases of healing (Yates et al. 2007), the aggregate data strongly suggest a significant role for CXCL10 in physiological vessel regression.

An important class of proteins, called matricellular proteins (Mosher and Adams 2012), which are capable of binding to the ECM microenvironment and acting upon resident ECs through specific receptors, are very likely to be instrumental in regulating physiological vessel involution (Fig. 2c). Two members of the Thrombospondin family, TSP-1 and TSP-2, are relatively well-studied potent anti-angiogenic factors. As matricellular proteins, they are capable of binding multiple ECM components and thus regulating critical cell–cell and cell–ECM interactions (Bornstein 2009). These molecules have been found to inhibit angiogenesis by downregulating EC proliferation and migration, inhibiting VEGF signaling, and initiating apoptosis (Lawler and Lawler 2012). Whereas TSP-1 is produced during the early phases of healing and likely functions to attenuate VEGF-mediated pro-angiogenic signals, TSP-2 is produced during the remodeling phase and is likely more involved in ECM remodeling-associated vessel regression (Kyriakides and Maclauchlan 2009). Integrating into the ECM microenvironment of the blood vessels ensures prime real estate for TSPs' spatio-temporal down-regulation of wound angiogenesis (Fig. 2c). However, a recent evaluation of wound healing in TSP-2-null mice did not find differences in EC apoptosis as compared to wild-type mice; instead, TSP-2 was implicated in the regulation of ECM remodeling in wounds (Maclauchlan et al. 2009), probably by tipping the matrix protease balance toward the creation of the angio-restrictive ECM scaffold of the resolving wounds.

Pigment Epithelium-Derived Factor (PEDF) is one of the most potent endogenous anti-angiogenic mediators, able to inhibit EC mitogenic behavior as well as induce EC apoptosis (Broadhead et al. 2010; Volpert et al. 2002). Similar to TSPs, PEDF is capable of binding multiple ECM components, especially Collagen Type I, a ubiquitous component of the resolving wound (Hosomichi et al. 2005) (Fig. 2c). Many of PEDF's properties, including its ability to specifically target neovessels without disrupting intact mature vasculature (Volpert et al. 2002), as well as its ability to reduce vasopermeability (Cai et al. 2011), imply an overarching role of promoting vascular homeostasis in many tissues (Broadhead et al. 2010). Though PEDF's in vivo functions have only been explored in models of pathological angiogenesis, such as in tumors and in ocular neovascularization, the aggregate data strongly suggest that PEDF also plays an important role in the process of physiological blood vessel regression during wound healing.

Other molecules likely involved in some way in the regulation of wound angiogenesis during wound repair are angiostatin and vasostatin (Nyberg et al. 2005). Vasostatin is a fragment of calreticulin, a multifunctional molecule in itself, and has been found to be anti-angiogenic when exogenously applied to excisional dermal wounds (Lange-Asschenfeldt et al. 2001), although it is unknown how its activity may integrate into the overall process of vessel regression. Angiostatin is a potent anti-angiogenic fragment of a ubiquitous protein present in the wound microenvironment, plasminogen (which functions to dissolve the fibrin clot) (Cao and Xue 2004), but its function has not yet been investigated in a model of physiological angiogenesis.

The discussion of angiostatin and vasostatin is valuable because it brings to light a potentially important role for the proteolytic processing of parent proteins into functional fragments in the downregulation of angiogenesis (Cao and Xue 2004). Besides the soluble and matricellular factors, an essential class of anti-angiogenic molecules are those derived from ECM components, generated when specific matrix proteases cleave large ECM proteins into bioactive peptides. Indeed, proteolytic processing of structural proteins is a characteristic component of the ECM remodeling that occurs during wound resolution and in coordination with vessel regression.

3.7 *Remodeling of the Extracellular Matrix: Dynamic Reciprocity in Vessel Regression*

The ECM microenvironment of blood vessels includes the immediate basement membrane, composed primarily of Collagen Type IV and laminin, and the surrounding connective tissue, composed of fibrillar collagens (Type I and III) and other ECM-associated molecules, like fibronectin (Eble and Niland 2009). The synthesis and remodeling of the various components of the ECM have been observed to drastically affect EC biology (Davis and Senger 2005), and, reciprocally, the resident ECs themselves participate in the regulation of their immediate ECM microenvironment (Schultz et al. 2010). During vessel growth, the EC basement membrane as well as the provisional matrix is dissolved ahead of the invading neovessel sprout. During vessel regression, the ECM as a whole undergoes remodeling from an angio-permissive scaffold into a more mature phenotype, which provides crucial anti-angiogenic cues, both molecular and biomechanical, that guide the vessels into a spatially distributed and functional network (Califano and Reinhart-King 2010; Egginton and Gaffney 2010).

The primary cell type responsible for the maintenance and remodeling of the ECM is the fibroblast. Multiple studies have found that fibroblasts may be important regulators of angiogenesis through the production and activation of specific soluble and matrix components (Hurley et al. 2010; Pollina et al. 2008; Sorrell et al. 2008). Gene expression profiles differ dramatically in proliferating/migrating *versus* quiescent fibroblasts, with activated fibroblasts shown to produce pro-angiogenic matrix proteases and soluble proteins, such as VEGF; in contrast, quiescent fibroblasts exhibit an elevated expression of multiple ECM precursors of potent anti-angiogenic peptides (to be discussed later), as well as anti-angiogenic matricellular proteins TSP2 and PEDF (Pollina et al. 2008). Co-culture studies of ECs and fibroblasts have shown a temporal regulation of EC function by fibroblasts both directly by expression of specific angiogenic factors like VEGF, and indirectly by altering the mechanical microenvironment via matrix disruption, deposition, and remodeling (Hurley et al. 2010). Finally, different fibroblast subpopulations may play a role in determining the fibroblast's pro- *versus* anti-angiogenic functions. Fibroblasts derived from the papillary dermis and co-cultured with ECs are angio-permissive, stimulating robust vessel growth, whereas reticular fibroblasts from the deeper tissue are angio-restrictive, presumably as a result of non-soluble factors such as the composition of the secreted ECM (Sorrell et al. 2008). It could be speculated from these in vitro studies that during wound healing, fibroblasts promote angiogenic growth in the proliferative phase by being themselves activated by the hypoxic environment. At the onset of the wound resolution phase, fibroblasts may switch to an anti-angiogenic phenotype—due to contact inhibition and normalizing oxygen levels—to regulate ECM remodeling, indirectly mediating vessel regression through their action upon the ECM microenvironment (Fig. 2c).

ECM remodeling by fibroblasts involves both the systematic breakdown and the deposition of new ECM components. The ECM itself has been demonstrated to be a rich source of endogenous anti-angiogenic mediators. Soluble proteases, called matrix metalloproteinases (MMPs), enzymatically cleave larger ECM components (Davis and Saunders 2006; Ghajar et al. 2008). Cleavage of the provisional ECM scaffold by MMPs may disrupt stable ECM contacts with EC integrins, causing a decrease in survival signaling and reduced pro-angiogenic integrin co-activation. Widespread proteolysis of various components of the ECM scaffold may also undermine the support for parts of the blood vessel network, leading to structural collapse and pruning (Davis and Saunders 2006) (Fig. 2c).

Importantly, cleavage of certain ECM components by specific MMPs may generate bioactive peptide fragments which have been shown to be anti-angiogenic (Mundel and Kalluri 2007; Nyberg et al. 2005). One of the better studied anti-angiogenic ECM-derived peptides is endostatin, a fragment of EC basement membrane component Collagen Type XVIII (Nyberg et al. 2005). Endostatin is readily cleaved from its parent matrix protein by multiple MMPs and is a potent inhibitor of EC proliferation and migration, also inducing EC apoptosis. Early studies on endostatin's angiogenic function during wound healing have produced conflicting results, and a recent study using Collagen Type XVIII-null and endostatin overexpressing mice found that endostatin's role in physiological wound angiogenesis may be indirect, causing changes in vessel quality rather than quantity (Seppinen et al. 2008). Another component of the basement membrane, Collagen Type IV, can be cleaved into several potent anti-angiogenic peptides, including arresten, canstatin, and tumstatin (Mundel and Kalluri 2007). Other ECM components that may be cleaved to anti-angiogenic peptides include fibronectin (yielding anastellin), Collagen Type VIII (yielding vastatin), and heparan sulfate proteoglycans (yielding endorepellin). Although the roles of these anti-angiogenic ECM-derived peptides in regulating physiological angiogenesis are yet to be investigated, the sheer number and variety suggests that their release into the neovessel microenvironment during the remodeling phase of wound healing may be an important mechanism for the spatial control of vessel regression (Fig. 2c).

Whether EC apoptosis is the direct cause of vessel pruning or happens as a result of EC dissociation from the existing network is still a matter of debate (Im and Kazlauskas 2006). For instance, CXCL10 has been shown to induce EC apoptosis only after its disruption of a key integrin that keeps the ECs anchored to the supporting ECM (Bodnar et al. 2009). Similar integrin-disrupting mechanisms have been observed in vitro for other endogenous anti-angiogenic mediators, particularly ECM-derived peptides such as endostatin, tumstatin, canstatin, and arresten (Nyberg et al. 2005). Importantly, it has been shown that interruption of stable cell-ECM connections leads to EC apoptosis (Cheresh and Stupack 2008). These data suggest a critical role for integrins in the resolution of angiogenesis (Fig. 2c).

During wound healing, vessel regression occurs concurrently with ECM remodeling. The process of ECM remodeling involves the cleavage of multiple ECM components by MMPs, leading to the simultaneous disruption of EC-ECM contacts (freeing integrins of stable scaffold-associated ligands) and the generation of ECM-derived anti-angiogenic peptides. This process may contribute to vessel regression in a variety of ways (Fig. 2c):

- Systematic disruption of EC connections to the matrix scaffold may cause structural collapse of the affected tubules via imbalance of transmitted forces, leading to dissociation from the neovessel network (Davis and Saunders 2006).
- The density and stiffness of the remodeled, more mature ECM has been shown to alter EC behavior toward an anti-angiogenic phenotype via biomechanical pathways (Califano and Reinhart-King 2010).
- The cleaved ECM-derived soluble anti-angiogenic peptides may competitively bind to the free integrin receptors on ECs, initiating anti-angiogenic signaling pathways (Nyberg et al. 2005).
- Since amplification of pro-angiogenic and pro-survival signaling pathways are often dependent on integrin co-activation (Eliceiri et al. 1998), loss of stable integrin-ECM contacts during remodeling may promote EC apoptosis.
- Activated ECs may find themselves in an ECM microenvironment that has already remodeled to a more mature phenotype, making their integrin repertoire inappropriate, and leading to insufficient integrin co-stimulation to prolong survival signals (Cheresh and Stupack 2008).

These concepts, while logical and supported by indirect evidence, need to be investigated in the context of physiological vessel regression.

3.8 *Scar Formation: Does Regression Impair Tissue Remodeling?*

The seemingly intimate relationship between the processes of ECM remodeling and vessel regression may help to explain the varieties of scarring outcomes between tissues. Oral mucosal and fetal dermal wounds are known to heal in a scarless fashion when compared to similar size wounds in the adult skin (Occleston et al. 2010). One remarkable common feature is the much reduced angiogenic response of both oral mucosal (Szpaderska et al. 2005) and fetal wounds (Wilgus et al. 2008) during the proliferative phase of healing as compared to dermal wounds in adult animals. Although the association between increased angiogenesis and scarring is notable, the potential mechanisms underlying this relationship are unknown. The reduced angiogenesis in non-scar forming wounds translates into a reduced necessity for vessel regression in these wounds. It is tempting to speculate that the mechanisms of vessel regression explored here contribute to less efficient ECM remodeling, leading to scarring rather than regeneration in adult dermal wounds. These largely unexplored questions underscore the implications for greater understanding of the resolving processes of physiological angiogenesis, which will likely have significant clinical applications to the field of tissue regeneration.

3.9 Endothelial Cell Apoptosis: Mechanisms at the Moment of Death

The method by which ECs are eliminated during vessel regression is programmed cell death or apoptosis (Dimmeler and Zeiher 2000). The characteristic morphological changes in cells undergoing apoptosis include cell shrinkage, chromatin condensation, membrane blebbing, and nuclear fragmentation (Stoneman and Bennett 2009). The proteases that perform these destructive actions are called caspases, which are present in cells in their inactive pro-enzyme form and must be activated in large enough numbers to promote apoptosis. Caspase activation results from signaling pathways that are induced by specific intra- or extracellular signals. Induction of apoptosis in ECs is mainly via the extrinsic pathway, which involves the expression of a death receptor, Fas, and the binding to it by Fas ligand (FasL). As discussed, activation by VEGF in hypoxic conditions has been found to increase the expression of Fas on ECs, thus sensitizing them to both mitogenic and apoptotic stimuli (Volpert et al. 2002). The dynamic ECM microenvironment is also an important regulator of EC fate, as it can promote either survival or death depending on the stability of the EC integrin-ECM connections (Cheresh and Stupack 2008). While many of the endogenous anti-angiogenic molecules discussed have been found to induce caspase-mediated EC apoptosis, they go about it in remarkably different ways, and what actually initiates apoptosis in ECs comprising the neovessel network during healing is likely very context-specific (Figs. 2c, 3).

The apoptotic signaling mechanisms that are initiated by the various anti-angiogenic stimuli are complex and diverse, and in-depth discussion of this topic is beyond the scope of this review. Here, we simply wish to give some examples to illustrate the incredible redundancy involved.

The proposed apoptotic pathway initiated by TSPs involves binding to the EC receptor CD36, propagating p38 Kinase and Jun Kinase (JNK) signals, leading to FasL production (Mirochnik et al. 2008). PEDF has been shown to promote the production of FasL via various signaling mechanisms in ECs, including PPAR-gamma, NF-κB, JNK and/or p53, depending on the context (Aurora et al. 2010; Ho et al. 2007; Konson et al. 2011; Volpert et al. 2002). The proposed mechanism by which CXCL10 causes apoptosis is different: binding to CXCR3 activates μ-calpain, which cleaves a crucial integrin, leading to EC dissociation from the vessel network and apoptosis induced by detachment from the ECM (Bodnar et al. 2009). Similar integrin-dependent apoptotic pathways have been implicated in cases where ECs lose stable contacts with the ECM during remodeling by MMPs (Cheresh and Stupack 2008). Furthermore, several of the ECM-derived anti-angiogenic peptides function by binding and blocking key EC integrins, presumably initiating apoptosis in a related fashion (Nyberg et al. 2005). It is also the case that loss of integrin-mediated survival signaling, both on the abluminal (via the ECM) and luminal (via shear stress caused by flow) sides of the EC, further sensitizes the cell to extrinsic apoptotic signals (Ando and Yamamoto 2009; Califano and Reinhart-King 2010; Stoneman and Bennett 2009).

Recent explorations into EC apoptosis have found that besides the canonical signaling pathways, unique EC-specific signaling proteins may be involved in vessel regression (Korn and Augustin 2012). One such protein, called FGD5 (FYVE, Rho guanine exchange factor, and pleckstrin homology domain containing 5), is expressed in ECs in a variety of tissues and promotes the activation of GTPase Cdc42 (Cheng et al. 2012). FGD5 activation of Cdc42 mediates vessel regression by two mechanisms: downregulation of pro-angiogenic VEGFR2 coincident with the upregulation of VEGFR1 and subsequent sequestration of VEGF, and by initiation of Hey-1/p53-dependent EC apoptosis. However, this mechanism appears to be context-specific, as FGD5 has previously been shown to promote tubule formation in the presence of VEGF (Korn and Augustin 2012). The actual response likely depends on a convergence of upstream signaling cascades initiated by anti-angiogenic stimuli leading up to FGD5, which are yet to be investigated. While the mechanistic data is robust, the study utilized two in vivo models of pathological angiogenesis to explore the anti-angiogenic function of FGD5, and it remains to be seen whether FGD5 is involved in physiological vessel regression.

3.10 Maturation: Bolstering the Essential Vessels

During the remodeling phase of healing, regression and maturation of neovessels occur simultaneously (Fig. 3). While the vast majority of the wound vasculature undergoes pruning by the mechanisms just discussed (Fig. 1d), a minority gets chosen for the opposite fate: survival and maturation (Fig. 1e). What are the determinants of vessel maturation in the complex neovessel network of the remodeling granulation tissue?

Vessel perfusion likely plays a critical role in the induction of vessel maturation (Ando and Yamamoto 2009; Califano and Reinhart-King 2010; Potente et al. 2011). Tubules that reconnect with a functional network re-establish a stable, laminar flow, leading to shear stress-mediated survival signaling (Figs. 1d, e, 2a). This signaling may give these vessels a competitive advantage over their non-perfused neighbors (Fig. 3). Furthermore, the microenvironment of the perfused vessel will over time be less hypoxic due to more efficient delivery of oxygenated blood, leading to the downregulation of pro-angiogenic stimuli. In these ECs, increased survival signaling, coupled with decreased mitogenic signaling, may promote their maturation into vessels that will make up the permanent blood vessel network of the healed tissue, a network that architecturally and functionally resembles that of the pre-wounded state (Fig. 1e).

Mechanisms of vessel maturation have been explored in some detail (reviewed in (Hellberg et al. 2010; Jain 2003; Murakami and Simons 2009)). The overarching processes involve the recruitment and attachment of network-stabilizing mural cells, such as pericytes (von Tell et al. 2006) and vascular smooth muscle cells (Korff et al. 2001), as well as the bolstering of EC–EC connections through VE-

Cadherins (Dejana et al. 2008). In a general sense, maturation aims at a return to a state of EC quiescence and vascular homeostasis (Fig. 2a).

The recruitment and differentiation of mural cells is largely mediated by Platelet-Derived Growth Factor (PDGF), which is produced by ECs undergoing maturation and binds to the PDGF receptor on nearby pericytes, promoting their migration toward and adhesion to vessel walls (Hellberg et al. 2010). The presence of pericytes stimulates the production by ECs of Transforming Growth Factor Beta (TGF-β), a multifunctional protein that attenuates EC response to pro-angiogenic stimuli and stimulates mural cell differentiation and activity (Murakami 2012). The sphingolipid metabolite, Sphingosine-1-phosphate (S1P), has been found to be important during vessel maturation. EC-derived S1P strengthens EC-pericyte interactions through the assembly of N-Cadherin between the EC and pericyte (Potente et al. 2011), and S1P promotes Akt-associated survival signaling as well as Rac1-mediated assembly of VE-Cadherin in ECs (Murakami 2012). Finally, Angiopoietin-1 produced by mural cells interacts with ECs through Tie-2 receptors and promotes vessel barrier integrity by attenuating VEGF-induced vasopermeability, stabilizing VE-Cadherin, and promoting pericyte adhesion (London et al. 2009; Thomas and Augustin 2009).

These and other mechanisms of vessel maturation promote the survival of a perfused and functional blood vessel network in the highly anti-angiogenic environment of the resolving wound. While protecting the ECs from pro- and anti-angiogenic stimuli, these mechanisms bolster EC–EC connections, reducing vascular leakage. Throughout wound resolution, while the majority of non-perfused and blind-ended neovessels undergo regression and are eliminated (Figs. 1d, 2c), the processes of vessel maturation promote the stabilization of the remaining vessel network, restoring original architecture and function and thus helping to return the tissue to homeostasis (Fig. 1e).

3.11 Tumor Angiogenesis: The Case of the Ever-Sprouting Blood Vessel

The exuberant angiogenesis that occurs in dermal wounds during the proliferative phase and results in the characteristic granulation tissue formation has been compared to the angiogenesis that occurs in solid tumors (Cuevas and Boudreau 2009; Dvorak 1986; Folkman 1974; Holash et al. 1999; Schafer and Werner 2008). In these tumors, there is a seemingly endless cycle of blood vessel growth and regression, with the tissue experiencing constant hypoxia and never quite being able to achieve homeostasis. That is, the tumor could be described as a wound that gets stuck in the proliferative phase, never fully transitioning into resolution and never quite adopting an anti-angiogenic phenotype as occurs in normal wounds. Based on our discussion of the microenvironmental and cellular changes that normally take place during the later phases of healing, several hypotheses can be made regarding the mechanisms that may be dysregulated in tumors as compared to wounds.

Attempts at treatments of tumors using anti-angiogenic therapy have yielded mixed results. Despite decades of rigorous research, no 'magic bullet' molecule has been found that effectively stops pathologic angiogenesis consistently and is able to starve the tumor in animal models and in the clinic. As we have seen, the biology inherent in the process of angiogenesis is extremely complex, with various cell types, countless redundant molecules and pathways, and cross-talk with a highly dynamic ECM microenvironment. A comprehensive understanding of not only the individual parts but how they come together may be necessary before we are able to predictably steer the angiogenic process in the desired direction. Importantly, this must include rigorous investigation into the mechanisms of vessel regression. By understanding how the organism deals with exuberant blood vessel growth in health, we may be able to recruit these natural resolving processes in cases of pathology.

Indeed, the field of cancer therapy appears to be approaching a similar conclusion. Recent observations have found that by normalizing tumor vasculature, via the induction of maturation mechanisms just discussed, the tumors respond better to therapy (Goel et al. 2011; Jain 2005; Sato 2011). The idea is that the leaky and non-perfused vasculature of tumors inefficiently delivers chemotherapeutic agents to the cancer cells. Furthermore, the implication of the dynamic reciprocity of the vasculature and its microenvironment is that active promotion of a more normal phenotype in vessels may induce normalization of the tumor tissue itself, halting the vicious cycle of hypoxia-driven growth. In light of these exciting concepts, it becomes clear that the field of cancer therapy may have much to learn from wound healing, particularly from more rigorous exploration of the mechanisms of wound resolution, and the tissue remodeling and vessel regression processes that naturally occur during healing.

4 Conclusions

Wound healing is a robust model for studying the physiological control of angiogenesis in vivo (Eming et al. 2007). During experimentally induced wound repair, the vascular network expands and regresses in a well-characterized and remarkably reproducible temporal pattern. Angiogenesis during wound repair is induced, controlled, and ultimately resolved to vascular and tissue homeostasis (Fig. 1). Spatio-temporal control of blood vessel growth and regression occurs through competing signals acting upon ECs, signals derived from the highly dynamic wound microenvironment. Pro- and anti-angiogenic stimuli are both molecular and biomechanical in nature, and the convergence of the induced signaling pathways determines EC fate: sprouting, regression, or maturation (Fig. 3).

Blood vessel regression likely occurs via various mechanisms (Fig. 2c). Throughout the process there is significant cross-talk between the different cells and their microenvironment. Blood vessels that are not perfused and experience low shear stress within the lumen are pruned. As the oxygen tension within the healing tissue normalizes, decreased hypoxia leads to a decrease in the production of pro-angiogenic factors. At the same time, endothelial cells that have been continually stimulated by pro-angiogenic factors begin to produce and activate inhibitors to pro-angiogenic signaling pathways in a negative feedback mechanism. The mounting intracellular inhibition makes these cells less sensitive to further pro-angiogenic stimulation while at the same time making them more sensitive to progressively accumulating anti-angiogenic stimuli, thus causing an 'anti-angiogenic switch'. Many direct anti- angiogenic mediators are produced in the microenvironment of the remodeling tissue. Soluble molecules, such as cytokines, as well as matricellular proteins, act directly upon ECs through specific receptors and downstream pathways. Often these then activate pro-apoptotic signaling pathways leading to systematic EC death. Another class of anti-angiogenic factors includes bioactive fragments of ECM components. These fragments are released into the microenvironment through the action of various matrix proteases that cleave ECM proteins at specific sites, and directly antagonize nearby ECs. In this process of matrix remodeling, support for blood vessels may become undermined, leading to their collapse and regression. Meanwhile, as the matrix remodels from a provisional to a more mature composition, the biomechanical properties of the matrix favor the anti-angiogenic phenotype. The net result is fine spatio-temporal control of blood vessel pruning occurring simultaneously and in cooperation with matrix remodeling, leading to an optimally distributed vessel network that adequately supplies the healing tissue.

The emerging picture is that of a superbly complex coordinated effort between multiple cell types and the dynamic wound microenvironment to bring about a resolution to the angiogenic process and the re-establishment of tissue homeostasis. The existence of such an active and deliberate anti-angiogenic phase has a number of critical implications.

First, it underscores the realization that nature leaves nothing to chance where vital biological processes are concerned. In the case of angiogenesis, mammals have evolved multilevel and redundant mechanisms to ensure the proper resolution of robust blood vessel growth. These mechanisms certainly include, but are not limited to, the simple withdrawal of pro-angiogenic signals. From negative feedback loops to matrix-embedded proteins, there are multiple molecular and biomechanical factors whose job it is to limit vessel sprouting and bring about the return to vascular homeostasis. The theme of competing yet complementary creative *versus* destructive pathways is most likely universal to other examples of tissue regeneration, and should inform our exploration of regenerative pathologies.

Second, the existence of an active anti-angiogenic phase highlights the limits of the prevailing scientific focus on the pro-angiogenic side of the equation. Indeed, this limited thinking extends to all areas of tissue regeneration and tissue engineering. Regeneration is *not* simply a matter of cellular reproduction and matrix expansion into its proper and final form. As our discussion of angiogenesis makes clear, the growth phase is merely the first step, and a crude one at that, whereby a temporary immature blood vessel network forms on a provisional matrix scaffold. In fact, overwhelming evidence suggests that this primary neovessel network is largely non-functional and architecturally does not come close to resembling the original structure. It is up to the anti-angiogenic side of the equation to remodel this network into its final, optimal form and function that are appropriate for tissue health. Without this side of the equation coming online, pathology invariably results, and cancer is the obvious example of this happening to the detriment of the entire organism.

Third, it is becoming abundantly clear that vessel regression and matrix remodeling are intimately linked processes. This is a great example of dynamic reciprocity at work: constant cross-talk between the cells and the dynamic microenvironment results in a remarkable coordination of functions that ensures a balanced resolution to tissue repair. Our brief discussion on the variability in regenerative outcomes between wounds from different anatomical and developmental origins and their remarkable association with angiogenic potential raises many intriguing questions about the nature of this relationship. Indeed, the study and more complete understanding of the processes involved in vessel pruning are bound to give critical insights into the determinants of tissue repair and scar formation *versus* regeneration.

Considering these implications for health and disease, greater research attention should be given to the understanding of biological processes from start to finish. Tunnel vision with pro-angiogenesis has seriously limited our perception of the angiogenic process as a whole, and it may have hindered our therapeutic abilities for the myriad of pathologies which feature angiogenic phenotypes. We need to explore physiological processes more closely, learn about the start *and* stop signals, and trust in the biological insights gained from millions of years of evolution. Only when we begin to understand biological processes in this way can we hope to have success in their optimization for our own ends.

References

Ando J, Yamamoto K (2009) Vascular mechanobiology: endothelial cell responses to fluid shear stress. Circ J 73:1983–1992

Aurora AB, Biyashev D, Mirochnik Y, Zaichuk TA, Sanchez-Martinez C, Renault MA, Losordo D, Volpert OV (2010) NF-kappaB balances vascular regression and angiogenesis via chromatin remodeling and NFAT displacement. Blood 116:475–484

Baffert F, Le T, Sennino B, Thurston G, Kuo CJ, Hu-Lowe D, McDonald DM (2006) Cellular changes in normal blood capillaries undergoing regression after inhibition of VEGF signaling. Am J Physiol Heart Circ Physiol 290:H547–H559

Bao P, Kodra A, Tomic-Canic M, Golinko MS, Ehrlich HP, Brem H (2009) The role of vascular endothelial growth factor in wound healing. J Surg Res 153:347–358

Barrientos S, Stojadinovic O, Golinko MS, Brem H, Tomic-Canic M (2008) Growth factors and cytokines in wound healing. Wound Repair Regen 16:585–601
Benjamin LE, Hemo I, Keshet E (1998) A plasticity window for blood vessel remodelling is defined by pericyte coverage of the preformed endothelial network and is regulated by PDGF-B and VEGF. Development 125:1591–1598
Bhadada SV, Goyal BR, Patel MM (2011) Angiogenic targets for potential disorders. Fundam Clin Pharmacol 25:29–47
Bloor CM (2005) Angiogenesis during exercise and training. Angiogenesis 8:263–271
Bluff JE, O'Ceallaigh S, O'Kane S, Ferguson MW, Ireland G (2006) The microcirculation in acute murine cutaneous incisional wounds shows a spatial and temporal variation in the functionality of vessels. Wound Repair Regen 14:434–442
Bodnar RJ, Yates CC, Rodgers ME, Du X, Wells A (2009) IP-10 induces dissociation of newly formed blood vessels. J Cell Sci 122:2064–2077
Bornstein P (2009) Thrombospondins function as regulators of angiogenesis. J Cell Commun Signal 3:189–200
Bou-Gharios G, Ponticos M, Rajkumar V, Abraham D (2004) Extra-cellular matrix in vascular networks. Cell Prolif 37:207–220
Broadhead ML, Becerra SP, Choong PF, Dass CR (2010) The applied biochemistry of PEDF and implications for tissue homeostasis. Growth Factors 28:280–285
Cabrita MA, Christofori G (2008) Sprouty proteins, masterminds of receptor tyrosine kinase signaling. Angiogenesis 11:53–62
Cai J, Wu L, Qi X, Li Calzi S, Caballero S, Shaw L, Ruan Q, Grant MB, Boulton ME (2011) PEDF regulates vascular permeability by a gamma-secretase-mediated pathway. PLoS ONE 6:e21164
Califano JP, Reinhart-King CA (2010) Exogenous and endogenous force regulation of endothelial cell behavior. J Biomech 43:79–86
Cao Y, Xue L (2004) Angiostatin. Semin Thromb Hemost 30:83–93
Chappell JC, Wiley DM, Bautch VL (2012) How blood vessel networks are made and measured. Cell Tiss Org 195:94–107
Chen RR, Silva EA, Yuen WW, Mooney DJ (2007) Spatio-temporal VEGF and PDGF delivery patterns blood vessel formation and maturation. Pharm Res 24:258–264
Cheng C, Haasdijk R, Tempel D, van de Kamp EH, Herpers R, Bos F, Den Dekker WK, Blonden LA, de Jong R, Burgisser PE, Chrifi I, Biessen EA, Dimmeler S, Schulte-Merker S, Duckers HJ (2012) Endothelial cell-specific FGD5 involvement in vascular pruning defines neovessel fate in mice. Circulation 125:3142–3159
Cheresh DA, Stupack DG (2008) Regulation of angiogenesis: apoptotic cues from the ECM. Oncogene 27:6285–6298
Cuevas I, Boudreau N (2009) Managing tumor angiogenesis: lessons from VEGF-resistant tumors and wounds. Adv Cancer Res 103:25–42
Davis GE, Saunders WB (2006) Molecular balance of capillary tube formation versus regression in wound repair: role of matrix metalloproteinases and their inhibitors. J Investig Dermatol Symp Proc 11:44–56
Davis GE, Senger DR (2005) Endothelial extracellular matrix: biosynthesis, remodeling, and functions during vascular morphogenesis and neovessel stabilization. Circ Res 97:1093–1107
Dejana E, Orsenigo F, Lampugnani MG (2008) The role of adherens junctions and VE-cadherin in the control of vascular permeability. J Cell Sci 121:2115–2122
Dimmeler S, Zeiher AM (2000) Endothelial cell apoptosis in angiogenesis and vessel regression. Circ Res 87:434–439
Distler JH, Hirth A, Kurowska-Stolarska M, Gay RE, Gay S, Distler O (2003) Angiogenic and angiostatic factors in the molecular control of angiogenesis. Q J Nucl Med 47:149–161
Dvorak HF (1986) Tumors: wounds that do not heal. Similarities between tumor stroma generation and wound healing. N Engl J Med 315:1650–1659
Eble JA, Niland S (2009) The extracellular matrix of blood vessels. Curr Pharm Des 15:1385–1400

Egginton S, Gaffney E (2010) Tissue capillary supply–it's quality not quantity that counts! Exp Physiol 95:971–979

Eliceiri BP, Klemke R, Stromblad S, Cheresh DA (1998) Integrin alphavbeta3 requirement for sustained mitogen-activated protein kinase activity during angiogenesis. J Cell Biol 140:1255–1263

Eming SA, Krieg T (2006) Molecular mechanisms of VEGF-A action during tissue repair. J Investig Dermatol Symp Proc 11:79–86

Eming SA, Brachvogel B, Odorisio T, Koch M (2007) Regulation of angiogenesis: wound healing as a model. Prog Histochem Cytochem 42:115–170

Folkman J (1974) Tumor angiogenesis. Adv Cancer Res 19:331–358

Fukumura D, Jain RK (2008) Imaging angiogenesis and the microenvironment. Apmis 116:695–715

Geudens I, Gerhardt H (2011) Coordinating cell behaviour during blood vessel formation. Development 138:4569–4583

Ghajar CM, George SC, Putnam AJ (2008) Matrix metalloproteinase control of capillary morphogenesis. Crit Rev Eukaryot Gene Expr 18:251–278

Goel S, Duda DG, Xu L, Munn LL, Boucher Y, Fukumura D, Jain RK (2011) Normalization of the vasculature for treatment of cancer and other diseases. Physiol Rev 91:1071–1121

Gosain A, Matthies AM, Dovi JV, Barbul A, Gamelli RL, DiPietro LA (2006) Exogenous pro-angiogenic stimuli cannot prevent physiologic vessel regression. J Surg Res 135:218–225

Gurtner GC, Werner S, Barrandon Y, Longaker MT (2008) Wound repair and regeneration. Nature 453:314–321

Hellberg C, Ostman A, Heldin CH (2010) PDGF and vessel maturation. Recent Results Cancer Res 180:103–114

Hickey MM, Simon MC (2006) Regulation of angiogenesis by hypoxia and hypoxia-inducible factors. Curr Top Dev Biol 76:217–257

Ho TC, Chen SL, Yang YC, Liao CL, Cheng HC, Tsao YP (2007) PEDF induces p53-mediated apoptosis through PPAR gamma signaling in human umbilical vein endothelial cells. Cardiovasc Res 76:213–223

Holash J, Maisonpierre PC, Compton D, Boland P, Alexander CR, Zagzag D, Yancopoulos GD, Wiegand SJ (1999) Vessel cooption, regression, and growth in tumors mediated by angiopoietins and VEGF. Science 284:1994–1998

Hosomichi J, Yasui N, Koide T, Soma K, Morita I (2005) Involvement of the collagen I-binding motif in the anti-angiogenic activity of pigment epithelium-derived factor. Biochem Biophys Res Commun 335:756–761

Hungerford JE, Little CD (1999) Developmental biology of the vascular smooth muscle cell: building a multilayered vessel wall. J Vasc Res 36:2–27

Hurley JR, Balaji S, Narmoneva DA (2010) Complex temporal regulation of capillary morphogenesis by fibroblasts. Am J Physiol Cell Physiol 299:C444–C453

Im E, Kazlauskas A (2006) New insights regarding vessel regression. Cell Cycle 5:2057–2059

Jain RK (2003) Molecular regulation of vessel maturation. Nat Med 9:685–693

Jain RK (2005) Normalization of tumor vasculature: an emerging concept in antiangiogenic therapy. Science 307:58–62

Kniazeva E, Putnam AJ (2009) Endothelial cell traction and ECM density influence both capillary morphogenesis and maintenance in 3-D. Am J Physiol Cell Physiol 297:C179–C187

Konson A, Pradeep S, D'Acunto CW, Seger R (2011) Pigment epithelium-derived factor and its phosphomimetic mutant induce JNK-dependent apoptosis and p38-mediated migration arrest. J Biol Chem 286:3540–3551

Korff T, Kimmina S, Martiny-Baron G, Augustin HG (2001) Blood vessel maturation in a 3-dimensional spheroidal coculture model: direct contact with smooth muscle cells regulates endothelial cell quiescence and abrogates VEGF responsiveness. Faseb J 15:447–457

Korn C, Augustin HG (2012) Born to die: blood vessel regression research coming of age. Circulation 125:3063–3065

Kyriakides TR, Maclauchlan S (2009) The role of thrombospondins in wound healing, ischemia, and the foreign body reaction. J Cell Commun Signal 3:215–225

Lange-Asschenfeldt B, Velasco P, Streit M, Hawighorst T, Pike SE, Tosato G, Detmar M (2001) The angiogenesis inhibitor vasostatin does not impair wound healing at tumor-inhibiting doses. J Invest Dermatol 117:1036–1041

Lasagni L, Francalanci M, Annunziato F, Lazzeri E, Giannini S, Cosmi L, Sagrinati C, Mazzinghi B, Orlando C, Maggi E, Marra F, Romagnani S, Serio M, Romagnani P (2003) An alternatively spliced variant of CXCR3 mediates the inhibition of endothelial cell growth induced by IP-10, Mig, and I-TAC, and acts as functional receptor for platelet factor 4. J Exp Med 197:1537–1549

Lawler PR, Lawler J (2012) Molecular basis for the regulation of angiogenesis by thrombospondin-1 and -2. Cold Spring Harb Perspect Med 2:a006627

London NR, Whitehead KJ, Li DY (2009) Endogenous endothelial cell signaling systems maintain vascular stability. Angiogenesis 12:149–158

Maclauchlan S, Skokos EA, Agah A, Zeng J, Tian W, Davidson JM, Bornstein P, Kyriakides TR (2009) Enhanced angiogenesis and reduced contraction in thrombospondin-2-null wounds is associated with increased levels of matrix metalloproteinases-2 and -9, and soluble VEGF. J Histochem Cytochem 57:301–313

Mirochnik Y, Kwiatek A, Volpert OV (2008) Thrombospondin and apoptosis: molecular mechanisms and use for design of complementation treatments. Curr Drug Targets 9:851–862

Modlich U, Kaup FJ, Augustin HG (1996) Cyclic angiogenesis and blood vessel regression in the ovary: blood vessel regression during luteolysis involves endothelial cell detachment and vessel occlusion. Lab Invest 74:771–780

Mosher DF, Adams JC (2012) Adhesion-modulating/matricellular ECM protein families: a structural, functional and evolutionary appraisal. Matrix Biol 31:155–161

Mundel TM, Kalluri R (2007) Type IV collagen-derived angiogenesis inhibitors. Microvasc Res 74:85–89

Murakami M (2012) Signaling required for blood vessel maintenance: molecular basis and pathological manifestations. Int J Vasc Med 2012:293641

Murakami M, Simons M (2009) Regulation of vascular integrity. J Mol Med (Berl) 87:571–582

Naldini A, Pucci A, Bernini C, Carraro F (2003) Regulation of angiogenesis by Th1- and Th2-type cytokines. Curr Pharm Des 9:511–519

Nissen NN, Polverini PJ, Koch AE, Volin MV, Gamelli RL, DiPietro LA (1998) Vascular endothelial growth factor mediates angiogenic activity during the proliferative phase of wound healing. Am J Pathol 152:1445–1452

Nyberg P, Xie L, Kalluri R (2005) Endogenous inhibitors of angiogenesis. Cancer Res 65:3967–3979

Occleston NL, Metcalfe AD, Boanas A, Burgoyne NJ, Nield K, O'Kane S, Ferguson MW (2010) Therapeutic improvement of scarring: mechanisms of scarless and scar-forming healing and approaches to the discovery of new treatments. Dermatol Res Pract

Olfert IM, Birot O (2011) Importance of anti-angiogenic factors in the regulation of skeletal muscle angiogenesis. Microcirculation 18:316–330

Owen MR, Alarcon T, Maini PK, Byrne HM (2009) Angiogenesis and vascular remodelling in normal and cancerous tissues. J Math Biol 58:689–721

Phelps EA, Garcia AJ (2010) Engineering more than a cell: vascularization strategies in tissue engineering. Curr Opin Biotechnol 21:704–709

Pollina EA, Legesse-Miller A, Haley EM, Goodpaster T, Randolph-Habecker J, Coller HA (2008) Regulating the angiogenic balance in tissues: a potential role for the proliferative state of fibroblasts. Cell Cycle 7:2056–2070

Potente M, Gerhardt H, Carmeliet P (2011) Basic and therapeutic aspects of angiogenesis. Cell 146:873–887

Ribatti D, Nico B, Crivellato E (2009) Morphological and molecular aspects of physiological vascular morphogenesis. Angiogenesis 12:101–111

Rodriguez PG, Felix FN, Woodley DT, Shim EK (2008) The role of oxygen in wound healing: a review of the literature. Dermatol Surg 34:1159–1169
Rogers PA (1996) Structure and function of endometrial blood vessels. Hum Reprod Update 2:57–62
Sakamaki K (2004) Regulation of endothelial cell death and its role in angiogenesis and vascular regression. Curr Neurovasc Res 1:305–315
Sato Y (2011) Persistent vascular normalization as an alternative goal of anti-angiogenic cancer therapy. Cancer Sci 102:1253–1256
Sato Y, Sonoda H (2007) The vasohibin family: a negative regulatory system of angiogenesis genetically programmed in endothelial cells. Arterioscler Thromb Vasc Biol 27:37–41
Schafer M, Werner S (2008) Cancer as an overhealing wound: an old hypothesis revisited. Nat Rev Mol Cell Biol 9:628–638
Schramm JC, Dinh T, Veves A (2006) Microvascular changes in the diabetic foot. Int J Low Extrem Wounds 5:149–159
Schultz GS, Davidson JM, Kirsner RS, Bornstein P, Herman IM (2010) Dynamic reciprocity in the wound microenvironment. Wound Repair Regen 19:134–148
Semenza GL (2010) Vascular responses to hypoxia and ischemia. Arterioscler Thromb Vasc Biol 30:648–652
Seppinen L, Sormunen R, Soini Y, Elamaa H, Heljasvaara R, Pihlajaniemi T (2008) Lack of collagen XVIII accelerates cutaneous wound healing, while overexpression of its endostatin domain leads to delayed healing. Matrix Biol 27:535–546
Shaterian A, Borboa A, Sawada R, Costantini T, Potenza B, Coimbra R, Baird A, Eliceiri BP (2009) Real-time analysis of the kinetics of angiogenesis and vascular permeability in an animal model of wound healing. Burns 35:811–817
Shaw TJ, Martin P (2009) Wound repair at a glance. J Cell Sci 122:3209–3213
Sorrell JM, Baber MA, Caplan AI (2008) Human dermal fibroblast subpopulations; differential interactions with vascular endothelial cells in coculture: nonsoluble factors in the extracellular matrix influence interactions. Wound Repair Regen 16:300–309
Staton CA, Reed MW, Brown NJ (2009) A critical analysis of current in vitro and in vivo angiogenesis assays. Int J Exp Pathol 90:195–221
Stoneman VE, Bennett MR (2009) Role of Fas/Fas-L in vascular cell apoptosis. J Cardiovasc Pharmacol 53:100–108
Swift ME, Kleinman HK, DiPietro LA (1999) Impaired wound repair and delayed angiogenesis in aged mice. Lab Invest 79:1479–1487
Szpaderska AM, Walsh CG, Steinberg MJ, DiPietro LA (2005) Distinct patterns of angiogenesis in oral and skin wounds. J Dent Res 84:309–314
Thomas M, Augustin HG (2009) The role of the Angiopoietins in vascular morphogenesis. Angiogenesis 12:125–137
Volpert OV, Zaichuk T, Zhou W, Reiher F, Ferguson TA, Stuart PM, Amin M, Bouck NP (2002) Inducer-stimulated Fas targets activated endothelium for destruction by anti-angiogenic thrombospondin-1 and pigment epithelium-derived factor. Nat Med 8:349–357
von Tell D, Armulik A, Betsholtz C (2006) Pericytes and vascular stability. Exp Cell Res 312:623–629
Wacker A, Gerhardt H (2011) Endothelial development taking shape. Curr Opin Cell Biol 23:676–685
Wietecha MS, Chen L, Ranzer MJ, Anderson K, Ying C, Patel TB, DiPietro LA (2012) Sprouty2 downregulates angiogenesis during mouse skin wound healing. Am J Physiol Heart Circ Physiol 300:H459–H467
Wilgus TA, Ferreira AM, Oberyszyn TM, Bergdall VK, Dipietro LA (2008) Regulation of scar formation by vascular endothelial growth factor. Lab Invest 88:579–590
Wong VW, Sorkin M, Glotzbach JP, Longaker MT, Gurtner GC (2011) Surgical approaches to create murine models of human wound healing. J Biomed Biotechnol 2011:969618

Yates CC, Whaley D, Kulasekeran P, Hancock WW, Lu B, Bodnar R, Newsome J, Hebda PA, Wells A (2007) Delayed and deficient dermal maturation in mice lacking the CXCR3 ELR-negative CXC chemokine receptor. Am J Pathol 171:484–495

Yuen WW, Du NR, Chan CH, Silva EA, Mooney DJ (2010) Mimicking nature by codelivery of stimulant and inhibitor to create temporally stable and spatially restricted angiogenic zones. Proc Natl Acad Sci USA 107:17933–17938

Wound Healing in Mammals and Amphibians: Toward Limb Regeneration in Mammals

Aiko Kawasumi, Natsume Sagawa, Shinichi Hayashi, Hitoshi Yokoyama and Koji Tamura

Abstract Mammalian fetal skin regenerates perfectly, but adult skin repairs by the formation of scar tissue. The cause of this imperfect repair by adult skin is not understood. In contrast, wounded adult amphibian (urodeles and anurans) skin is like mammalian fetal skin in that it repairs by regeneration, not scarring. Scar-free wound repair in adult *Xenopus* is associated with expression of the paired homeobox transcription factor *Prx1* by mesenchymal cells of the wound, a feature shared by mesenchymal cells of the regeneration blastema of the axolotl limb. Furthermore, mesenchymal cells of *Xenopus* skin wounds that harbor the mouse *Prx1*-limb-enhancer as a transgene exhibit activation of the enhancer despite the fact that they are *Xenopus* cells, suggesting that the mouse *Prx1* enhancer possesses all elements required for its activation in skin wound healing, even though activation of the same enhancer in the mouse is not seen in the wounded skin of an adult mouse. Elucidation of the role of the *Prx1* gene in amphibian skin wound healing will help to clarify the molecular mechanisms of scarless wound healing. Shifting the molecular mechanism of wound repair in mammals to that of amphibians, including reactivation of the *Prx1*-limb-enhancer, will be an important clue to stimulate scarless wound repair in mammalian adult skin. Finding or creating *Prx1*-positive stem cells in adult mammal skin by activating the *Prx1*-limb-enhancer may be a fast and reliable way to provide for scarless skin wound repair, and even directly lead to limb regeneration in mammals.

A. Kawasumi · N. Sagawa · S. Hayashi · H. Yokoyama · K. Tamura (✉)
Department of Developmental Biology and Neurosciences,
Graduate School of Life Sciences, Tohoku University, Aobayama Aoba-ku,
Sendai 980-8578, Japan
e-mail: tam@m.tohoku.ac.jp

Current Topics in Microbiology and Immunology (2013) 367: 33–49
DOI: 10.1007/82_2012_305
© Springer-Verlag Berlin Heidelberg 2012
Published Online: 23 December 2012

Abbreviations

AEC	Apical epithelial cap
D	Dermis
E	Epidermis
ECM	Extracellular matrix
EPC	Endothelial progenitor cell
H	Hypodermis
MIF	Migration inhibitory factor
MMP	Matrix metalloproteinase
PDGF	Platelet-derived growth factor
TGF	Transforming growth factor
VEGF	Vascular endothelial growth factor

Contents

1 Structure of Mammalian Adult Skin

The skin in mammals is the largest organ of the body and is composed of two layers, the epidermis and dermis. In addition to these two layers, the hypodermis, which is sometimes considered a layer of the skin, lies beneath the dermis and is composed mainly of adipose tissue. It is noteworthy that there are significant differences in the anatomy and physiology of each skin layer between species such as humans and mice (Wong et al. 2011), and principle structures based on human skin are overviewed here (Fig. 1).

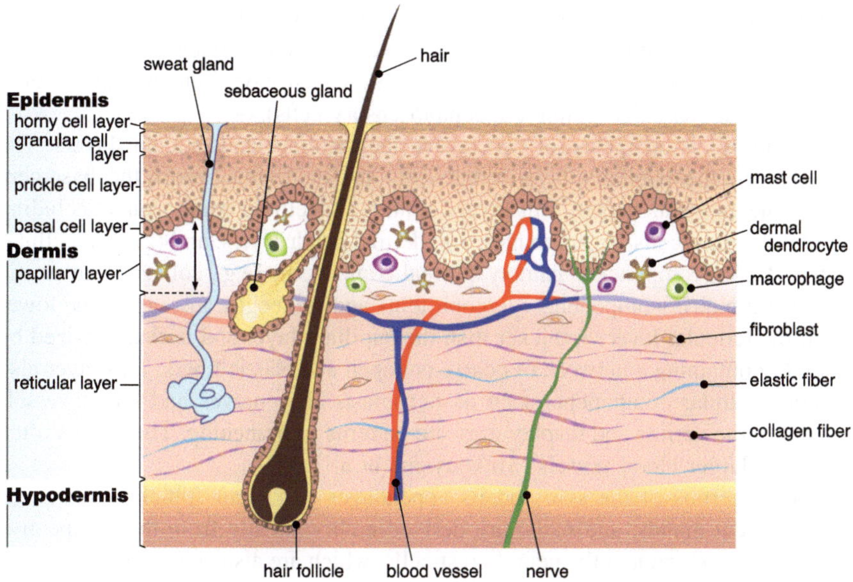

Fig. 1 Adult human skin is a layered organ consisting of an epidermis and a dermis. The epidermis is composed of four distinct layers: horny, granular, prickle, and basal cell layers (from top to bottom). The dermis is a highly elastic, tough, and flexible tissue made up of a meshwork of collagenous, reticular, and elastic fibers. It is divided into two functional layers, the papillary dermis and reticular dermis. These two layers are separated by a vascular plexus, which is fed by another vascular plexus located at the base of the reticular dermis (not shown). The papillary dermis contains a higher density and a greater variety of cells than does the reticular dermis. Fibroblasts, the main cells in the dermis, are essentially located in the papillary layer and are found only in very small numbers in the reticular dermis. They play an important part in production of the extracellular matrix. The skin also contains hair follicles, glands (A sweat gland and a sebaceous gland are shown here), and nerve endings responsible for the sense of touch and pain.

The epidermis, which is a terminally differentiated and stratified squamous epithelium, is the most superficial layer of the skin and is classified into several layers, including the basal layer at the bottom. The basal cell layer, the innermost cell layer of the epidermis, is a single layer consisting of basal cells including the epidermal stem cell subpopulation. The stem cell population in the basal layer gives rise to distinguishable layers, such as prickle, granular, and horny cell layers, as the cells from the basal layer move outwards and progressively differentiate into them. The prickle cell layer, the suprabasal cell layer, is composed of five to ten layers that appear connected to each other by prickle-like structures. The granular cell layer is composed of two or three layers of cells containing flattened nuclei and many granules. The main component of the granules is released as corneum lipid into the intercellular space of the horny cell layer. The horny cell layer is the outermost cell layer where keratinocytes are enucleated and become corneocytes as a terminal differentiation. Corneocytes consist of a stabilized array of keratin filaments contained within a covalently cross-linked protein envelope and serve as a protective surface (Matoltsy et al. 1968). Additionally, there are several special cells in the epidermis, including melanocytes, which provide pigment to the keratinocytes, Langerhans, and dendritic cells, which have immunological functions, and Merkel cells (putative mechanosensory cells).

The dermis is composed of two layers, the papillary and reticular layers. The upper layer, the papillary layer, edges into the epidermis across the basement membrane and nourishes it. This layer contains (1) cellular components, including fibroblasts, mast cells, macrophages, and dermal dendrocytes, and (2) extracellular matrix (ECM) components, namely stromal components (e.g., collagen and elastic fibers) and matrix components (glycoproteins and proteoglycans, etc.). The lower layer, the reticular layer, is thicker than the papillary layer and is characterized by an ECM containing a network of coarse collagen and elastic fibers. This layer also contains cellular components such as fibroblasts. The nerves and blood vessels come into the dermis and display intricate patterns of branching. The nerves often run along larger blood vessels (Mukouyama et al. 2002).

The epidermis of the skin and its appendages (e.g., hair follicles, sebaceous glands, sweat glands, and nails) are derived embryonically from the prospective epidermal ectoderm and the neural crest cells, which are also ectodermal in origin. The skin appendages have their roots in the dermis or even in the hypodermis, both of which are derived from embryonic mesoderm such as somites and lateral plate.

2 Phases of Repair After Skin Injury

The skin is an intricate structure as described above, and it functions as armor for protection of the body from external environments and as an anatomical barrier from pathogens and physical damage, water resistance, UV protection, and so on. The armor of the vertebrate body is always exposed to irritants and assailants and is therefore sometimes injured. When the skin is injured, the skin initiates a complex process of events, namely wound healing that involves inflammation as well as formation and remodeling of new tissue, which require orchestrated regulation of different cell types. The wound healing process results in reconstitution of injured skin, although the process in adult mammals is imperfect and less restorative, and the wounds heal with a scar (Fig. 2). In fact, skin wounds deep through the basement membrane trigger scar formation and, therefore, the wounded dermis is repaired with scar formation.

2.1 Hemostasis

Wound healing needs hemostasis for the first step. At the onset of skin injury, capillary blood enters the wound bed, and this is followed by a requirement to reduce continuous blood loss. Humoral and cellular components, such as fibrinogen and

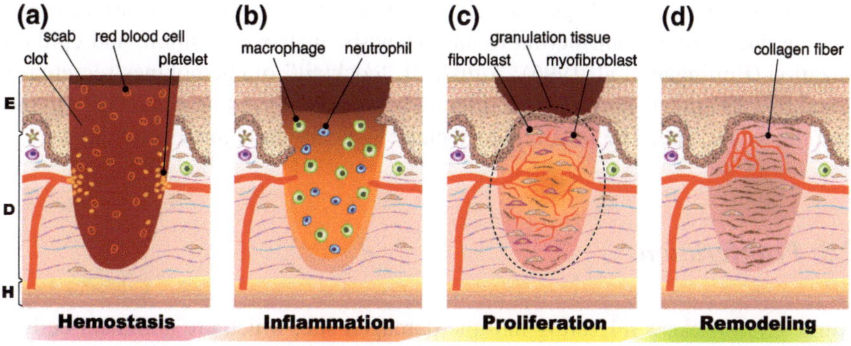

Fig. 2 Wound healing is a complex process encompassing a number of overlapping phases, including hemostasis, inflammation, proliferation, and remodeling. **a** Platelets adhere to damaged blood vessel walls. At this point, the platelets act to initiate blood clotting by activating coagulation factors. The resulting clot consists of platelets, red blood cells, and extracellular matrix molecules. **b** The inflammatory phase begins immediately and lasts for a few hours to a few days in acute wounds. Inflammatory cells such as phagocytic neutrophils and macrophages invade the clot, phagocytizing and triggering inflammatory response. Macrophages play an important role in subsequent angiogenesis, matrix deposition, reepithelialization, and fibroblast migration by secreting chemokines. **c** Migration and proliferation of fibroblasts and endothelial cells, which accompany reepithelialization, result in the formation of granulation tissue. Fibroblasts gradually replace the provisional matrix with a collagen-rich matrix and transform into myofibroblasts. **d** The transition from granulation to scar tissue occurs, leaving a collagen-rich scar tissue that is slowly remodeled in the following months under the wound surface that has been completely covered with a neoepidermis

platelets, stem blood loss and additionally provide signals that contribute to the earliest phases of wound healing. For arrest of bleeding, fibrinogen is activated in response to injured epithelium to form fibrin meshes that trap platelets, which adhere to the ruptures of blood vessels, preventing further blood loss. In addition, as platelets come into contact with damaged ECM components, they release coagulation factors, leading to formation of blood clots on the injured tissue. Various growth factors and chemokines such as platelet-derived growth factor (PDGF), vascular endothelial growth factor (VEGF), and transforming growth factor beta (TGF-beta), all of which have pivotal roles in the following phases, are rapidly released (Abe et al. 2001). This earliest phase is called hemostasis, and it occurs over a period of minutes to hours under normal circumstances.

2.2 *Inflammation*

The subsequent inflammatory phase is characterized by invasion of phagocytic neutrophils, the most abundant type of white blood cells in mammals, in the wound area (Enoch et al. 2006), usually within the first 24 h after injury. Neutrophils release chemotactic factors that attract monocytes to egress from blood vessels and enter injured tissue, in addition to phagocytosing foreign particles, bacteria and necrotic cell components. Macrophages as well as monocytes are essential for further wound healing due to their contribution to angiogenesis, matrix deposition, and epithelialization (Bellingan et al. 1996). A distinct "cocktail" of chemokines secreted by macrophages attracts fibroblasts to migrate from the surrounding undamaged tissues into the wound site. The inflammatory phase typically lasts for several days.

2.3 Proliferation

Attracted fibroblasts migrating into the damaged site, which are dominant cells in the wound area, start growing. Fibroblast proliferation is oxygen-dependent, and thus a sufficient peripheral oxygen partial pressure is necessary for adequate wound healing (Gordillo and Sen 2003). Extension of blood vessels into the injured tissue allows oxygen to be transported further into the wound region, and the extension is achieved as follows.

The blood vessel network in the wound region is formed through angiogenesis and vasculogenesis of endothelial cells. For the first step of angiogenesis, in which new blood vessels originate from preexisting ones, growth factors such as VEGF and TGF-beta stimulate endothelial cells in the nearby healthy tissue to release proteases such as matrix metalloproteinases (MMPs) into the wound area (Urlich et al. 2005). MMPs digest the basement membrane, allowing endothelial cells to escape from their parent vessel, elongate, and form a new capillary sprout extending away from the original vessel. Newly formed blood vessels enter the collagenous network, which is a major component of the ECM produced by fibroblasts and fibrocytes (circulating cells that rapidly enter sites of tissue injury and exhibit potent immune-stimulatory activities) at the wounded site, and form highly vascularized "granulation tissue".

Vasculogenesis, in contrast, is defined as the formation of new blood vessels by differentiation of endothelial progenitor cells (EPCs) usually circulating in the blood and also contributes to enhanced blood vessel growth during the wound healing process (Tepper et al. 2005; Velazquez 2007). Some distinct factors have been proposed to play a major role in mobilization and recruitment of EPCs to hypoxic and injured sites. These factors include cognate and non-cognate ligands to CXCR4 (a chemokine receptor), that is, CXCL12 and macrophage migration inhibitory factor (MIF) (Ceradini et al. 2004; Grieb et al. 2010).

Nutrients and oxygen provided through blood vessels extending into the injured tissue promote fibroblast proliferation. Fibroblasts generally differentiate further into myofibroblasts that generate a tensile strength across the wound, leading to contraction and further closure of the wound. Fibroblasts continue to secrete growth factors such as TGF-beta, PDGF and VEGF which activate the migration and proliferation of keratinocytes in the epidermis, creating the epithelial layer that covers the top of the wound. This process is accompanied by a constant transition from the proliferative phase into the next phase of remodeling in the wound-healing process.

2.4 Remodeling

The granulation tissue is remodeled into a relatively acellular fibrous scar tissue. Eventually, granulation tissue is converted into a scar. The scar differs from normal dermis: (1) the number of elastic fibers is reduced in scar tissue and (2) type I collagen fibers in the scar are broken by MMPs, and growth factors influence the amount of collagen available for cross-linking but do not affect the cross-linking process itself (Mast 1992). As the scar matures, the density of blood vessels returns to normal and the number of fibroblasts is reduced by apoptosis (Mast 1992). The remodeling phase can last for several months or even years. During this period, type I collagen fibers in the scar are crosslinked by the enzyme lysyl oxidase into thick bundles oriented parallel to the surface of the wound and the ECM is remodeled into a more mature structure with greater integrity. Due to myofibroblast contraction, complete wound closure occurs and the wound strength is increased from 20 % of normal tensile strength at 3 weeks after injury to about 80 % within 2 years in human skin (Levenson et al. 1965).

3 Wound Healing in the Mammalian Fetus

It is well known that skin wounds in early mammalian embryos can heal perfectly with no sign of scar tissue formation and complete restitution of the normal skin architecture (Whitby and Ferguson 1991). There are many differences between the healing process of embryonic and adult wounds. However, most of those differences do not appear to be the reason for embryonic scarless healing. Most differences merely stem from the differences in development itself. However, some differences, as overviewed below, have been shown to be involved in fetal scarless wound healing.

3.1 Inflammation

One of the most important differences between embryonic and adult wound-healing processes is the inflammatory response. The embryonic response in the number of inflammatory cells recruited to the wound is different from that of the adult response, and the embryo skin has less activated cells at the wound site and longer duration of inflammatory cells at the wound site (Hopkinson-Woolley et al. 1994; Cowin et al. 1998; Wulff et al. 2012). Therefore, the possibility that recruitment of inflammatory cells to a wound site affects scar formation prompted researchers to manipulate the recruitment of these cells. For example, sheep embryos that were artificially stimulated to produce an inflammatory response showed an adult-like response with scar formation (Ozturk et al. 2001). On the other hand, PU.1 null mice, which lack macrophages and functioning neutrophils, took more time to repair skin wounds than did wild-type siblings and could not repair the skin scarlessly (Martin et al. 2003). These studies suggest that an appropriate balance of inflammatory response is required for the wound healing.

3.2 Reepithelialization

After fetal skin is injured, surrounding epidermal cells rapidly move to fill the wound and to reepithelialize it. In this process, these cells are pulled forward by the contraction of actin fibers that draw the wound edges together as the opening of a purse is closed by a purse string (Martin and Lewis 1992). This process is unique to embryonic skin and different from that of adult wounds in which crawling of periwound epidermal cells resurface wounds, and the cells before crawling undergo retraction of intracellular tonofilaments and dissolution of most of the intercellular desmosomes that provide physical connections between the cells.

Epidermal cells are adherent to the extracellular matrix, which can change the speed of keratinocyte migration. Cell adhesion molecules such as fibronectin and tenascin and cell surface receptors such as integrins emerge earlier in healing fetal wounds than in adult wounds (Whitby and Ferguson 1991; Cass et al. 1998; Whitby et al. 1991). This can alter the phenotype of epidermal cells moving to fill in the wound, including composition of the underlying extracellular matrix can modify the speed of keratinocyte migration in the presence of growth factors (Nickoloff et al. 1988; Putnins et al. 1991).

3.3 Fibroblast Migration and Reorganization of the ECM

Fetal fibroblasts migrate to the wound site by chemoattractants derived from macrophages and neutrophils as adult fibroblasts do (see Sect. 2.2). Interestingly, fetal fibroblasts show a greater ability than adult fibroblasts to migrate, and it has been suggested that this is because fetal fibroblasts have more surface receptors for hyaluronic acid, which serves to enhance fibroblast migration (Chen et al. 1989). Their increased migration velocity during wound repair likely affects collagen deposition and distribution patterns. Fetal fibroblasts synthesize more type III and IV collagen than do their adult counterparts in vitro (Lorenz and Adzick 1993). Collagen type III fibers may allow a more reticular pattern of fiber deposition because they are smaller and finer than type I fibers, which predominate and are the principal component of both adult and fetal ECM. Collagen synthesis is slower in adult wounds than in fetal wounds and, therefore, adult fibroblasts probably have difficulty in synthesizing collagen in parallel with proliferation, whereas fetal fibroblasts simultaneously proliferate and synthesize collagen (Clark 1996). This difference in reorganization of the ECM gives rise to different tissue organization at the wound site that is involved in contraction and degree of scarring. In the adult wound site, granulation tissue, which plays a considerable role in wound contraction, is mainly composed of myofibroblasts that is characterized by smooth muscle features acquired in fibroblasts in granulation tissue. Myofibroblasts can also be detected temporarily in fetal wounds at earlier time points than in scarring wounds that have progressively more active myofibroblasts (Cass et al. 1997; Ellis and Schor 1996).

3.4 Growth Factors for Embryonic Wound Healing without Scar Formation

Growth factors and their receptors play a pivotal role in wound healing and sometimes lead to a number of aberrations associated with abnormal wound healing such as pathological scarring. Embryonic skin contains much higher levels of morphogenetic growth factors than does adult skin because an embryo is rapidly developing and growing with a considerable expansion of skin volume. Differences in inflammatory responses between an embryo and an adult also give rise to differences in levels of the growth factors derived from these cells. Consequently, the growth factor profile at a fetal wound site is very different from that at an adult wound site in quality, quantity and duration. For example, the TGF-beta family, which contains at least three isoforms known as TGF-beta-1, TGF-beta-2 and TGF-beta-3, is multifunctional and is believed important in both tissue repair and scarring. In an embryonic wound site, the levels of TGF-beta-1 and TGF-beta-2 were lower and TGF-beta-3 was higher than in an adult wound site (Martin et al. 1993; Cowin et al. 2001; Soo et al. 2003). It was shown that blocking TGF-beta-1 and TGF-beta-2 or the addition of exogenous TGF-beta-3 in an adult wound site can reduce scar formation (Shah et al. 1994 and 1995). However, in other studies using a different animal model, TGF-beta-3 had no effect in reducing scar formation (Wu et al. 1997). Wound healing, particularly that of the epidermis has long been studied as a part of limb regeneration because it is recognized as an indispensable process for limb regeneration (Werner and Richard 2003).

4 Relationship Between Wound Healing and Limb Regeneration in Amphibians

Amphibians, in particular urodele amphibians such as newts and salamanders, exhibit perfect regeneration of many organs, including the skin, and this remarkable ability would save the animals from their traumas over the course of their life. A newt, for example, can completely regenerate its tail, limbs, jaws and ocular tissues such as the lens. Unlike urodele amphibians, anuran amphibians such as frogs and toads have a limited regenerative ability that sometimes depends on their developmental stage. Regarding limb regeneration, *Xenopus laevis* can completely regenerate its developing limb buds before metamorphosis, but this regenerative capacity declines as metamorphosis proceeds (Dent 1962; Muneoka et al. 1986) The decline of regenerative capacity is due to intrinsic properties of limb cells rather than extrinsic properties such as neurotrophic factors or growth hormones (Sessions and Bryant 1988; reviewed by Suzuki et al. 2006). Despite the declined capacity of limb regeneration in the adult *Xenopus*, they can perfectly heal wounded skin, and the healing process from cutaneous traumas appears similar histologically to that in urodele amphibians (Suzuki et al. 2005; Yokoyama and Maruoka et al. 2011). We call the perfect wound healing in amphibians "skin regeneration" hereafter in this review.

The wound healing has long been studied as a part of limb regeneration because it can be recognized as a crucial and necessary event for limb regeneration.

4.1 Limb Regeneration in Amphibians

4.1.1 Immediate Reepithelialization and Subsequent Dedifferentiation

A hemostatic response occurs within a few seconds after amputation of a limb. Then the wound surface at the amputated site is covered with epidermal cells migrating from the circumference of the stump. Within 12 h after amputation, this wound-healing phase by the epidermis is thought to be achieved by cell movement with little cell division (Repesh and Oberpriller 1978; Carlson et al. 1998).

Over the next few days, this thin layer of epithelial cells thickens into a multilayered wound epithelium, called the apical epithelial cap (AEC). Underneath the AEC, mesenchymal cells around the wound site that have been saved from cell death enter the cell cycle and execute dedifferentiation. In this earliest process of limb regeneration, matrix metalloproteinases (MMPs) as well as AEC factors promote the conversion of mesenchymal cells to an undifferentiated state, so-called blastema cells (Yang and Bryant 1994). By the end of this process, a cone-shaped blastema, a cluster of blastema cells plus wound epithelium including the AEC, is formed, and this is the origin of tissue restoration, which induces and maintains limb regeneration (Muneoka and Sassoon 1992).

4.1.2 Redevelopment and Positional Memory for Repatterning

From 3 to 7 weeks after amputation, the basic pattern and main components of the limb are restored by the redevelopment phase, which includes re-differentiation and proliferation of the participating cell population and re-patterning by molecular mechanisms similar to those for developmental limb formation (Muneoka et al. 1992).

An important aspect of the re-development process is that each cell at the amputated site remembers its own position and what it should restore to regenerate a complete replica of the original limb structure. During limb development, for example, the transcriptional condition for a specific combination of gene expression in a cell successively progresses to form a skeletal pattern along the axes (reviewed by Tamura et al. 2010). The final state of the transcriptional condition in the genome is thought to be fixed and memorized but masked in a cell (probably by epigenetic regulation, reviewed by Yakushiji et al. 2009). Stimulation by limb amputation unveils the mask on the transcriptional condition and rewinds it into the final state that has been fixed in the cell. The program for patterning reprogresses from the middle (at the final state) to re-finish the development, restoring only the lost part of the pattern. To achieve this re-patterning from the amputated point, the genomic condition on position and pattern should not be erased after limb amputation, although the cell condition is initialized into an undifferentiated state. The molecular nature of this model remains unresolved, and it is essential to clarify the molecular mechanism of positional memory in amphibians for the purpose of successful organ regeneration in mammals.

4.1.3 Nerve Dependency of Blastema Formation

Nerve dependency is a characteristic of limb regeneration in amphibians. Nerve axons in the limb secrete neurotrophic factors into the blastema that are essential for its formation and growth, and surgical removal of axons in the limb therefore inhibits proper growth of the blastema, gives rise to no regenerate, and results in simple restoration of wounded skin (Stocum 2011).

Simple skin removal with little injury to nerve axons ends in wound healing, but if nerves are deviated into the site of the wound, outgrowth of a blastema-like structure can be induced (Bodemer 1959; Lheureux 1977; Reynolds et al. 1983; Maden and Holder 1984). Endo et al. (2004) clearly demonstrated that skin grafting from a different position of the limb with nerve deviation induces limb formation, suggesting a close relationship between wound healing and limb regeneration mediated by nerve signals in amphibians.

4.2 Perfect Wound Healing in Amphibian Skin

Amphibian skin is composed of two layers, a layered outer epidermis and a spongy inner dermis as in mammalian skin. The epidermis of an adult frog has a typical stratified squamous epithelium composed of germinative basal, spinous, granular, and cornified cells (Yoshizato et al. 2007). Many aspects of amphibian skin wound healing remain unclear, but some studies have suggested that the initial phase of wound healing in urodeles shares mechanisms with that in mammals.

Despite histological homologies of amphibian skin to mammalian skin and the common process of early healing, regenerative ability differs remarkably; mammalian wound healing results in a scar formation as described above (Ferguson and O'Kane 2004), while amphibian wound healing leads to a perfect restoration of tissue architecture and function (Levesque et al. 2010; Yokoyama and Maruoka et al. 2011; Seifert et al. 2012), although there is an exception (Poll 2009). The processes of wound healing in the metamorphosed *Xenopus* and axolotl are outlined here.

4.2.1 Immediate Hemostasis and Subsequent Reepithelialization

When the skin is injured, hemostasis begins within a few seconds as seen after limb amputation. Then the wound site is closed by epidermal tissue composed of two or three layers, referred to as wound epidermis, within several hours to one day by migration of epidermal cells using pseudopodial projections. Therefore, the epidermis is in direct contact with the subcutaneous muscle and connective tissue underneath (Levesque et al. 2010; Yokoyama and Maruoka et al. 2011). It should be noted that this reepithelialization in amphibians is very fast; a wound with a diameter of 1.5 mm in axolotl skin can be fully reepithelialized within 8 h (Levesque et al. 2010).

4.2.2 Dedifferentiation and Proliferation

Within 24 h after injury, in the case of metamorphosed *Xenopus*, the subcutaneous musculature underneath the wound epidermis is disrupted and a few mononuclear cells emerge there. Then, within 4 days after injury, the number of mononuclear cells greatly increase (Yokoyama and Maruoka et al. 2011). Similar mononuclear cells can also be seen in the surrounding dermal layer, and these appear to be the source of mesenchymal cells that repair the wound. These cells are eosin-negative, highly proliferative and express *Prx1* and *Tbx5*, reliable markers of blastema cells in limb regeneration (Suzuki et al. 2005; Satoh et al. 2007; Yokoyama and Maruoka et al. 2011). Furthermore, these mononuclear cells harboring the mouse *Prx1*-limb-enhancer as a transgene exhibit activation of the enhancer despite the fact that they are *Xenopus* cells, suggesting that the mouse *Prx1* enhancer possesses all elements required for its activation in skin wound healing. However, this activation of the same enhancer in the mouse is not seen at skin wound site of an adult mouse (Yokoyama and Maruoka et al. 2011).

4.2.3 Skin Regeneration

Within 10 days after injury, an almost normal dermis and well-organized muscle begin reappearing (Levesque et al. 2010; Yokoyama and Maruoka et al. 2011). In the case of the *Xenopus*, immature exocrine glands appear within this dermis, suggesting that skin derivatives are regenerated at this stage (Yokoyama and Maruoka et al. 2011). In the case of the axolotl, however, it takes more than 45 days for recovery of the basement membrane in the healing skin (Levesque et al. 2010). Eventually, within a couple of months, the wound is indistinguishable from the surrounding skin, and no scar forms, indicating that the skin wound healing is perfect in amphibians.

5 Perspectives on Perfect Wound Healing in Mammalian Adult Skin

As repeatedly discussed in this review, injured mammalian adult skin can be healed, but the wound healing is imperfect with scar/cicatrix formation and few skin derivatives. Although many researches have revealed cellular events and molecular mechanisms involved in the process of imperfect wound healing, the nature or cause of the imperfectness remains unclear. In contrast to the situation in the adult stage, embryonic skin can repair the wound without a scar. However, the repair appears not to be completely equivalent to that in the adult skin because the embryonic skin structure before wounding is immature, and embryonic wound healing therefore must include recovery of the immature skin and subsequent development into mature skin. Despite its nonequivalence, embryonic skin wound healing that does not form a scar is highly suggestive for successful perfect wound healing in the adult skin. Elucidating differences in molecular characteristics between embryonic and adult skin wound healing will help to ascertain the condition of scarless wound healing and regeneration of skin derivatives such as hairs and secretion glands.

On the other hand, in amphibians, wound healing of the skin is perfect, and the skin even in metamorphosed adults of both urodeles (Seifert et al. 2012) and anurans (Yokoyama and Maruoka et al. 2011) repairs the wound without scar. Wound healing must be a trigger of limb regeneration because an ectopic limb can be generated from a wound on the side of a urodele limb. Moreover, mesenchymal cells for skin wound healing and limb blastema mesenchymal cells share some molecular mechanisms, including *Prx1* activation, suggesting their close relationship. Although nerve signals are the key to connect wound healing to limb regeneration, *Prx1* activation is a nerve-independent event (though its maintenance is nerve-dependent) (Suzuki et al. 2007). Thus, regeneration of amphibian skin shares cellular and molecular features of the early events of limb regeneration but regenerates only skin in a nerve-independent manner.

The Prx1 molecule itself may contribute to scarless wound healing. In fact, Prx1 can directly activate tenascin-C expression in cultured fibroblasts (McKean et al. 2003). Since a high level of tenascin-C protein was observed in an axolotl skin wound, which can be healed in a scarless manner (Seifert et al. 2012), *Prx1* may regulate scarless skin wound healing through activation of tenascin-C expression. Elucidation of the role of the *Prx1* gene itself in amphibian skin wound healing will help to clarify the molecular mechanisms of scarless wound healing. Mouse adult skin cannot reactivate *Prx1*-limb-enhancer after being wounded, although the enhancer contains all elements required for skin regeneration because the mouse sequence is sufficient for reactivation during amphibian wound healing (Yokoyama and Maruoka et al. 2011). It is possible that fibrosis, which fulfills the injured skin region including the dermal layer in adult mammals, uses a distinct molecular mechanism for the initial step, which is totally different from that in amphibians. However, Levesque et al. (2010) showed that a fibrosis-like response can be induced also in axolotl skin by bleomycin treatment as in mammal skin. This finding suggests that amphibian skin is a comparable model to skin in adult mammals for studying fibriosis or scar formation. Shift of the molecular mechanism to that in amphibians, including reactivation of *Prx1*-limb-enhancer, will be an important clue for successful perfect wound healing in mammalian adult skin. In this sense, it would be interesting to examine the *Prx1*-enhancer activity in embryonic skin wound healing in mammals. Finding or creating *Prx1*-positive stem cells for skin regeneration in the adult mammal skin that have activated *Prx1*-limb-enhancer may be a fast and secure way for perfect wound healing, and this will directly lead to limb regeneration in mammals.

References

Abe R, Donnelly SC, Peng T, Bucala R, Metz CN (2001) Peripheral blood fibrocytes: differntiation pathway and migration to wound sites. J Immunol 166:7556–7562

Bellingan GJ, Caldwell H, Howie SE, Dransfield I, Haslett C (1996) In vivo fate of the inflammatory macrophage during the resolution of inflammation: inflammatory macrophages do not die locally, but emigrate to the draining lymph nodes. J Immunol 157:2577–2585

Bodemer CW (1959) Observations on the mechanism of induction of supernumerary limbs in adult Triturus viridescens. J Exp Zool 140:79–99

Carlson MR, Bryant SV, Gardiner DM (1998) Expression of Msx-2 during development, regeneration, and wound healing in axolotl limbs. J Exp Zool 282:715–723

Cass DL, Sylvester KG, Yang EY, Crombleholme TM, Adzick NS (1997) Myofibroblast persistence in fetal sheep wounds is associated with scar formation. J Pediatr Surg 32:1017–1021

Cass DL, Bullard KM, Sylvester KG, Yang EY, Sheppard D, Herlyn M, Adzick NS (1998) Epidermal integrin expression is upregulated rapidly in human fetal wound repair. J Pediatr Surg 33:312–316

Ceradini DJ, Kulkami AR, Callaghan MJ, Tepper OM, Bastidas N, Kleinman ME, Capla JM, Galiano RD, Levine JP, Gurtner GC (2004) Progenitor cell trafficking is regulated by hypoxic gradients through HIF-1 induction of SDF-1. Nat Med 10:858–864

Chen WY, Grant ME, Schor AM, Schor SL (1989) Differences between adult and foetal fibroblasts in the regulation of hyaluronate synthesis: correlation with migratory activity. J Cell Sci 94:577–584

Clark RAF (1996) The molecular and cellular biology of wound repair, 2nd edn. Plenum Press, New York, pp 427–474

Cowin AJ, Brosnan MP, Holmes TM, Ferguson MW (1998) Endogenous inflammatory response to dermal wound healing in the fetal and adult mouse. Dev Dyn 212:385–393

Cowin AJ, Holmes TM, Brosnan P, Ferguson MW (2001) Expression of TGF-beta and its receptors in murine fetal and adult dermal wounds. Eur J Dermatol 11:424–431
Dent JN (1962) Limb regeneration in larvae and metamorphosing individuals of the South African clawed toad. J Morphol 110:61–77
Ellis IR, Schor SL (1996) Differential effects of TGF-beta1 on hyaluronan synthesis by fetal and adult skin fibroblasts: implications for cell migration and wound healing. Exp Cell Res 228:326–333
Endo T, Bryant SV, Gardiner DM (2004) A stepwise model system for limb regeneration. Dev Biol 270:135–145
Enoch S, Grey JE, Harding KG (2006) Recent advances and emerging treatments. Br Med J 332:962–965
Ferguson MW, O'Kane S (2004) Scar-free healing: from embryonic mechanisms to adult therapeutic intervention. Philos Trans R Soc Lond B Biol Sci 359:839–850
Gordillo GM, Sen CK (2003) Revisiting the essential role of oxygen in wound healing. Am J Surg 186:259–263
Grieb G, Piatkowski A, Simons D, Hormann N, Dewor M, Steffans G, Bernhagen J, Pallua N (2010) Macrophage migration inhibitory factor is a potential inducer of endothelial progenitor cell mobilization after flap operation. Surgery 151:268–277
Hopkinson-Woolley J, Hughes D, Gordon S, Martin P (1994) Macrophage recruitment during limb development and wound healing in the embryonic and foetal mouse. J Cell Sci 107:1159–1167
Levenson SM, Geever EF, Vrowley LV, Oates JF 3rd, Berard CW, Rosen H (1965) The healing of rat skin wounds. Ann Surg 161:293–308
Lévesque M, Villiard E, Roy S (2010) Skin wound healing in axolotls: a scarless process. J Exp Zool B Mol Dev Evol 314:684–697
Lheureux E (1977) Importance of limb tissue associations in the development of nerve-induced supernumerary limbs in the newt Pleurodeles waltlii Michah. J Embryol Exp Morphol 38:151–173
Lorenz HP, Adzick NS (1993) Scarless skin wound repair in the fetus. West J Med 159:350–355
Malcolm M, Holder N (1984) Axial characteristics of nerve induced supernumerary limbs in the axolotl. Roux's Arch.\ Dev Biol 193:394–401
Martin P, Lewis J (1992) Actin cables and epidermal movement in embryonic wound healing. Nature 360:179–183
Martin P, Dickson MC, Millan FA, Akhurst RJ (1993) Rapid induction and clearance of TGF beta 1 is an early response to wounding in the mouse embryo. Dev Genet 14:225–238
Martin P, D'Souza D, Martin J, Grose R, Cooper L, Maki R, McKercher SR (2003) Wound healing in the PU.1 null mouse–tissue repair is not dependent on inflammatory cells. Curr Biol 13:1122–1128
Mast BA (1992) The skin. In: Cohen IK, Diegelmann RF, Lindblad WJ (eds) Wound Healing: Biochemical and Clinical Aspects. WB Saunders, Philadelphia, pp 344–355
Matoltsy AG, Downes AM, Sweeney TM (1968) Studies of the epidermal water barrier. II investigation of the chemical nature of the water barrier. J Invest Dermatol 50:19–26
McKean DM, Sisbarro L, Ilic D, Kaplan-Alburquerque N, Nemenoff R, Weiser-Evans M, Kern MJ, Jones PL (2003) FAK induces expression of Prx1 to promote tenascin-C-dependent fibroblast migration. J Cell Biol 161:393–402
Mukouyama YS, Shin D, Britsch S, Taniguchi M, Anderson DJ (2002) Sensory nerves determine the pattern of arterial differentiation and blood vessel branching in the skin. Cell 109:693–705
Muneoka K, Holler-Dinsmore G, Bryant SV (1986) Intrinsic control of regenerative loss in Xenopus laevis limbs. J Exp Zool 240:47–54
Muneoka K, Sassoon D (1992) Molecular aspects of regeneration in developing vertebrate limbs. Dev Biol 152:37–49
Nickoloff BJ, Mitra RS, Riser BL, Dixit VM, Varani J (1988) Modulation of keratinocyte motility. Correlation with production of extracellular matrix molecules in response to growth promoting and antiproliferative factors. Am J Pathol 132:543–551

Oztürk S, Deveci M, Sengezer M, Günhan O (2001) Results of artificial inflammation in scarless foetal wound healing: an experimental study in foetal lambs. Br J Plast Surg 54:47–52

Poll CP (2009) Wound Management in Amphibians: etiology and Treatment of Cutaneous Lesions. J Exotic Pet Med 18:20–35

Putnins EE, Firth JD, Lohachitranont A, Uitto VJ, Larjava H (1999) Keratinocyte growth factor (KGF) promotes keratinocyte cell attachment and migration on collagen and fibronectin. Cell Adhes Commun 7:211–221

Repesh LA, Oberpriller JC (1978) Scanning electron microscopy of epidermal cell migration in wound healing during limb regeneration in the adult newt, Notophthalmus viridescens. Am J Anat 151:539–555

Reynolds S, Holder N, Fernandes M (1983) The form and structure of supernumerary hindlimbs formed following skin grafting and nerve deviation in the newt Triturus cristatus. J Embryol Exp Morphol 77:221–241

Satoh A, Gardiner DM, Bryant SV, Endo T (2007) Nerve-induced ectopic limb blastemas in the axolotl are equivalent to amputation-induced blastemas. Dev Biol 312:231–244

Seifert AW, Monaghan JR, Voss SR, Maden M (2012) Skin regeneration in adult axolotls: a blueprint for scar-free healing in vertebrates. PLoS ONE 7(4):e32875

Sessions SK, Bryant SV (1988) Evidence that regenerative ability is an intrinsic property of limb cells in Xenopus. J Exp Zool 247:39–44

Shah M, Foreman DM, Ferguson MW (1994) Neutralising antibody to TGF-beta 1,2 reduces cutaneous scarring in adult rodents. J Cell Sci 107:1137–1157

Shah M, Foreman DM, Ferguson MW (1995) Neutralisation of TGF-beta 1 and TGF-beta 2 or exogenous addition of TGF-beta 3 to cutaneous rat wounds reduces scarring. J Cell Sci 108:985–1002

Soo C, Beanes SR, Hu FY, Zhang X, Dang C, Chang G, Wang Y, Nishimura I, Freymiller E, Longaker MT, Lorenz HP, Ting K (2003) Ontogenetic transition in fetal wound transforming growth factor-beta regulation correlates with collagen organization. Am J Pathol 163:2459–2476

Stocum DL (2011) The role of peripheral nerves in urodele limb regeneration. Eur J Neurosci 34:908–916

Suzuki M, Satoh A, Ide H, Tamura K (2005) Nerve-dependent and -independent events in blastema formation during Xenopus froglet limb regeneration. Dev Biol 286:361–375

Suzuki M, Yakushiji N, Nakada Y, Satoh A, Ide H, Tamura K (2006) Limb regeneration in Xenopus laevis froglet. Sci World J 6:26–37

Suzuki M, Satoh A, Ide H, Tamura K (2007) Transgenic Xenopus with prx1 limb enhancer reveals crucial contribution of MEK/ERK and PI3 K/AKT pathways in blastema formation during limb regeneration. Dev Biol 304:675–686

Tamura K, Ohgo S, Yokoyama H (2010) Limb blastema cell: a stem cell for morphological regeneration. Dev Growth Differ 52:89–99

Tepper OM, Capla JM, Galiano RD, Ceradini DJ, Callaghan MJ, Kleinman ME, Gurtner GC (2005) Adult vasculogenesis occurs through in situ recruitment, proliferatioon, and tubulization of circulating bone marrow-derived cells. Blood 105:1068–1077

Ulrich D, Lichtenegger F, Unglaub F, Smeets R, Pallua N (2005) Effect of chronic wound exudates and MMP-2/-9 inhibitor on angiogenesis in vitro. Plast Reconstr Surg 116:539–545

Velazquez OC (2007) Angiogenesis and vasculogenesis: inducing the growth of new blood vessels and wound healing by stimulation of bone marrow-derived progenitor cell mobilization and homing. J Vasc Surg 45:A39–A47

Werner S, Richard G (2003) Regulation of wound healing by growth factors and cytokines. Physiol Rev 83:835–870

Whitby DJ, Ferguson MW (1991) The extracellular matrix of lip wounds in fetal, neonatal and adult mice. Development 122:651–668

Whitby DJ, Longaker MT, Harrison MR, Adzick NS, Ferguson MW (1991) Rapid epithelial-isation of fetal wounds is associated with the early deposition of tenascin. J Cell Sci 99:583–586

Wong VW, Sorkin M, Glotzbach JP, Longaker MT, Gurtner GC (2011) Surgical approaches to create murine models of human wound healing. J Biomed Biotechnol 969618. Epub 1 Dec 2010
Wu L, Siddiqui A, Morris DE, Cox DA, Roth SI, Mustoe TA (1977) Transforming growth factor beta 3 (TGF beta 3) accelerates wound healing without alteration of scar prominence. Histologic and competitive reverse-transcription-polymerase chain reaction studies. Arch Surg 132:753–760
Wulff BC, Parent AE, Meleski MA, DiPietro LA, Schrementi ME, Wilgus TA (2012) Mast cells contribute to scar formation during fetal wound healing. J Invest Dermatol 132:458–465
Yang EV, Bryant SV (1994) Developmental regulation of a matrix metalloproteinase during regeneration of axolotl appendages. Dev Biol 166:696–703
Yakushiji N, Yokoyama H, Tamura K (2009) Repatterning in amphibian limb regeneration: a model for study of genetic and epigenetic control of organ regeneration. Semin Cell Dev Biol 20:565–574
Yokoyama H, Maruoka T, Aruga A, Amano T, Ohgo S, Shiroishi T, Tamura K (2011) Prx-1 expression in Xenopus laevis scarless skin-wound healing and its resemblance to epimorphic regeneration. J Invest Dermatol 131:2477–2485
Yoshizato K (2007) Molecular mechanism and evolutional significance of epithelial-mesenchymal interactions in the body- and tail-dependent metamorphic transformation of anuran larval skin. Int Rev Cytol 260:213–260

Part II
Limb and Lens Regeneration

Cellular Plasticity During Vertebrate Appendage Regeneration

James R. Monaghan and Malcolm Maden

Abstract Many vertebrates have the amazing ability to regenerate all or portions of appendages including limbs, tails, fins, and digits. Unfortunately, our understanding of the cellular and molecular basis of appendage regeneration is severely lacking. However, recent technological advances that facilitate the tracking of cell lineages in vivo through space and time are allowing us to address the unknowns of regeneration, such as characterizing the cells that contribute to regeneration and identifying the tissues these cells differentiate into during regeneration. Here, we describe the experiments and the surprisingly uniform results that have emerged across diverse vertebrate species when specific cell lineages have been tracked during vertebrate appendage regeneration. These investigations show that vertebrates, from zebrafish to salamanders to mammals, utilize a limited amount of cellular plasticity to regenerate missing appendages. The universal approach to appendage regeneration is not to generate pluripotent cells that then differentiate into the new organ, but instead to generate lineage-restricted cells that are propagated in a progenitor-like state. Lessons learned from these natural cases of complex tissue regeneration might inform regenerative medicine on the best approach for re-growing complex tissues.

J. R. Monaghan (✉)
Department of Biology, Northeastern University, Boston, MA 02115, USA
e-mail: j.monaghan@neu.edu

M. Maden (✉)
Department of Biology and UF Genetics Institute, University of Florida, Gainesville, FL 32610, USA
e-mail: malcmaden@ufl.edu

Current Topics in Microbiology and Immunology (2013) 367: 53–74
DOI: 10.1007/82_2012_288

© Springer-Verlag Berlin Heidelberg 2012
Published Online: 13 December 2012

Contents

1 Introduction

The aim of regenerative medicine is to replace diseased, amputated, or damaged organs by therapeutic intervention. The future success of regenerative medicine will be fueled by regenerative biology, a field that studies tissue growth at its most fundamental levels, merging the disciplines of stem cell biology, developmental biology, cell biology, and bioengineering. Specifically, regenerative biology aims to combine the knowledge gained across these disciplines, providing the means to regenerate tissues and organs by directing cellular growth and differentiation in a controlled manner. Despite considerable progress in each individual field, our understanding of how to integrate this information to provide fundamental insights into the cellular and molecular basis of organ regeneration is still severely lacking. However, one recent technological advance that is beginning to change this is the ability to track and manipulate cells in vivo through space and time. Here, we describe the experiments and the surprisingly uniform results that have emerged across diverse vertebrate species when this technology is applied to the study of vertebrate appendage regeneration.

Vertebrate appendages such as limbs, tails, and fins are a particularly attractive model for understanding complex tissue regeneration because of their accessibility. Multiple vertebrates are capable of regenerating appendages, which is surprising considering that they are in direct contact with the harsh external environment, can be relatively large compared to the size of the animal, and are morphologically complex both in tissue type and morphology (Fig. 1). Here, we review the recent flurry of investigations that have attacked the problem of cellular plasticity during vertebrate appendage regeneration, while keeping the simple question in mind: do the cellular processes that regulate natural examples of

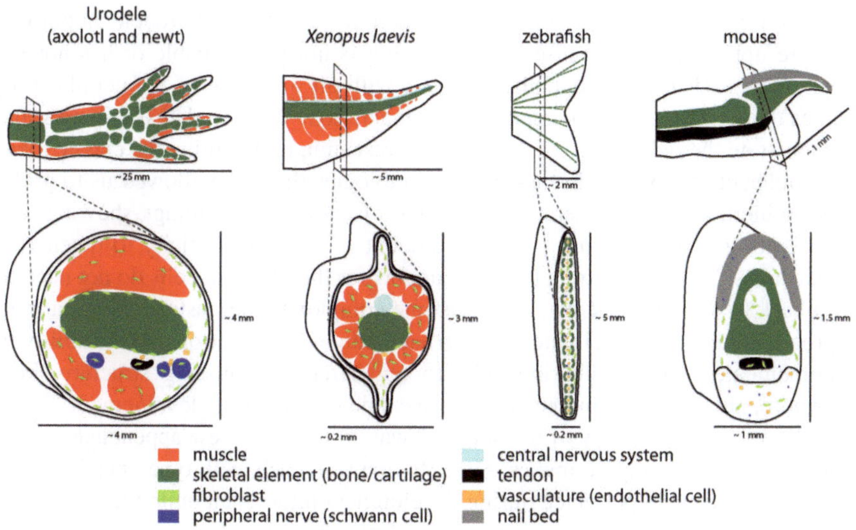

Fig. 1 Images of vertebrate appendages capable of regeneration and have techniques available to study cellular lineage. Each animal group's appendage is represented by a longitudinal cartoon and cross-section of the amputation stump, highlighting the tissues and cell types present in each appendage. The approximate sizes of appendages are presented to give an indication of the capabilities of tissue regeneration capable in each animal group

appendage regeneration deviate from the development of that appendage? We focus specifically on examples of vertebrate appendage regeneration because of their direct relevancy to the goals of regenerative medicine, highlighting our current understanding of cellular plasticity, and how this information provides a unified model of vertebrate appendage regeneration.

Conceptually, there are multiple ways to regrow an appendage. It may take place by the activation of stem cells localized throughout the body, which are called into action upon appendage amputation and then travel to the site of damage. Alternatively, cells located near the injury sight may become activated and liberate their potential to regrow the appendage. It has been known for decades that the latter is how vertebrate appendages regenerate, but our understanding of how these cells go from a resting state to a re-growing state is unclear. For example, do appendages regenerate using local pluripotent stem cells or can all cell types dedifferentiate and take part in regeneration? How much cellular plasticity is needed to regrow an appendage? Is appendage regeneration the same as appendage development?

We focus on three animal groups capable of appendage regeneration, all of which have modern techniques available to track cell lineage through space and time: the amphibians (limbs and tails), teleost fish (caudal fins), and mammals (digit tips) (Fig. 1). Unfortunately, important examples of vertebrate appendage regeneration like the lizard tail, catfish barbell, deer antler, and urodele amphibian jaw are not discussed because lineage tracing is not yet available or has not yet been addressed in these animals. We start with an overview of the cellular processes that typify vertebrate appendage regeneration. We include experiments that demonstrate the rich history of regeneration research, highlighting the classic ideas of cellular plasticity and discuss the methodologies that have allowed us to probe this problem during regeneration. We then discuss recent findings showing that regenerating vertebrates use a limited amount of cellular plasticity to regenerate appendages. There is a strong basis of support for the idea that cells do not need to turn back to a pluripotent state to regenerate complex tissue, instead they return to a progenitor state that facilitates an increased rate of proliferation and the ability to communicate with neighboring cells to regenerate the missing structure. Furthermore, animals do so by reusing the rules laid down during development. When one thinks parsimoniously of the simplest way to regrow a new appendage from a complex tissue stump it makes sense to expand each cell type locally using mechanisms already perfected during development rather than build the new organ *de novo* from a homogeneous population of pluripotent cells.

2 Appendage Regeneration Across Vertebrates

Natural cases of regeneration can be found in two general forms, physiological regeneration and reparative regeneration. Physiological regeneration is the continuous replacement of old or damaged tissues throughout life, which is widespread across vertebrates. Examples include the continuous turnover by proliferation of resident progenitor cells found in the skin, gut epithelium, blood cells, and bone. Reparative regeneration, in contrast, is the replacement of organs due to damage or disease. Reparative regeneration is less common across vertebrates and the mechanisms underlying this process are less well understood. Nevertheless, there are some striking examples of appendage regeneration among vertebrates. For example, many amphibians can regenerate limbs and tails after amputation, teleost fish can regenerate portions of their caudal fins, lizards can regenerate their tail, deer antlers regenerate yearly, and even humans can regenerate the distal portion of their digit tip (Goss 1969). The burning question that needs to be addressed is whether there are common mechanisms that regulate each of these examples of regeneration.

Appendage regeneration is so striking because limbs, tails, and jaws are externally visible for all to see and because the progress of regeneration can be very simply visualized and recorded in real time. Curious onlookers have described regeneration, since the time of antiquity (Dinsmore 1991) and over the centuries have come to the conclusion that several common events occur in all vertebrates capable of regeneration. First, amputation is followed by wound closure and some investigators have observed that wounds close more rapidly in animals capable of regeneration compared to non-regenerating animals; e.g. zebrafish (<24 h)(Akimenko et al. 2003), Urodele amphibians (<24 h), and mammalian fetal wound healing are examples of rapid wound closure in regenerating animals compared to much slower closure in human wounds. Rapid wound closure must not be necessary for regeneration though, because many animals capable of regeneration close wounds at a slow pace, including the gecko tail (3–8 days; McLean and Vickaryous 2011), Anolis lizard tail (7–9 days; Cox 1969), adult mice digit tips (6–8 days; Han et al. 2003), and macaque monkey digit tips (7–10 days; Singer et al. 1987). It is possible, however, that rapid wound closure may regulate regeneration rate. Seifert et al. showed that flank wounds inflicted in metamorphosed axolotl salamanders re-epithelialize within 72 h compared to 24 h in aquatic axolotls, which correlated with a slower regeneration rate in metamorphosed animals (Seifert et al. 2012). It is clear that a functional wound epidermis is necessary for appendage regeneration across vertebrates because regeneration is inhibited if the epidermis is removed during salamander limb regeneration (Goss 1956a; Stocum and Cameron 2011) and if the clot at the end of the regenerating mouse digit is removed (Lehoczky et al. 2011). Furthermore, zebrafish fin regeneration is inhibited if the signaling pathways involved for epithelial-mesenchymal communication are blocked including fibroblast growth factor (Poss et al. 2000), insulin growth factor (Chablais and Jazwinska 2010), retinoic acid signaling (Blum and Begemann 2012), fibroblast growth factor 20 (Whitehead et al. 2005), and sdf-1 (Dufourcq and Vriz 2006).

Second, regenerating appendages have a specialized structure similar to a developing limb or tail bud called a blastema, which is an accumulation of proliferating mesenchymal cells located beneath the wound epidermis. The blastema is made up of locally derived cells that are generated from dedifferentiation of mature cell types or from activation of local stem cells. The blastema is the "functional unit" of the regenerating appendage, containing all the cell types necessary to generate the missing appendage, although it is unclear whether the cellular make-up of blastemas is equivalent across vertebrates. The best characterized blastema, that of the salamander limb, was shown to be a self-organizing structure because transplantation of blastemas to neutral locations on the body generated entire limbs from the graft tissue in the ectopic location (reviewed by Stocum 1984). Because it is crucial for us to know how a blastema is generated, the last century of appendage regeneration research has been dominated by investigations attempting to understand where blastemal cells come from, how they communicate with one another to organize morphogenesis, and what they turn into in the regenerated structure (for review Stocum and Cameron 2011). All three of these problems have yet to be solved in any model of appendage regeneration, although the recent major advances in the origin of blastemal cells are described below and when we do, the question remains whether it will be possible to mimic nature's strategy of appendage regeneration for human applications.

3 Amphibian Limb and Tail Regeneration

3.1 Blastemal Cells Are Locally Derived and Not Pluripotent

Amphibian appendage regeneration is the classic model for vertebrate regeneration and has undergone detailed scrutiny for well over a century (reviewed by Wallace 1981; Dinsmore 1991). Because of this rich history, our understanding of the origin of blastemal cells and what they become during regeneration is well supported in the literature. Experiments up to the mid-twentieth century were confined to a limited arsenal of methods: namely histological analysis, extirpation of tissues, irradiation to block cell proliferation, grafting of polyploid or haploid cells between donors and hosts, and lineage labeling of dividing cells using H^3 thymidine. Even within these confines, much was learned before the advent of modern molecular biology about the source and plasticity of cells during regeneration.

At the beginning of experimental biology, the preformationist Charles Bonnet, argued that all animals, including those that can regenerate, contained "germs" that circulate through the body, and that these must therefore give rise to the regenerated structures (Liversage 1991). As early as 1766, in a correspondence from Lazarro Spallanzani to Bonnet, Spallanzani cast doubt on the role of "germs" in regeneration because new tissues such as vasculature, intestines, muscles, and nerves looked like extensions of the old version (translation by Tsonis and Fox 2009). The first experimental evidence for the cell source of blastemal cells being locally derived was provided over a century later, when Hertwig grafted haploid limbs to diploid hosts and observed that the blastemas formed after amputation of the grafts contained only haploid cells (Hertwig 1927). The final blow against "germ cells" was provided by Butler, who used the technique of irradiation to block cell proliferation in reciprocal experiments. In one design, a narrow strip of the knee region of the salamander hindlimb was irradiated and the rest of the body was shielded from irradiation. In the converse design, the narrow knee strip was shielded and the rest of the body irradiated. When amputation was performed through the irradiated portions of the limb, regeneration did not take place, while the limb regenerated when amputated through the non-irradiated regions even though this was surprisingly narrow. This clearly demonstrated that all the cells necessary to regenerate a limb are located approximately 1–2 mm proximal to the amputation surface and discounted the theory that remote cell sources are necessary for limb regeneration (Butler 1933, 1935; Butler and O'Brien 1942). More recently, transgenics have been used to show that grafts of GFP-positive blood anlage to non-transgenic axolotl host embryos do not contribute circulating blood cells to local regenerating tissues (Sobkow et al. 2006). Similar results were found during axolotl tail regeneration, showing that the cell source for spinal cord regeneration is within 500 μm of the amputation plane (McHedlishvili et al. 2007). A multitude of detailed histological studies supported these findings, concluding that amphibian blastemas arise from local tissues including muscle, cartilage, connective tissue, spinal cord tissue (tail only), Schwann cells, and epidermis (Manner 1953; Chalkey 1954; Hay 1959; Maden 1977).

At the gross histological level, the blastema appears to be a homogeneous population of cells, morphologically similar to embryonic cells. Based upon the similarity to embryonic cells, it is often suggested that blastemal cells have a heightened cellular plasticity, possibly to an embryonic state. This issue is important in terms of regenerative medicine, because the current strategy for repairing human tissue is to generate pluripotent cells as a cell source for regeneration and then differentiating them either in vivo or in vitro to the required cell type. If this assumption is true, blastema cells should act like blastula/gastrula-stage cells, which are capable of generating cloned animals following nuclear injection into enucleated host eggs (Briggs and King 1952; Signoret et al. 1962). To test this hypothesis, Burgess injected nuclei of gastrula-stage endoderm cells, limb blastemal cells, or differentiated epidermal cells into enucleated, activated host *Xenopus laevis* eggs. She showed that 29.1 % of transplanted gastrula-stage endoderm nuclei generated feeding stage tadpoles, while transplanted blastemal cell nuclei collected from the regenerating hindlimb as well as differentiated epidermal cell nuclei did not give rise to cloned animals (Burgess 1967). Similar results were obtained in axolotls, showing that the ability to generate cloned animals significantly decreases during development and a blastemal cell's potential to become totipotent is no higher than other somatic cells (Briggs et al. 1964; Dasgupta 1970; Gurdon and Byrne 2003). Therefore, blastemal cells are not fundamentally different from other somatic cells in the body except that they can proliferate, pattern, and differentiate into a new organ.

3.2 Are Blastemal Cells Multipotent-Epidermis and Cartilage/Fibroblasts?

Although blastemal cells may not be equivalent to embryonic stem cells or germ cells, these studies did not address the question of whether blastemal cells become multipotent (rather than pluripotent) during regeneration. The major problem in conducting the original cellular plasticity experiments was in generating a pure source of tissue for grafting and so the question was originally posed only for epidermis and cartilage, which can be dissected clean, whereas muscle, for example, cannot and contains many different cell types. With regard to the epidermis which covers the blastema, the answer to metaplasia is unequivocally no. Hay and Fischman were the first to conclusively show that the epidermis does not transdifferentiate into mesodermal tissues during normal limb regeneration by tracking H^3 thymidine-labeled epidermis during regeneration (Hay and Fischmann 1961). Furthermore, triploid mesodermal tissues grafted into haploid irradiated limbs could not be detected in the epidermis clearly showing no metaplasia between mesoderm and epidermis (Hay and Fischmann 1961; Namenwirth 1974).

With regard to the fibroblast/cartilage mesodermal lineage, however, there is a clear but limited degree of plasticity. Experiments to test this concept have been

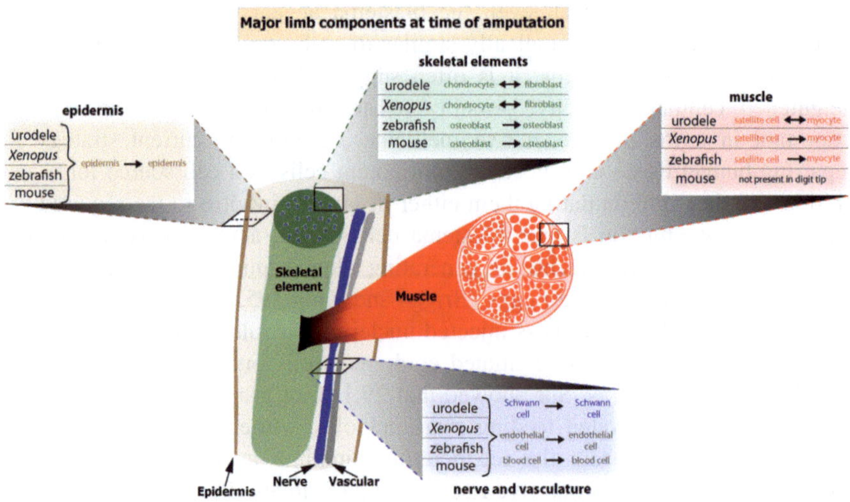

Fig. 2 Tissue lineage restrictions during appendage regeneration in urodeles, *Xenopus*, zebrafish, and mice. Notice the high degree of similarity between each animal group. Skeleton, muscle, nerve fibers, vasculature, and epidermis are represented in the cartoon. A *double-sided arrow* indicates that transdifferentiation can take place between cell types. A *single-sided arrow* indicates that differentiation has been observed only in one direction. Note that the zebrafish caudal fin does not contain muscle fibers (Fig. 1), although Rodrigues et al. showed that satellite cells generate new muscle rather than dedifferentiation if amputations are performed in a muscle-containing region proximal to the caudal fin. Also, mouse digit tips do not contain muscle fibers, although it is known that satellite cells are the sole progenitors of myocytes rather than dedifferentiation (Le Grand and Rudnicki 2007)

performed in two ways. First, during *normal* regeneration by grafting in replacement tissues labeled in some fashion to address what normally happens during regeneration. Second, in a more demanding regime whereby the stump is irradiated or individual tissue types are removed from the stump and labeled grafts are inserted to ask whether plasticity can be forced upon blastemal cells. Using triploid or radioactively labeled limb cartilage grafted into regenerating diploid limbs, both Patrick and Briggs (1964) and Steen (1968) showed that cartilage almost exclusively differentiates into new cartilage during normal regeneration (93 % of triploid cells were chondrocytes and the rest were 'undifferentiated' or 'other cell types'). When triploid cartilage is grafted into irradiated limbs there is no obvious increase in the degree of plasticity shown even though some triploid cells are now seen in the joints and connective tissue fibroblasts (Namenwirth 1974). Therefore, during either normal or 'forced' limb regeneration, little metaplasia of cartilage occurs, and when it does occur, it is into the fibroblastic lineage (Fig. 2).

Based upon these studies, it was therefore very surprising to find that when resident skeletal elements are completely extirpated before amputation, a perfect skeleton nevertheless regenerates from the depleted amputation plane, suggesting a huge amount of metaplasia from a non-skeletal source (Bischler 1926; Thornton 1938b; Goss 1956b; Steen 1968; Foret 1970). A similar result was found in the anuran *X. laevis*, which regenerates a cartilaginous spike after forelimb amputation. In these studies, it was suggested that non-chondrogenic cells, possibly fibroblasts differentiated into cartilage during amphibian appendage regeneration (Korneluk and Liversage 1984; Inoue et al. 1988). The potential of fibroblasts to generate cartilage was the most parsimonious explanation for the universal finding that irradiated limbs could generate perfect regenerates containing all the cartilage elements when mixed tissues such as muscle, dermis or even nerves were used for grafting and this conclusion was verified in triploid experiments (Namenwirth 1974; Dunis and Namenwirth 1977; Lheureux 1983). Indeed, during normal regeneration there is an overwhelming contribution of dermal fibroblasts to the blastema. On average, 43 % of blastemal cells arise from the dermal fibroblasts, but they only represent 19 % of the cells in the stump, whereas only 2 % of blastemal cells arise from the cartilage (Muneoka et al. 1986).

These conclusions concerning the metaplasia of fibroblasts have now been fully confirmed with the latest transgenic lines. Using a transgenic line of GFP-expressing axolotls (Sobkow et al. 2006), Tanaka's group showed that during normal limb regeneration muscle, Schwann cells, and epidermis are lineage-restricted tissues, whereas connective tissue fibroblasts can transdifferentiate into cartilage (Kragl et al. 2009). They further revealed that many erroneous results obtained from previous salamander grafting studies were due to impure grafts containing fibroblasts. Identical results were obtained using the axolotl accessory limb model, where an additional limb is induced on the flank of an intact limb by deviating a nerve and skin graft from the contralateral side of a GFP donor limb (Hirata et al. 2010). In this study, a GFP-labeled skin graft gave rise to epidermis, cartilage, tendons, and connective tissues, but did not generate muscle or Schwann cells.

3.3 Are Blastemal Cells Multipotent-Muscle?

The limitations of the triploid approach, and the mixed nature of a muscle graft left open the possibility of transdifferentiation between muscle and cartilage as a way of explaining why complete limbs could be generated from a muscle graft onto an irradiated stump. However, the converse transformation, fibroblasts to muscle, clearly does not happen because skin grafted onto an irradiated limb generates muscle-less regenerates (Holder 1989). Not until transgenics became available in amphibians was this problem resolved. Ryffell et al. generated and crossed a muscle-specific *X. laevis* cre driver line to a fluorescent cre recombination reporter line to show that no transdifferentiation took place between muscle and other tissues during larval tail regeneration (Ryffel et al. 2003). This work was corroborated by grafting tissue primordia between GFP-expressing transgenic *X. laevis* embryos to wild-type hosts to generate animals with only GFP-positive spinal cord, muscle, or skeleton. After tail amputation, each GFP tissue gave rise to only their particular cell type, demonstrating no metaplasia between tissues (Gargioli and Slack 2004). Similarly, when axolotls are created with GFP muscle in their limbs by grafting presomitic mesoderm at embryonic stages between GFP and white animals, no GFP- positive cells are seen either in the cartilage or epidermis of the regenerate (Kragl et al. 2009). With regard to muscle, therefore, it is clear that this is a lineage-restricted tissue (Fig. 2).

3.4 Similarity to Development

The studies described above reveal that during normal amphibian appendage regeneration there is no transdifferentiation of muscle and a limited degree of transdifferentiation between fibroblasts and cartilage which can be increased if, for example, cartilage is removed from the stump prior to regeneration. Furthermore, these fibroblasts contain all the patterning information needed to generate a limb. When one thinks of the origin of tissues during limb development, these findings are not so surprising as at first sight. The developing limb bud arises from the lateral plate mesoderm as a mass of mesenchymal tissue, which subsequently differentiates into patterned skeletal units through endochondral ossification (differentiation into cartilage and then bone). Therefore, transdifferentiation of fibroblasts to limb cartilage is very similar to skeletal formation during development. The limb musculature, however, arises from an entirely different cellular lineage. It differentiates after the limb bud has formed, and enters the limb as migrating muscle progenitors that originate in the hypaxial dermomyotome of the somites (Gros et al. 2005; Relaix et al. 2005), and responds to patterning signals emanating from the limb bud (Bladt et al. 1995; Hayashi and Ozawa 1995). Similarly, Schwann cells and melanocytes are derived from neural crest migrating from the dorsal neural tube and endothelial cells migrate from the vasculature in the trunk, both arising from primordia outside the limb. Therefore, switching from muscle or Schwann cells to connective tissue/skeletal tissue would greatly deviate from normal limb development.

Although strong evidence described above suggests that most cells are lineage-restricted during amphibian appendage regeneration, it is important to mention that multiple investigations have concluded that transdifferentiation occurs between connective tissue, muscle, cartilage, and epidermis. Oberpriller (1967) and Tuchkova (1973) concluded that in newts, intestinal epithelium transdifferentiates into limb cartilage after they grafted H^3 thymidine-labeled intestinal blastemal cells into unlabeled limb blastemas. Others have made similar conclusions after grafting labeled peripheral nerves (Wallace 1972; Wallace and Wallace 1973), muscle (Thornton 1942; Steen 1968), tail fin connective tissue (Steen 1970), and heart tissue (Laube et al. 2006) into the limb blastema. Considering that all of these tissues contain connective tissue fibroblasts, it is likely that the fibroblasts, and not the muscle, Schwann cells, or epidermis are giving rise to connective tissue and cartilage in the host limbs. Similarly, others have electroporated GFP-expressing DNA into the axolotl spinal cord and concluded that neural cells give rise to large numbers of cartilage and muscle cells during tail regeneration (Echeverri and

Tanaka 2002), although more recent studies using transgenic methods have shown that little metaplasia takes place between spinal cord cells and muscle or cartilage (McHedlishvili et al. 2007, 2012). It is possible that electroporation may change certain cell properties that increase cell fusion events, cellular plasticity, or cell death (leading to phagocytosis), giving the impression of metaplasia. Lastly, it has been suggested in newts that muscle satellite cells can contribute to cartilage (and even epidermis) by labeling cultured myofibers or the cells that bud off of myofibers with fluorescent dextran (Lo et al. 1993), BrdU (Morrison et al. 2006), and adenovirus (Morrison et al. 2010).

Discrepancies between cellular plasticity studies may be due to a combination of factors. The surprising result of transdifferentiation of BrdU-labeled muscle fibers into epidermis (Morrison et al. 2006) could be due to the epidermis phagocytosing BrdU-labeled cells, a well recognized property of the amphibian wound epidermis (Singer and Salpeter 1961). It has also been suggested that discrepancies in the plasticity of muscle fibers between axolotls (Kragl et al. 2009) and newts (Morrison et al. 2010) is due to differences between the axolotl and newt, but this is unlikely because two near-identical studies showed that irradiated limbs can regenerate into muscle-less limbs after grafting of non-irradiated skin (containing dermal fibroblasts and epidermis) in axolotl (Dunis and Namenwirth 1977) and Spanish ribbed newt (Lheureux 1983), suggesting that connective tissue generated the new limb and did not generate new muscle cells in both species. A similar controversy exists in the mammalian satellite cell literature, where multiple investigations have suggested that muscle satellite cells can give rise to adipogenic lineages (Asakura et al. 2001; Csete et al. 2001; Shefer et al. 2004; Rossi et al. 2010), osteogenic lineages (Katagiri et al. 1994; Asakura et al. 2001; Wada et al. 2002), and fibroblasts (Brack et al. 2007). Recently, this idea has been challenged by investigations showing that non-satellite cells located in the muscle interstitium have the potential to give rise to adipocytes, osteogenic cells, and fibroblasts (Lounev et al. 2009; Joe et al. 2010; Uezumi et al. 2010, 2011). Furthermore, MyoD-cre knock-in mice have been used to show that muscle satellite cells are uni-potent, giving rise only to the myogenic lineage in culture (Starkey et al. 2011). Overall, it is clear that non-myogenic progenitors are present in the mammalian muscle interstitium and it is possible that similar non-myogenic cell populations are present in newt muscle fibers. Transgenic approaches in a newt species should provide further insight into this controversy.

4 Teleost Fish Fin Regeneration

Fin regeneration occurs in a variety of teleost fish including tilapia, minnows, goldfish, trout, and zebrafish, among others (Akimenko et al. 2003). Compared to amphibians, zebrafish entered the field of regenerative biology relatively late (Géraudie et al. 1993; White et al. 1994), but its rise as a powerful genetic model has elevated it to the predominant regeneration model in recent years. Zebrafish caudal fins are made up of repeating dermally derived fin rays, growing along with the animal by terminal addition of new rays (lepidotrichia) (Fig. 1). Amputation along the rays leads to regeneration within 2 weeks throughout their 3-year life-span (Azevedo et al. 2011; Itou et al. 2012), although others have observed some regenerative decline with age (Tsai et al. 2007; Anchelin et al. 2011). The caudal fin consists of a limited number of cell types including epidermis, fibroblasts, vasculature, glia associated with nerve fibers, and lepidotrichia osteoblasts (Tu and Johnson 2011). Histological observations and ideas based upon studies from the regenerating salamander limb suggested that the fin blastema is made up of a homogeneous population of dedifferentiated cells. Lineage restriction during fin regeneration was first hinted at by Goss and Stagg (1957), who showed that the original bony ray in *Fundulus heteroclitus* is necessary to regenerate a new ray. Excising half of a duplicated bony ray followed by amputation led to a regenerated fin containing half a ray, missing the half that would arise from the excised ray. More recently, similar results have been obtained in zebrafish, suggesting a conserved mechanism across fish species (Murciano et al. 2007). Furthermore, implanting extra rays into undamaged fins also gave rise to corresponding extra rays in the regenerate, demonstrating that the source of cells for new rays arise locally, most likely from within the resident ray (Goss and Stagg 1957). Surprisingly, these results suggested the existence of even tighter bone/cartilage lineage restrictions than in the Urodele limb where cartilage removal in the stump does not lead to deficiencies in the regenerate.

Recent developments in transgenics have allowed readdressing of this issue using genetic lineage labeling. Several studies have clearly demonstrated that caudal fin regeneration proceeds through the expansion of lineage-restricted progenitor cells generated by dedifferentiation of mature fin cells rather than multipotent stem cells. Using promoters to the preosteoblast marker *runx2*, the intermediate bone marker *osterix* (*sp7*), and the late differentiation marker *osteocalcin*, Knopf et al. (2011) demonstrated osteoblast dedifferentiation to a progenitor-like state by showing that mature osteoblasts located near the amputation plane lose expression of *osterix* and *osteocalcin* and up-regulate *runx2* along with the immature bone extracellular protein, Tenascin C. Furthermore, *osterix*-driven cre-mediated lineage- labeling was used to permanently switch mature osteoblasts from green to red fluorescence, showing that osteoblasts only give rise to other osteoblasts in the regenerated fin. Simultaneously, Sousa et al. showed that mature osteoblasts dedifferentiate and are lineage restricted using gene expression and permanent fluorescent lineage tracking (Sousa et al. 2011).

Others extended the list of lineage-restricted cell types beyond osteoblasts to all fin cell types by generating over 1000 mosaic transgenic zebrafish that fluoresced in distinct lineages of the caudal fin (Tu and Johnson 2011). This study showed that the caudal fin is made up of nine distinct lineage classes including four neuroectodermal lineages (melanocyte/xanthophores, iridophores, intraray glia, and lateral line), an osteoblast lineage, dermal fibroblast lineage, vascular endothelium lineage, epithelial cell lineage, and circulating blood lineage with no transdifferentiation ever occurring between lineages during regeneration. A second study (Stewart and Stankunas 2012) came to the same conclusions and added a putative tissue-specific macrophage lineage to the list of lineage-restricted cells by generating inducible cre-lox transgenic animals that allowed permanent fluorescent lineage labeling of small, distinct populations of cells. Together with the work of Goss and others, these studies conclusively show that normal zebrafish caudal fin regeneration proceeds by dedifferentiation of mature fin cells to a progenitor-like state, retaining their lineage and positional memory during fin regeneration (Fig. 2).

A question left open from these studies was whether dedifferentiation of mature osteoblasts is absolutely necessary for bone replacement during fin regeneration. This question is important in regards of regenerative medicine because the answer informs us whether organs can be generated using multiple cell sources (i.e. iPS cells, eSCs, or transdifferentiation) or whether dedifferentiation of lineage-restricted progenitors is the only option for re-growing an appendage. Using inducible cell ablation and cre-lox lineage labeling in resident osteoblasts, Singh et al. killed all osterix-expressing osteoblasts before amputation,and saw that new osteoblasts regenerated *de novo* from some other cell source (Singh et al. 2012). This study confirmed that resident osteoblasts generate only osteoblasts during normal fin regeneration, but can arise from a different cell source if needed, which is precisely the case in the Urodele limb. Based upon the original work of Goss showing that the new ray arises from the old ray, the most likely other cell type is the intraray fibroblasts located between the bony elements. Indeed, intraray fibroblasts express the skeletal markers, *sox9* and *col2a*, and some fibroblasts had osterix reporter activity after osteoblast ablation. It will be important to address this issue of whether intraray fibroblasts are the true source cells for ablated osteoblasts by specific lineage labeling of fibroblasts.

The zebrafish, *X. laevis*, and urodele data together thus make a strong case for these animals regenerating using very similar mechanisms (Fig. 2). First, each animal seems to use progenitor cells located just proximal to the injury. Second, little transdifferentiation takes place between cellular lineages during regeneration. Lastly, the existence of transdifferentiation of fibroblasts into skeletal elements as a natural process seems to be conserved between each animal, which may be similar to the development of the skeleton in each animal. Unfortunately, the exact developmental origin of the lepidotrichia in the zebrafish caudal fin is not yet clear, although it is believed to be the fibroblasts. If this is indeed the case, like amphibians, the origin of this cell type during regeneration would be the same as that which occurred during development.

5 Mammalian Digit Tip Regeneration

Neonatal and adult mice, rats, monkeys, and humans have the capacity to regenerate a near replicate of amputated digits if the amputation is distal to the terminal phalanx (Illingworth 1974; Borgens 1982; Singer et al. 1987; Said et al.

2004; Fernando et al. 2011). One of the most studied examples of digit regeneration is the adult mouse, which has a distal phalange containing an internal bone marrow compartment encased in dense cortical bone, surrounded by loose connective tissue, which is covered by an epidermis associated with epidermal derivatives (nail matrix, nail bed, and nail plate). Proximal to the nail is a pair of dorsal and ventral tendons and a ventrally located fat pad containing sweat glands, melanocytes, nerves with associated glia, dermal fibroblasts, and epidermis (Fig. 1). Regeneration of these components takes place only when amputation is performed at the distal portion of the fat pad, past the interphalangeal joint and not through more proximal phalanges (Neufeld and Zhao 1993, 1995; Fernando et al. 2011; Rinkevich et al. 2011). This is a much more limited model of appendage regeneration since there is no muscle present in this terminal phalange but, being mammalian, represents an excellent model for the same phenomenon which occurs in humans (Illingworth 1974).

Histological studies of digit tip regeneration in mice (Borgens 1982) and rhesus monkeys (Singer et al. 1987) suggest that digit regeneration does not form a blastema-like population of cells similar to the Urodele amphibian. On the other hand, molecular analyses show some similarities to Urodele limb regeneration including *msx1* expression (Hanna et al. 2007) and *bmp4* expression (Han et al. 2003), suggesting that similar molecular processes may be underlying both examples of regeneration. Two groups have recently set out to address whether digit regeneration involves transdifferentiation of cells or uses lineage-restricted progenitors like amphibians and zebrafish. Regenerated mouse digit tips were imaged following labeling of specific tissues using cre/lox mediated recombination and demonstrating that, like lower vertebrates, no transdifferentiation takes place between ectodermal, mesodermal, and endothelial tissues during both fetal and adult digit tip regeneration (Lehoczky et al. 2011; Rinkevich et al. 2011). Furthermore, parabiosis of genetically marked mice showed that cells from distant sources did not contribute to any regenerated tissue, demonstrating that all the cells necessary for digit regeneration are locally derived, similar to the case in amphibians and zebrafish.

Amphibian and zebrafish studies have demonstrated that during normal regeneration, skeletal elements give rise to new skeletal elements, yet fibroblasts may become skeletal tissue if the skeletal element is removed. Permanent osteoblast lineage labeling using *sp7* driven cre-mediated recombination showed that osteoblasts only give rise to new osteoblasts thus revealing skeletal tissue mainly gives rise to new skeletal tissue during appendage regeneration across vertebrates. The important question remains whether mice have the ability to use an alternative source for new osteoblasts when they are removed from the stump, as is the case in Urodeles and most likely zebrafish, thus revealing another universal vertebrate appendage principle.

An important issue is whether the distal phalanx regenerates using the same mechanisms used in development. The distal phalanx bone in the mouse originally develops through endochondral ossification and elongates by direct ossification at the distal tip (Casanova and Sanz-Ezquerro 2007; Han et al. 2008). During regeneration, bone is regenerated only by direct ossification from pre-existing osteoblasts (Lehoczky et al. 2011), which has been argued to be different from development (Fernando et al. 2011). Given that direct ossification occurs during outgrowth of the developing distal phalanx, regeneration does not deviate from mechanisms of distal phalanx development. Therefore, mammalian digit regeneration does not use novel mechanisms that have not been previously used during development of the organ, which is similar to results found in amphibian and teleost fish appendage regeneration.

6 Dedifferentiation: What Does It Really Mean?

The studies discussed above strongly support a model of appendage regeneration whereby cells retain their particular lineage that was established during appendage development (Fig. 2). The question remains though, where do these lineage-restricted progenitors come from and can all cell types contribute to the lineage-restricted blastema? Almost 80 years ago, Butler (1933, 1935) and Thornton (1938a) used the term dedifferentiation to describe the observation that cells near the amputation plane of salamander limbs lose their differentiated morphological phenotype. This observation was supported by studying sections of regenerating limbs using electron microscopy, which showed that muscle fibers lose their striated arrangement (sarcomeres), and break up their basement membranes, while rounding up their nuclei (Hay 1959). More recently, stronger evidence for the cellularization of salamander muscle fibers was provided by tracking the cellular progeny of myotubes or myofibers in vivo after labeling with fluorescent dextran (Lo et al. 1993; Echeverri et al. 2001), retrovirus (Kumar et al. 2000), or cell tracking dye (Kumar et al. 2004). These investigations confirmed that the myofibers transition from a multinucleate syncytium to mononucleate cells, which proliferate and contribute to the salamander limb and tail blastema. Others have shown that satellite cells can also contribute to the blastema, suggesting that two sources of muscle progenitors contribute to the blastema (Cameron et al. 1986; Morrison et al. 2006, 2010). As described above, cells that are not of the myogenic lineage are likely present in the myofiber interstitium as is the case in mammals (Paylor et al. 2011), so it will be important to determine the relative contribution of satellite cells, non-myogenic cells and cells derived from myofiber dedifferentiation to the blastema. Furthermore, our understanding of dedifferentiation of other tissues such as cartilage, peripheral nerves, and fibroblasts is presently non-existent. The availability of transgenics in urodeles (Khattak et al. 2009; Casco-Robles et al. 2011) should provide the means to address these issues.

Unlike Urodeles, however, transgenics have provided the means to investigate the contribution of satellite cells versus muscle dedifferentiation to the blastema in *X. laevis* and zebrafish. Ryfell et al. labeled muscle fibers using cre-lox techniques to show that muscle actin-expressing cells located within muscle fibers do *not* participate in *X. laevis* tail regeneration (Ryffel et al. 2003). A similar cre-lox approach showed that muscle fibers do not fragment or re-enter the cell cycle during *X. laevis* tail regeneration. Interestingly, these investigators also showed that no muscle dedifferentiation takes place after amputation of the zebrafish caudal fin as well, suggesting that muscle fiber dedifferentiation may be less common during appendage regeneration than previously appreciated (Rodrigues et al. 2012). Gargioli and Slack (Gargioli and Slack 2004) used grafts from GFP-positive transgenic *X. laevis* embryos to wild-type hosts to specifically label satellite cells or muscle fibers. Following tail amputation, new muscle arose only from GFP-positive satellite cells and not muscle fibers suggesting that muscle regeneration occurs entirely from proliferation of satellite cells. Furthermore, expression of a dominant-negative form of the satellite cell gene, *pax7*, killed satellite cells and reduced muscle fiber regeneration in the tail (Chen et al. 2006). Together, these results show that muscle regeneration in the *X. laevis* tail and zebrafish caudal fin proceeds by proliferating satellite cells rather than dedifferentiation of myofibers, highlighting an apparent difference between these species and Urodeles.

These results are significant because they suggest that muscle regeneration in some lower vertebrates is similar to mammalian muscle regeneration, which occurs exclusively by proliferation of satellite cells, and not myocyte dedifferentiation (Le Grand and Rudnicki 2007). Therefore, dedifferentiation of myocytes may not be necessary in order to regenerate complex mammalian tissues. In the future, it will be critical to understand the role of muscle dedifferentiation during urodele regeneration.

7 Future Perspectives

Contemporary concepts in regenerative medicine are based on first, the generation of pluripotent stem cells either from the embryo or by the induction of pluripotency in adult cells, second, inducing their differentiation into the desired cell type, and third, grafting the differentiated cells into patients or using the cells to grow organs ex vivo (Robinton and Daley 2012). Although the promise of patient-specific, made-to-order organs may someday be possible, this strategy is currently limited by our ability to differentiate enough stem cells into the desired cell type at high frequency and obtaining enough engraftment in the damaged or diseased organ to make a significant recovery of function, while at the same time not unleashing the teratoma or cancer forming potential of these cells. In addition, we know almost nothing about how to control the morphogenesis of communities of cells, so that an organ of the appropriate form and structure can be re-grown. This is what combinatorial investigation of animal development and animal regeneration will tell us—are signaling pathways that were originally used during development re-used again to regenerate organs? For this we need to ask basic questions of those animals that can already regenerate organs such as appendages, a behavior found frequently in fish and widespread in amphibians as described above. By studying morphogenesis and basic questions of the characteristics of cell behavior during regeneration we may not only provide information to the regenerative medicine strategy outlined above, but also may be possible to directly mimic the strategy that these animals employ for human applications.

There is one obvious contrast between this induction of complete pluripotency that regenerative medicine seeks to induce and the recent work on lineage specification during appendage regeneration that we have reviewed here. The universal approach to appendage regeneration in fish and amphibians and to digit tip regeneration in mammals is not to generate pluripotent blastemal cells which can then be differentiated into the new organ, but instead to generate lineage-restricted cells which are only partially dedifferentiated and which are then propagated in a progenitor-like state (Fig. 2). In other words, blastemal cells are not fully pluripotent as IPS or ES cells, but retain memories of their historical past as bone, muscle, epidermis, or fibroblasts. Thus the most critical question that needs to be addressed in regeneration is to identify the molecular mechanisms of this partial, lineage-specific dedifferentiation. This will surely provide a major impetus to regenerative medicine by answering the question of how to stimulate dedifferentiation in human cells.

References

Akimenko M-A, Marí-Beffa M, Becerra J, Géraudie J (2003) Old questions, new tools, and some answers to the mystery of fin regeneration. Dev Dyn 226:190–201

Anchelin M, Murcia L, Alcaraz-Pérez F, García-Navarro EM, Cayuela ML (2011) Behaviour of telomere and telomerase during aging and regeneration in zebrafish. PLoS ONE 6:e16955

Asakura A, Komaki M, Rudnicki M (2001) Muscle satellite cells are multipotential stem cells that exhibit myogenic, osteogenic, and adipogenic differentiation. Differentiation 68:245–253

Azevedo AS, Grotek B, Jacinto A, Weidinger G, Saúde L (2011) The regenerative capacity of the zebrafish caudal fin is not affected by repeated amputations. PLoS ONE 6:e22820

Bischler V (1926) L'influence du squelette de Dans la régénération, et les potentialités des plongeurs Territoires du membre chez Triton cristatus. RSZ 33:431–560

Bladt F, Riethmacher D, Isenmann S, Aguzzi A, Birchmeier C (1995) Essential role for the c-met receptor in the migration of myogenic precursor cells into the limb bud. Nature 376:768–771

Blum N, Begemann G (2012) Retinoic acid signaling controls the formation, proliferation and survival of the blastema during adult zebrafish fin regeneration. Development 139:107–116

Borgens RB (1982) Mice regrow the tips of their foretoes. Science 217:747–750

Brack AS, Conboy MJ, Roy S, Lee M, Kuo CJ, Keller C, Rando TA (2007) Increased Wnt signaling during aging alters muscle stem cell fate and increases fibrosis. Science 317: 807–810

Briggs R, King TJ (1952) Transplantation of living nuclei from blastula cells into enucleated frogs' eggs. Proc Natl Acad Sci U S A 38:455–463

Briggs R, Signoret J, Humphrey RR (1964) Transplantation of nuclei of various cell types from neurulae of the Mexican axolotl (*Ambystoma mexicanum*). Dev Biol 10:233–246

Burgess AM (1967) The developmental potentialities of regeneration blastema cell nuclei as determined by nuclear transplantation. J Embryol Exp Morphol 18:27–41

Butler EG (1933) The effects of X-radiation on the regeneration of the fore limb of Amblystoma larvae. J Exp Zool 65:271–315

Butler EG (1935) Studies on limb regeneration in X-rayed amblystoma larvae. Anat Rec 62: 295–307
Butler EG, O'Brien JP (1942) Effects of localized x-radiation on regeneration of the urodele limb. Anat Rec 84:407–413
Cameron JA, Hilgers AR, Hinterberger TJ (1986) Evidence that reserve cells are a source of regenerated adult newt muscle in vitro. Nature 321:607–610. doi:10.1038/321607a0. Accesed 5 June 1986
Casanova JC, Sanz-Ezquerro JJ (2007) Digit morphogenesis: is the tip different? Dev Growth Differ 49:479–491
Casco-Robles MM, Yamada S, Miura T, Nakamura K, Haynes T, Maki N, Del Rio-Tsonis K, Tsonis PA, Chiba C (2011) Expressing exogenous genes in newts by transgenesis. Nat Protoc 6:600–608
Chablais F, Jazwinska A (2010) IGF signaling between blastema and wound epidermis is required for fin regeneration. Development 137:871–879
Chalkey D (1954) A quantitative histological analysis of forelimb regeneration in Triturus viridescens. J Morphol 94:21–70
Chen Y, Lin G, Slack JMW (2006) Control of muscle regeneration in the Xenopus tadpole tail by Pax7. Development 133:2303–2313
Cox PG (1969) Some aspects of tail regeneration in the lizard, *Anolis carolinensis*. II the role of the peripheral nerves. J Exp Zool 171:151–159
Csete M, Walikonis J, Slawny N, Wei Y, Korsnes S, Doyle JC, Wold B (2001) Oxygen-mediated regulation of skeletal muscle satellite cell proliferation and adipogenesis in culture. J Cell Physiol 189:189–196
Dasgupta S (1970) Developmental potentialities of blastema cell nuclei of the Mexican Axolotl. J Exp Zool 175:141–147
Dinsmore C (1991) A history of regeneration research: milestones in the evolution of a science. Press Syndicate of the University of Cambridge, New York
Dufourcq P, Vriz S (2006) The chemokine SDF-1 regulates blastema formation during zebrafish fin regeneration. Dev Genes Evol 216:635–639
Dunis DA, Namenwirth M (1977) The role of grafted skin in the regeneration of x-irradiated axolotl limbs. Dev Biol 56:97–109
Echeverri K, Clarke JD, Tanaka EM (2001) In vivo imaging indicates muscle fiber dedifferentiation is a major contributor to the regenerating tail blastema. Dev Biol 236:151–164
Echeverri K, Tanaka EM (2002) Ectoderm to mesoderm lineage switching during axolotl tail regeneration. Science 298:1993–1996
Fernando WA, Leininger E, Simkin J, Li N, Malcom CA, Sathyamoorthi S, Han M, Muneoka K (2011) Wound healing and blastema formation in regenerating digit tips of adult mice. Dev Biol 350:301–310
Foret JE (1970) Regeneration of larval urodele limbs containing homoplastic transplants. J Exp Zool 175:297–321
Gargioli C, Slack JMW (2004) Cell lineage tracing during Xenopus tail regeneration. Development 131:2669–2679
Géraudie J, Monnot MJ, Ridet A, Thorogood P, Ferretti P (1993) Is exogenous retinoic acid necessary to alter positional information during regeneration of the fin in zebrafish? Prog Clin Biol Res 383B:803–814
Goss R (1969) Principles of regeneration. Academic Press, New York
Goss RJ (1956a) Regenerative inhibition following limb amputation and immediate insertion into the body cavity. Anat Rec 126:15–27
Goss RJ (1956b) The relation of bone to the histogenesis of cartilage in regenerating forelimbs and tails of adult Triturus viridescens. J Morphol 98:89–123
Goss RJ, Stagg MW (1957) The regeneration of fins and fin rays in Fundulus heteroclitus. J Exp Zool 136:487–507

Gros J, Manceau M, Thomé V, Marcelle C (2005) A common somitic origin for embryonic muscle progenitors and satellite cells. Nature 435:954–958

Gurdon JB, Byrne JA (2003) The first half-century of nuclear transplantation. PNAS 100: 8048–8052

Han M, Yang X, Farrington JE, Muneoka K (2003) Digit regeneration is regulated by Msx1 and BMP4 in fetal mice. Development 130:5123–5132

Han M, Yang X, Lee J, Allan CH, Muneoka K (2008) Development and regeneration of the neonatal digit tip in mice. Dev Biol 315:125–135

Hanna J, Wernig M, Markoulaki S, Sun C-W, Meissner A, Cassady JP, Beard C, Brambrink T, Wu L-C, Townes TM, Jaenisch R (2007) Treatment of sickle cell anemia mouse model with iPS cells generated from autologous skin. Science 318:1920–1923

Hay E, Fischmann D (1961) Origin of the blastema in regenerating limbs of the newt Triturus viridescens. Dev Biol 3:26–59

Hay ED (1959) Electron microscopic observations of muscle dedifferentiation in regenerating Amblystoma limbs. Dev Biol 1:555–585

Hayashi K, Ozawa E (1995) Myogenic cell migration from somites is induced by tissue contact with medial region of the presumptive limb mesoderm in chick embryos. Development 121:661–669

Hertwig G (1927) Experimentelle untersuchungen uber die Herkunft des Regenerationsblastemas. Anat Anz 63:90–96

Hirata A, Gardiner DM, Satoh A (2010) Dermal fibroblasts contribute to multiple tissues in the accessory limb model. Dev Growth Differ 52:343–350

Holder N (1989) Organization of connective tissue patterns by dermal fibroblasts in the regenerating axolotl limb. Development 105:585–593

Illingworth CM (1974) Trapped fingers and amputated finger tips in children. J Pediatr Surg 9:853–858

Inoue S, Shimoda Y, Fujikura K (1988) Cartilage differentiation in the bone-removed and amputated Xenopus forelimb with or without innervation. In: Regeneration and development: the proceedings of the 6th international marcus singer symposium. Okada Pub Co, Maebashi, pp 271–285

Itou J, Kawakami H, Burgoyne T, Kawakami Y (2012) Life-long preservation of the regenerative capacity in the fin and heart in zebrafish. Biol Open 1:739–746

Joe AWB, Yi L, Natarajan A, Le Grand F, So L, Wang J, Rudnicki MA, Rossi FMV (2010) Muscle injury activates resident fibro/adipogenic progenitors that facilitate myogenesis. Nat Cell Biol 12:153–163

Katagiri T, Yamaguchi A, Komaki M, Abe E, Takahashi N, Ikeda T, Rosen V, Wozney JM, Fujisawa-Sehara A, Suda T (1994) Bone morphogenetic protein-2 converts the differentiation pathway of C2C12 myoblasts into the osteoblast lineage. J Cell Biol 127:1755–1766

Khattak S, Richter T, Tanaka EM (2009) Generation of transgenic axolotls (Ambystoma mexicanum). Cold Spring Harb Protoc 2009, pdb.prot5264

Knopf F, Hammond C, Chekuru A, Kurth T, Hans S, Weber CW, Mahatma G, Fisher S, Brand M, Schulte-Merker S, Weidinger G (2011) Bone regenerates via dedifferentiation of osteoblasts in the zebrafish fin. Dev Cell 20:713–724

Korneluk RG, Liversage RA (1984) Effects of radius–ulna removal on forelimb regeneration in Xenopus laevis froglets. J Embryol Exp Morphol 82:9–24

Kragl M, Knapp D, Nacu E, Khattak S, Maden M, Epperlein HH, Tanaka EM (2009) Cells keep a memory of their tissue origin during axolotl limb regeneration. Nature 460:60–65

Kumar A, Velloso CP, Imokawa Y, Brockes JP (2000) Plasticity of retrovirus-labelled myotubes in the newt limb regeneration blastema. Dev Biol 218:125–136

Kumar A, Velloso CP, Imokawa Y, Brockes JP (2004) The regenerative plasticity of isolated urodele myofibers and its dependence on MSX1. PLoS Biol 2:E218

Laube F, Heister M, Scholz C, Borchardt T, Braun T (2006) Re-programming of newt cardiomyocytes is induced by tissue regeneration. J Cell Sci 119:4719–4729

Le Grand F, Rudnicki MA (2007) Skeletal muscle satellite cells and adult myogenesis. Curr Opin Cell Biol 19:628–633

Lehoczky JA, Robert B, Tabin CJ (2011) Mouse digit tip regeneration is mediated by fate-restricted progenitor cells. Proc Natl Acad Sci U S A 108:20609–20614

Lheureux E (1983) The origin of tissues in the X-irradiated regenerating limb of the newt. In: Liss A (ed) In limb development and regeneration. Progress in clinical and biological research. Wiley-Liss, New York, pp 455–465

Liversage R (1991) Origin of the blastema cells in epimorphic regeneration of urodele appendages: A history of ideas. In: Dinsmore CE (ed) A history of regeneration research: milestones in the evolution of science. Cambridge University Press, Cambridge, pp 179–199

Lo DC, Allen F, Brockes JP (1993) Reversal of muscle differentiation during urodele limb regeneration. Proc Natl Acad Sci U S A 90:7230–7234

Lounev VY, Ramachandran R, Wosczyna MN, Yamamoto M, Maidment ADA, Shore EM, Glaser DL, Goldhamer DJ, Kaplan FS (2009) Identification of progenitor cells that contribute to heterotopic skeletogenesis. J Bone Joint Surg Am 91:652–663

Maden M (1977) The role of Schwann cells in paradoxical regeneration in the axolotl. J Embryol Exp Morphol 41:1–13

Manner HW (1953) The origin of the blastema and of new tissues in regenerating forelimbs of adult Triturus viridescens viridescens (Rafinesque). J Exp Zool 122:229–257

McHedlishvili L, Epperlein HH, Telzerow A, Tanaka EM (2007) A clonal analysis of neural progenitors during axolotl spinal cord regeneration reveals evidence for both spatially restricted and multipotent progenitors. Development 134:2083–2093

McHedlishvili L, Mazurov V, Grassme KS, Goehler K, Robl B, Tazaki A, Roensch K, Duemmler A, Tanaka EM (2012) Reconstitution of the central and peripheral nervous system during salamander tail regeneration. Proc Natl Acad Sci 109(34):E2258–2266

McLean KE, Vickaryous MK (2011) A novel amniote model of epimorphic regeneration: the leopard gecko Eublepharis macularius. BMC Dev Biol 11:50

Morrison JI, Borg P, Simon A (2010) Plasticity and recovery of skeletal muscle satellite cells during limb regeneration. FASEB J 24:750–756

Morrison JI, Lööf S, He P, Simon A (2006) Salamander limb regeneration involves the activation of a multipotent skeletal muscle satellite cell population. J Cell Biol 172:433–440

Muneoka K, Fox WF, Bryant SV (1986) Cellular contribution from dermis and cartilage to the regenerating limb blastema in axolotls. Dev Biol 116:256–260

Murciano C, Pérez-Claros J, Smith A, Avaron F, Fernández TD, Durán I, Ruiz-Sánchez J, García F, Becerra J, Akimenko M-A, Marí-Beffa M (2007) Position dependence of hemiray morphogenesis during tail fin regeneration in Danio rerio. Dev Biol 312:272–283

Namenwirth M (1974) The inheritance of cell differentiation during limb regeneration in the axolotl. Dev Biol 41(1):42–56

Neufeld DA, Zhao W (1993) Phalangeal regrowth in rodents: postamputational bone regrowth depends upon the level of amputation. Prog Clin Biol Res 383A:243–252

Neufeld DA, Zhao W (1995) Bone regrowth after digit tip amputation in mice is equivalent in adults and neonates. Wound Repair Regen 3:461–466

Oberpriller J (1967) A radioautographic analysis of the potency of blastemal cells in the adult newt, Diemictylus viridescens. Growth 31:251–296

Patrick J, Briggs R (1964) Fate of cartilage cells in limb regeneration in the axolotl (Ambystoma mexicanum). Experientia 20:431–432

Paylor B, Natarajan A, Zhang R-H, Rossi F (2011) Nonmyogenic cells in skeletal muscle regeneration. Curr Top Dev Biol 96:139–165

Poss KD, Shen J, Nechiporuk A, McMahon G, Thisse B, Thisse C, Keating MT (2000) Roles for Fgf signaling during zebrafish fin regeneration. Dev Biol 222:347–358

Relaix F, Rocancourt D, Mansouri A, Buckingham M (2005) A Pax3/Pax7-dependent population of skeletal muscle progenitor cells. Nature 435:948–953

Rinkevich Y, Lindau P, Ueno H, Longaker MT, Weissman IL (2011) Germ-layer and lineage-restricted stem/progenitors regenerate the mouse digit tip. Nature 476:409–413

Robinton DA, Daley GQ (2012) The promise of induced pluripotent stem cells in research and therapy. Nature 481:295–305

Rodrigues AMC, Christen B, Martí M, Izpisúa Belmonte JC (2012) Skeletal muscle regeneration in Xenopus tadpoles and zebrafish larvae. BMC Dev Biol 12:9

Rossi CA, Pozzobon M, Ditadi A, Archacka K, Gastaldello A, Sanna M, Franzin C, Malerba A, Milan G, Cananzi M, Schiaffino S, Campanella M, Vettor R, De Coppi P (2010) Clonal characterization of rat muscle satellite cells: proliferation, metabolism and differentiation define an intrinsic heterogeneity. PLoS ONE 5:e8523

Ryffel GU, Werdien D, Turan G, Gerhards A, Goosses S, Senkel S (2003) Tagging muscle cell lineages in development and tail regeneration using Cre recombinase in transgenic Xenopus. Nucleic Acids Res 31:e44

Said S, Parke W, Neufeld DA (2004) Vascular supplies differ in regenerating and nonregenerating amputated rodent digits. Anat Rec A Discov Mol Cell Evol Biol 278:443–449

Seifert AW, Monaghan JR, Voss SR, Maden M (2012) Skin regeneration in adult axolotls: a blueprint for scar-free healing in vertebrates. PLoS ONE 7:e32875

Shefer G, Wleklinski-Lee M, Yablonka-Reuveni Z (2004) Skeletal muscle satellite cells can spontaneously enter an alternative mesenchymal pathway. J Cell Sci 117:5393–5404

Signoret J, Briggs R, Humphrey RR (1962) Nuclear transplantation in the axolotl. Dev Biol 4:134–164

Singer M, Salpeter M (1961) Regeneration in vertebrates: the role of the wound epithelium. In: Zarrow MX, Beevers H (eds) Growth in living systems. Basic Books, Inc., New York, pp 277–311

Singer M, Weckesser EC, Géraudie J, Maier CE, Singer J (1987) Open finger tip healing and replacement after distal amputation in rhesus monkey with comparison to limb regeneration in lower vertebrates. Anat Embryol 177:29–36

Singh SP, Holdway JE, Poss KD (2012) Regeneration of amputated zebrafish fin rays from de novo osteoblasts. Dev Cell 22:879–886

Sobkow L, Epperlein H–H, Herklotz S, Straube WL, Tanaka EM (2006) A germline GFP transgenic axolotl and its use to track cell fate: dual origin of the fin mesenchyme during development and the fate of blood cells during regeneration. Dev Biol 290:386–397

Sousa S, Afonso N, Bensimon-Brito A, Fonseca M, Simões M, Leon J, Roehl H, Cancela ML, Jacinto A (2011) Differentiated skeletal cells contribute to blastema formation during zebrafish fin regeneration. Development 138:3897–3905

Starkey JD, Yamamoto M, Yamamoto S, Goldhamer DJ (2011) Skeletal muscle satellite cells are committed to myogenesis and do not spontaneously adopt nonmyogenic fates. J Histochem Cytochem 59:33–46

Steen TP (1968) Stability of chondrocyte differentiation and contribution of muscle to cartilage during limb regeneration in the axolotl (Siredon mexicanum). J Exp Zool 167:49–78

Steen TP (1970) Origin and differentiative capacities of cells in the blastema of the regenerating salamander limb

Stewart S, Stankunas K (2012) Limited dedifferentiation provides replacement tissue during zebrafish fin regeneration. Dev Biol 365(2):339–49

Stocum DL (1984) The urodele limb regeneration blastema, determination and organization of the morphogenetic field. Differentiation 27:13–28

Stocum DL, Cameron JA (2011) Looking proximally and distally: 100 years of limb regeneration and beyond. Dev Dyn 240:943–968

Thornton C (1938a) The histogenesis of muscle in the regenerating forelimb of larval Amblystoma punctatum. J Morphol 62:17–47

Thornton CS (1938b) The histogenesis of the regenerating fore limb of larval Amblystoma after exarticulation of the humerus. J Morphol 62:219–241

Thornton CS (1942) Studies on the origin of the regeneration blastema in Triturus viridescens. J Exp Zool 89:375–389

Tsai SB, Tucci V, Uchiyama J, Fabian NJ, Lin MC, Bayliss PE, Neuberg DS, Zhdanova IV, Kishi S (2007) Differential effects of genotoxic stress on both concurrent body growth and gradual senescence in the adult zebrafish. Aging Cell 6:209–224

Tsonis PA, Fox TP (2009) Regeneration according to Spallanzani. Dev Dyn 238:2357–2363

Tu S, Johnson SL (2011) Fate Restriction in the growing and regenerating zebrafish fin. Dev Cell 20:725–732

Tuchkova SI (1973) Study of the potencies of the regenerated blastoma by transplantation and autoradiographic methods. Dokl Akad Nauk SSSR 213:1217–1220

Uezumi A, Fukada S, Yamamoto N, Takeda S, Tsuchida K (2010) Mesenchymal progenitors distinct from satellite cells contribute to ectopic fat cell formation in skeletal muscle. Nat Cell Biol 12:143–152

Uezumi A, Ito T, Morikawa D, Shimizu N, Yoneda T, Segawa M, Yamaguchi M, Ogawa R, Matev MM, Miyagoe-Suzuki Y, Takeda S, Tsujikawa K, Tsuchida K, Yamamoto H, Fukada S (2011) Fibrosis and adipogenesis originate from a common mesenchymal progenitor in skeletal muscle. J Cell Sci 124:3654–3664

Wada MR, Inagawa-Ogashiwa M, Shimizu S, Yasumoto S, Hashimoto N (2002) Generation of different fates from multipotent muscle stem cells. Development 129:2987–2995

Wallace BM, Wallace H (1973) Participation of grafted nerves in amphibian limb regeneration. J Embryol Exp Morphol 29:559–570

Wallace H (1972) The components of regrowing nerves which support the regeneration of irradiated salamander limbs. J Embryol Exp Morphol 28:419–435

Wallace H (1981) Vertebrate limb regeneration. Wiley, Chichester

White JA, Boffa MB, Jones B, Petkovich M (1994) A zebrafish retinoic acid receptor expressed in the regenerating caudal fin. Development 120:1861–1872

Whitehead GG, Makino S, Lien C-L, Keating MT (2005) fgf20 is essential for initiating zebrafish fin regeneration. Science 310:1957–1960

Cell Signaling Pathways in Vertebrate Lens Regeneration

Jonathan J. Henry, Alvin G. Thomas, Paul W. Hamilton, Lisa Moore and Kimberly J. Perry

Abstract Certain vertebrates are capable of regenerating parts of the eye, including the lens. Depending on the species, two principal forms of in vivo lens regeneration have been described wherein the new lens arises from either the pigmented epithelium of the dorsal iris or the cornea epithelium. These forms of lens regeneration are triggered by retinal factors present in the eye. Studies have begun to illuminate the nature of the signals that support lens regeneration. This review describes evidence for the involvement of specific signaling pathways in lens regeneration, including the FGF, retinoic acid, TGF-beta, Wnt, and Hedgehog pathways.

Abbreviations

BMP	Bone morphogenetic protein
cDNA	Complementary DNA
EGF	Epidermal growth factor
EGFR	Epidermal growth factor receptor
EST	Expressed sequence tag
HH	Hedgehog
FGF	Fibroblast growth factor
FGFR	Fibroblast growth factor receptor
IGF	Insulin like growth factor
mRNA	Messenger RNA
PECs	Pigmented epithelial cells
PCP	Planar cell polarity
RA	Retinoic acid

J. J. Henry (✉) · A. G. Thomas · P. W. Hamilton · L. Moore · K. J. Perry
Department of Cell and Developmental Biology, University of Illinois, 601 S. Goodwin Avenue, Urbana, IL 61801, USA
e-mail: j-henry4@life.illinois.edu

Current Topics in Microbiology and Immunology (2013) 367: 75–98
DOI: 10.1007/82_2012_289

© Springer-Verlag Berlin Heidelberg 2012
Published Online: 8 December 2012

RAREs Retinoic acid response elements
TGFβ Transforming growth factor-beta
VEGF Vascular endothelial growth factor

Contents

1 Introduction: Phenomena of Vertebrate Lens Regeneration

Virtually all multicellular organisms possess the ability to repair damaged tissues. On the other hand, some animals can replace entire organs or even complex body parts. One remarkable example of these phenomena is that of complete lens regeneration, which takes place within the eyes of certain vertebrates, including newts and salamanders, a cobitid fish, as well as frogs in the genus *Xenopus* (examples are shown in Fig. 1a–c).

One form of lens regeneration, first observed over 200 years ago, is that of Wolffian lens regeneration that takes place in some adult newts, salamanders, and the cobitid fish *Misgurnus anguillicaudatus* (for examples see Fig. 1a–b, reviewed by Henry 2003; Roddy and Tsonis 2008; Henry and Tsonis 2010). In these animals, a new lens arises from the dorsal edge of the iris (specifically the pigmented iris epithelium) once the original lens is removed. This process is diagrammed in Fig. 2a–d. Wolffian lens regeneration takes place via transdifferentiation, which begins with dedifferentiation of the pigmented iris epithelial cells (PECs), followed by their re-differentiation and patterning to form the new lens. This process is triggered by signals provided by the neural retina once the original lens is removed (reviewed by Henry 2003; Roddy and Tsonis 2008). The presence of the lens plays a role in normally preventing Wolffian lens regeneration, though the mechanistic basis of this is not fully understood (Henry 2003; Roddy and Tsonis 2008; Henry and Tsonis 2010). In vivo, only the dorsal iris pigmented epithelium can support complete lens regeneration, although the initial changes associated with dedifferentiation (including depigmentation) do appear to be initiated in ventral iris tissues following lens removal (Roddy and Tsonis 2008).

A different form of lens regeneration occurs in species of frogs in the genus *Xenopus* (i.e., *X. laevis, X. tropicalis* and *X. borealis*; see Freeman 1963; Henry and Elkins 2001; Filoni et al. 2006, and reviewed by Henry 2003; Henry et al. 2008; Roddy and Tsonis 2008; Henry and Tsonis 2010). This process has also been documented in the salamander *Hynobius unnangso* (Ikeda 1936a, b, 1939). In these cases, a new lens arises from cells of the inner or basal layer of the cornea epithelium, and takes place during larval stages. Cornea lens regeneration is diagrammed in Fig. 3e–h. Although this process has historically been described as one involving transdifferentiation of cornea epithelial cells, new evidence suggests that this could involve the deployment of uncommitted, basal cornea epithelial stem cells (Perry and Henry 2012). As is the case in Wolffian lens regeneration, the retina provides the key signal(s) that trigger lens regeneration in the cornea (see Henry 2003). These factors are normally prevented from reaching the cornea by the presence of the lens and the inner cornea endothelium, which act as simple blocks to the diffusion of these key retinal factors. Interestingly, only the cornea and the pericorneal epithelium that immediately surrounds the eye are capable of responding to these signals (Freeman 1963). Other ectoderm outside of these regions cannot participate in lens regeneration, even if these tissues are implanted directly inside the eye where they can be continuously exposed to the retinal factors.

Together, these data reveal that the capacity to undergo lens regeneration is highly restricted to specific tissues (i.e., lens regeneration competent tissues). Furthermore, the retina provides a key source of paracrine signals that support the process of lens regeneration, which may be involved in initiating this process and/or supporting subsequent stages of lens differentiation, polarization (patterning) and growth. Once the lens begins to form, it may also provide autocrine signals that regulate its own formation. This review addresses the question as to what

Fig. 1 Examples of various animals capable of regenerating the lens of the eye. **a**. The newt *Cynops pyrrhogaster* (the Japanese Fire Belly Newt). **b**. The cobitid fish, *Misgurnus anguillicaudatus* (the Japanese Weather Loach) **c**. The tadpole larvae of the frog *Xenopus laevis* (the South African Clawed Frog). See text for further details. *Scale bar* equals 7 mm for a, 15 mm for b and 6 mm for c

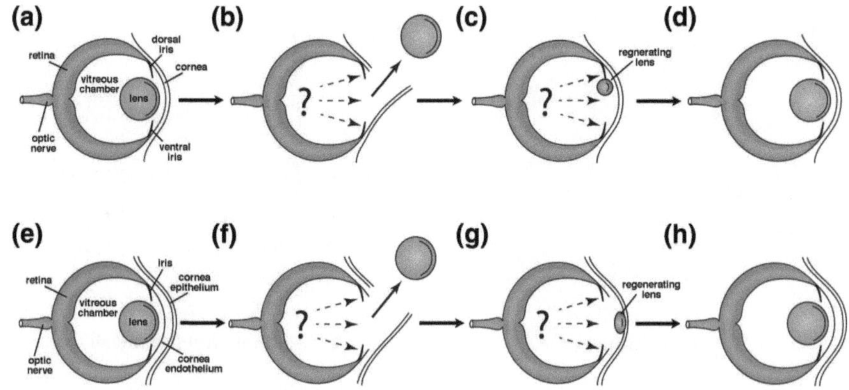

Fig. 2 Diagrams illustrating the two main processes of lens regeneration. **a–d** Steps involved in Wolffian lens regeneration that take place in certain newts and salamanders. This process takes place within the dorsal iris of the adult eye following removal of the lens (**b**) (i.e., the new lens is derived from PECs that undergo transdifferentiation). *Dashed arrows* in *b* and *c* indicate key retinal signals that support the process of lens regeneration. **e–h** Steps involved in cornea lens regeneration that takes place in frogs of the genus *Xenopus* and the salamander *Hynobius unnangso*. The process takes place within cells of the basal layer of the larval cornea epithelium following removal of the lens and perforation of the cornea endothelium (**f**). *Dashed arrows* in *f* and *g* indicate key retinal signals that support the process of lens regeneration

signaling pathways play important roles in regulating these processes of lens regeneration, and whether these may be shared among the different forms of lens regeneration found in these systems.

2 Fibroblast Growth Factor Signaling

2.1 The FGF Signaling Pathway

The fibroblast growth factor (FGF) protein family members, historically identified through their affinity for heparin, are divided into seven evolutionarily conserved subfamilies (reviewed in Itoh 2007; Itoh and Ornitz 2011). With the exception of one subfamily (FGF11-14), FGFs bind and activate specific subsets of their corresponding receptor isoforms (FGFRs), causing autophosphorylation of FGFRs and downstream signaling (Ornitz et al. 1996; Zhang et al. 2006). The activated FGFRs signal through three main downstream pathways: the phospholipase C-gamma (PLCγ), phosphoinositide 3-kinase (PI3K), and mitogen-activated protein kinase (MAPK) pathways (reviewed in Dailey et al. 2005; Mason 2007; Dorey and Amaya 2010). Of these three pathways, only the phosphoinositide 3-kinase (PI3K), and MAPK pathways have been directly implicated in lens development and regeneration (Fig. 3a).

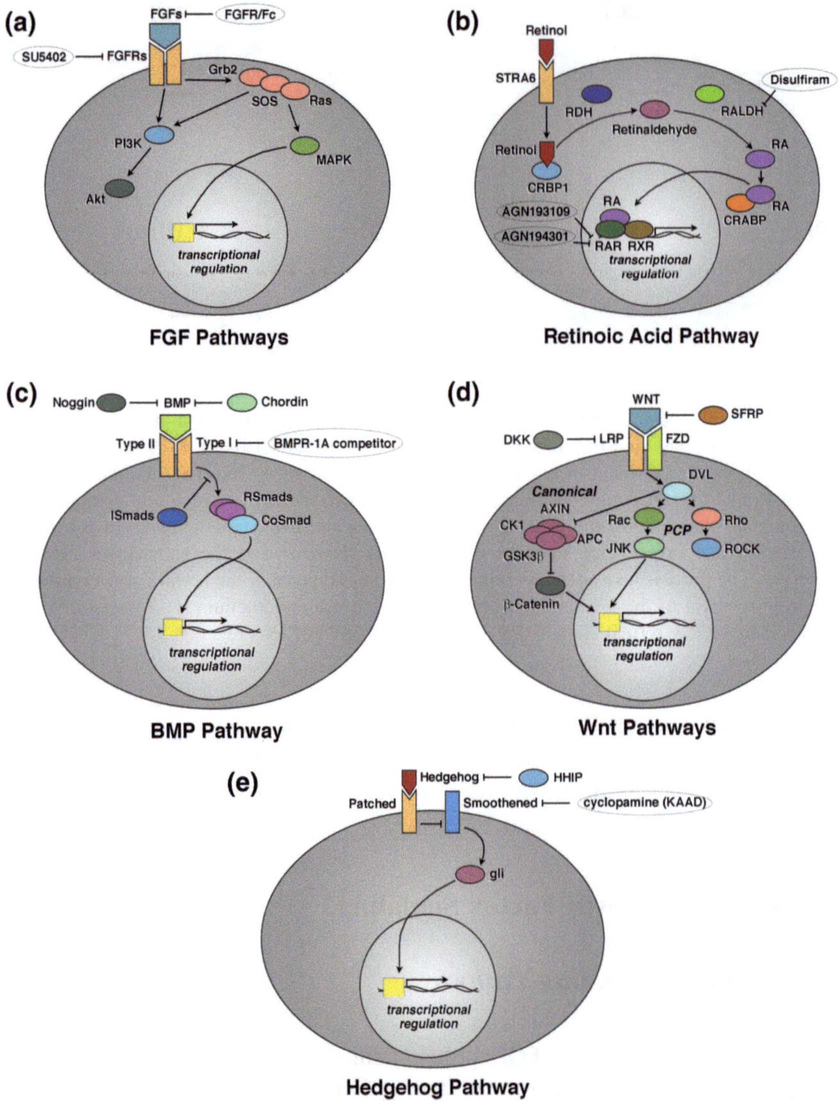

Fig. 3 Highly simplified summary diagrams of various signaling pathways known to play roles during vertebrate lens regeneration. Various pathways and pathway members are as labeled. Not all known elements are included here for the sake of simplicity. Specific activators and inhibitors used to manipulate these pathways are also included, and specific synthetic agents are enclosed in ovals. See text for further details

2.2 FGF Signaling in Lens Development

The FGF pathway has been identified to play important roles in embryonic lens development (reviewed in Donner et al. 2006; Robinson 2006; Gunhaga 2011). In early lens development, the FGF pathway participates in establishing lens-forming competence, bias, and specification. FGF signaling is essential for cell survival in the lens placode, as conditional deletion of FGFR1 and FGFR2 in the mouse lens placode increases apoptosis and prevents lens formation (Garcia et al. 2011). During later eye development, FGF is essential for inducing lens fiber development and establishing proper lens structure (Robinson 2006). A greater concentration of FGF is present in the vitreous humor than in the aqueous humor (Schulz et al. 1993), and normal lens polarity depends upon this FGF concentration gradient (Chamberlain and McAvoy 1997). For instance, perturbation of this gradient through lens-specific overexpression of FGFs in transgenic mice results in disorganized lens development (Stolen et al. 1997; Lovicu and Overbeek 1998; Robinson et al. 1998). In a fascinating recent study, FGF2 addition was required for inducing human embryonic stem cells to differentiate into lentoid bodies, mirroring the requirement of FGF signaling during embryonic lens development (Yang et al. 2010).

2.3 FGF Signaling in Wolffian Lens Regeneration

Members of the FGF family and their receptors are expressed in iris tissues during Wolffian regeneration. *fgf1* mRNA is present in the dedifferentiating dorsal iris and regenerating lens (Del Rio-Tsonis et al. 1997), and *fgf2* mRNA levels increase in both the dorsal and ventral iris regions following lens removal (Hayashi et al. 2004). As for the receptors, FGFR1, FGFR2, and FGFR3 are present in the dedifferentiating dorsal iris and early lens vesicle during lens regeneration (Del Rio-Tsonis et al. 1997; McDevitt et al. 1997; Del Rio-Tsonis et al. 1998). Interestingly, FGFR1 is present in the dorsal lens-forming iris but not in the regeneration incompetent ventral iris (Del Rio-Tsonis et al. 1998), and similarly, FGFR3 is present more in the dorsal iris than the ventral iris (McDevitt et al. 1997). The presence of these specific factors in the dorsal iris points to a potential role in lens regeneration.

In fact, the FGF pathway is necessary for Wolffian lens regeneration and proper morphology of the regenerated lens. Application of the small molecule inhibitor of tyrosine kinase activity (SU5402) inhibits FGFRs and lens regeneration, showing the necessity of FGFR function in this process (Fig. 3a; Del Rio-Tsonis et al. 1998). Furthermore, the ligands for FGFR2 isoform IIIc appear to be necessary for lens regeneration, as daily injection of a competitive recombinant FGFR2 isoform IIIc (FGFR2/Fc) inhibited lens regeneration (Hayashi et al. 2004). In contrast, injection of soluble recombinant FGFR2 (IIIb) did not inhibit regeneration. These results suggest that FGFR2 (IIIc) receptor function itself may be essential for lens regeneration, and point to the importance of the ligand specificity of FGFR isoforms (Hayashi et al. 2004). Furthermore, FGFs appear to play a role in lens fiber differentiation and lens structure, similar to the case in development. Application of exogenous recombinant FGF1 or FGF4 after lens removal perturbs regenerated lens polarity and structure, and may result in the formation of multiple lenses (Del Rio-Tsonis et al. 1997; Yang et al. 2005).

Specific FGFs appear to be the key ligands sufficient for inducing Wolffian lens regeneration. An in vitro study of newt dorsal iris epithelial cell cultures noted that the addition of FGF2 or FGF4 to culture media was sufficient to induce lens development akin to Wolffian lens regeneration in vivo, whereas FGF8, FGF10, EGF, IGF1, and VEGF were not found to be sufficient (Hayashi et al. 2002). This effect by FGF2 and FGF4 was completely inhibited by the FGFR inhibitor SU5402; therefore, these FGFs are likely acting as the key ligand(s) for specific FGF receptors (Hayashi et al. 2002). Likewise, Hayashi et al. (2004) showed the sufficiency of FGF2 for in vivo Wolffian lens regeneration. In that study, the functions of members of the FGF family (1, 2, 4, 7, 8, 9, 10), as well as EGF, IGF, and VEGF were examined. Each growth factor was introduced into the anterior chamber of the eye without lens removal, and of those factors tested, only FGF2 was sufficient to induce secondary lens formation from the dorsal iris tissue. Overall, the results above suggest that the FGF signaling plays a key role in initiating Wolffian lens formation. Furthermore, a precise gradient of FGF is necessary for establishing proper lens polarity during subsequent stages of regeneration.

2.4 FGF Signaling in Cornea Lens Regeneration

fgf and *fgfr* expression have been observed in eye tissues during *Xenopus* cornea lens regeneration (Fukui and Henry 2011). Though many FGF family members are expressed, a few FGFs (*fgf1, 8,* and *9*) are more highly expressed in the retina than in the cornea, and thus could be candidate signals originating from the neural retina that induce lens regeneration. In a microarray analysis comparing differences in expression between lenses, regenerating corneas, and sham operated corneas (where the inner cornea and lens were left intact), *fgf8b* expression was identified to be up-regulated in regenerating corneas relative to lens and sham operated corneas (Day and Beck 2011). This raises the possibility of additional autocrine or paracrine FGF8 signaling within the cornea after it has received an initial signal from the neural retina. As for the receptors, *fgfr1*, *fgfr2*, and *fgfr3* mRNA are present in larval corneas, suggesting that any of these could act as the key receptor for induction of lens formation from corneas (Fukui and Henry 2011). Further, from a subtracted cDNA library enriched for genes expressed during lens regeneration, *fgfr3* expression was found to be up-regulated in regenerating corneas (Henry et al. 2002; Malloch et al. 2009). In addition, FGFR2 expression has been correlated with lens-forming competence in *Xenopus* epidermal tissues (Arresta et al. 2005). Specifically, FGFR2 (*bek* isoform) was detected by immunohistochemistry in the cornea and pericorneal ectoderm, but not in regeneration incompetent flank epidermis (Arresta et al. 2005). Other downstream FGF pathway markers are also up-regulated in regenerating corneas. In particular, *mek2* (a member of the MAPK pathway), *grb2* (an upstream member of the MAPK and PI3 K pathways), and *spry2* (a negative regulator of FGFRs) have been identified in the regenerating cornea (Malloch et al. 2009). Together, these expression data suggest that FGF1, 8, and/or 9 may interact with FGFR1, 2, and/or 3 in the cornea to induce lens regeneration.

There is also functional evidence that the FGF pathway is both necessary and sufficient for cornea lens regeneration. Similar to observations made for Wolffian lens regeneration above, FGFR function also appears to be necessary for cornea lens regeneration, as addition of SU5402 inhibited lens regeneration (Fig. 3a; Fukui and Henry 2011). Furthermore, exogenous FGF1 appears to be sufficient for inducing lens cell formation within cultured *Xenopus* corneas (Bosco et al. 1994, 1997).

3 Retinoic Acid Signaling

3.1 The RA Signaling Pathway

Retinol (Vitamin A) is the biological source of retinoids in animals, which is delivered to a transmembrane protein located on the cell surface, STRA6 (Stimulated by retinoic acid (RA), Fig. 3b). In the cytosol, retinol is bound to Cytoplasmic Retinol Binding Protein (CRBP1), and is oxidized to retinaldehyde by a suite of dehydrogenases such as RDH10, and is subsequently oxidized to RA by tissue-specific Retinaldehyde Dehydrogenases RALDH1/2/3 (Moloktov et al. 2002a, b). RA is bound by Cellular Retinoic Acid Binding Protein (CRABP) and ultimately transported to the nucleus where it is released. RA exerts its influence by binding to nuclear Retinoic Acid Receptors (RAR$\alpha/\beta/\gamma$), which dimerize with other RARs or Retinoid X Receptors (RXR$\alpha/\beta/\gamma$), These complexes, in turn, bind to specific genomic elements known as Retinoic Acid Response Elements (RAREs), where they act as transcriptional activators or repressors. In the absence of ligand, the aporeceptors generally act to recruit chromatin condensing histone deacetylases, and when RA is bound to the receptors the chromatin is remodeled to enable transcription, via the recruitment of histone acetyl transferases (reviewed by Niederreither and Dolle 2008). RA may also form a complex with CRABP and exit the cell, enabling RA signaling to act in both autocrine and paracrine fashions (reviewed by Cvekl and Wang 2009).

3.2 RA Signaling in Lens Development

RA signaling plays important roles in the development of various eye tissues, including the retina, lens, and cornea (Kastner et al. 1994; Enwright and Grainger 2000; Wagner et al. 2000; see also review by Cvekl and Wang 2009). For instance, proper eye morphogenesis itself depends on RA (Hyatt et al. 1996; Molotkov et al. 2006), and studies using mice have shown that RA induces crystallin expression in lens cells (Gopal-Srivastava et al. 1998).

3.3 RA Signaling in Wolffian Lens Regeneration

RA has been shown to play a necessary role in Wolffian lens regeneration through the use of small molecule inhibitors such as disulfiram, which inhibits RA-synthesis enzymes (retinaldehyde dehydrogenases), as well as inhibitors of RAR nuclear receptors (Fig. 3b). While there is almost certainly a redundancy of function of RAR receptor isotypes within the regenerating lens, RARα was specifically implicated when the use of a RARα specific inhibitor, AGN194301, stunted the regenerative capacity of the dorsal iris (Fig. 3b; Tsonis et al. 2000, 2002). In many cases there was a failure to form a dedifferentiated lens vesicle from the dorsal iris, suggesting a role for RA in the earliest events of regeneration. Furthermore, it is interesting to note that the use of a pan-RAR antagonist (AGN193109, Fig. 3b) resulted in some cases of ectopic lens formation, including one case of lens regeneration from the cornea, reminiscent of that which takes place in *Xenopus*. This effect was not seen when the RARα specific antagonist was used, suggesting that whatever function RARα might have within the regenerating tissue, it does not involve defining the site of lens formation. *rara* (RARα) is not detected by in situ analysis in unoperated lenses, but it is expressed in the lens epithelial cells and the fiber cells within the regenerating lens, as well as the dedifferentiated lens vesicle that initially buds from the dorsal iris at the start of regeneration (Tsonis et al. 2002). The same expression pattern is true for *RARδ* (*RARγ*), but its functional role has not been specifically investigated in the regenerating lens (Tsonis et al. 2000). Work is needed to reveal what cellular or molecular mechanisms the various RA nuclear receptors control in the context of vertebrate lens regeneration.

Grogg et al. (2005) demonstrated that the normally lens-incompetent ventral iris could be induced to regenerate lenses when transfected with a cDNA expression construct encoding the transcription factor *six-3*, but only when these samples were co-treated with RA. These treatments forced the ventral iris to adopt patterns of gene expression that are seen only in the dorsal iris, and the authors suggest that it is this "dorsalization" that renders the ventral iris capable of transdifferentiation. Since RA was necessary for this transformation to take place, it can be reasoned that gene regulation by RA signaling is critical for conferring lens-competence upon lens-incompetent tissue. Taken together with the earlier studies described above, it appears that RA is important in all phases of Wolffian regeneration: acquiring/maintaining lens-competence, dedifferentiation of the iris, and terminal differentiation of lens fiber cells. Investigations of which gene loci are under the control of RA-mediated chromatin remodeling within the regenerating tissue will cast light upon currently unknown genes or regulatory regions that are a critical part of lens regeneration.

3.4 RA Signaling in Cornea Lens Regeneration

The role of RA in *Xenopus* cornea lens regeneration is less understood. Malloch et al. (2009) identified *rxrg* (encoding RXRγ) from the subtracted cDNA library, but there is a lack of data regarding either function or expression of any RA nuclear receptors within the larval eye.

4 Transforming Growth Factor β Signaling

4.1 The TGFβ/BMP Signaling Pathways

Signaling via the Transforming Growth Factor β (TGFβ) superfamily of proteins is one of the major pathways of signal transduction in cells, and plays key roles in cell renewal and differentiation, as well as in the formation of complex tissue patterns and animal body plans. The superfamily encompasses many signaling pathways, including TGFβ, Bone Morphogenetic Protein (BMP), Growth Differentiation Factor (GDF), Activin, Anti-Muellerian Hormone (AMH), and others. Signaling begins at the cell surface when a ligand, such as BMP, binds and activates Type I and Type II receptors (Fig. 3c). This leads to the subsequent phosphorylation of the transducers of the signal, the Smad proteins. Receptor Smads (RSmads, i.e. Smads 1/2/3/5/8) form a complex with Co-Smad (Smad4), which then translocate to the nucleus to effect transcriptional changes. Other Inhibitory Smads (ISmads, i.e. Smads 6/7) may act to negatively regulate these pathways (reviewed by Massagué et al. 2005; Moustakas and Heldin 2009).

4.2 TGFβ/BMP Signaling in Lens Development

BMP signaling has a well established role in eye development, including establishing lens-competence, lens placode development and induction (Luo et al. 1995; Furuta and Hogan 1998; Wawersik et al. 1999), and the differentiation of lens fiber cells (Belecky-Adams et al. 2002; Faber et al. 2002). The TGFβ ligand is found in the aqueous humor of many animals including humans (Jampel et al. 1990; Cousins et al. 1991; Granstein et al. 1990), and it is known to inhibit lens epithelial cell proliferation in vitro (Kurosaka and Nagamoto 1994). It is also important for the repair of wounded lens epithelia (Saika et al. 2001, 2004), but can induce cataract-like effects in explanted lenses (Liu et al. 1994).

4.3 TGFβ/BMP Signaling in Wolffian Lens Regeneration

Among the many members of the superfamily, only BMP signaling has been functionally characterized in the context of lens regeneration. A very striking involvement of BMP signaling in Wolffian regeneration was established when Grogg et al. (2005) enabled the ventral iris to transdifferentiate by treating the eyes with BMP inhibitors (Fig. 3c). Specifically, treatment with either Chordin or a BMPR-IA competitor was sufficient to enable lens regeneration in the ventral iris, even in the absence of *six3* and RA, as mentioned earlier. Moreover, treating the normally lens-competent dorsal iris with BMP activators, such as BMP-4 and BMP-7, significantly diminished their capacity to regenerate. These results are beautifully consistent with what is known about the role of BMPs in establishing a ventral identity (DeRobertis and Kuroda 2004), and the authors suggest that the ventralizing action of active BMP signaling within the ventral iris prohibits transdifferentiation. While BMP inhibition drove the lens-incompetent ventral iris of the newt to transdifferentiate, it did not have this effect on a related salamander (the axolotl), which is normally not capable of lens regeneration (Grogg et al. 2005). This suggests that the barrier to lens regeneration imposed by ventralizing factors may be a situation unique to the newt. A detailed understanding of the molecular differences in how the newt and axolotl eye structures develop and maintain their irides will provide insight as to why BMPs have different effects on the same tissue in related species. Our current understanding of TGFβ superfamily activity is further complicated by a recent study, where several members of the BMP and TGFβ pathways were identified among genes expressed in the dorsal iris during the dedifferentiation stage of Wolffian regeneration (Maki et al. 2010).

4.4 TGFβ/BMP in Cornea Lens Regeneration

BMP-5 and a protein known to inhibit BMP signaling, Sclerostin domain-containing protein 1 (SOSTDC1), were both implicated in *Xenopus* cornea lens regeneration when cDNAs encoding these proteins were identified in a subtracted cDNA library representing genes up-regulated in the course of regeneration (Henry et al. 2002; Malloch et al. 2009). The same study implicated the TGFβ type III receptor (encoded by *tgfbr3*), also known as Betaglycan. This member of the

TGFβ pathway is a ligand-presenting proteoglycan and can have stimulatory effects on TGFβ signaling (Blobe et al. 2001; Santander and Brandan 2006).

In the case of cornea lens regeneration in *Xenopus*, BMP inhibition has the diametrically opposite consequence on regeneration as compared to the newt. When BMP signaling was inhibited in the larval *Xenopus* cornea, it drastically lost its ability to regenerate lenses (Day and Beck 2011). Through the use of transgenic animals that inducibly express the BMP signaling inhibitor Noggin Day, and Beck demonstrated that sustained Noggin activity for the first 48 h of regeneration greatly reduced successful regeneration (Fig. 3c). It was shown further that the first stage of lens regeneration (Freeman Stage 1, see Freeman 1963), is not affected by Noggin expression, as the cornea epithelial cells still transform from a squamous to a cuboidal configuration as they do in wildtype regenerating animals, but subsequently these reverted to a sqauamous state rather than develop a differentiated lens. In light of these findings, it is possible that BMP signaling governs later events specific to the transformation/differentiation of lens cell fates, similar to those in the newt, but direct molecular evidence of this in *Xenopus* is currently lacking.

5 Wnt Signaling

5.1 The Wnt Signaling Pathways

Wnt signaling is typically divided into the canonical Wnt/β-catenin signaling pathway and the non-canonical Wnt/Planar Cell Polarity (PCP), and Wnt/Calcium pathways (Fig. 3d). The canonical Wnt/β-catenin pathway is centered on the regulation of β-catenin (reviewed in Logan and Nusse 2004; Clevers 2006; MacDonald et al. 2009). In the absence of a Wnt ligand, the β-catenin degradation complex, comprised of Glycogen synthase kinase 3 (GSK3), Casein kinase 1 (CK1), Adenomatous polyposis coli (APC), and the scaffolding protein Axin, form a complex to phosphorylate β-catenin, leading to its degradation via the proteasome. However, when Wnt binds to the Frizzled (FZD) receptor and the Low density lipoprotein receptor-related protein (LRP) co-receptor, signaling is activated. This ultimately inhibits the ability of the degradation complex to phosphorylate β-catenin, allowing it to accumulate and translocate into the nucleus. Once in the nucleus, β-catenin is able to activate transcription through T cell factor/Lymphoid enhancer factor (TCF/LEF) complexes.

Wnt signaling can also activate two non-canonical pathways (Fig. 3d). In the Wnt/PCP pathway, Wnt signaling leads to the activation of either Rho or Rac. These small GTPases lead to cytoskeletal changes through Rho-associated protein kinase (ROCK) or gene transcription through the Jun N-terminal kinase (JNK) cascade (reviewed in Gao and Chen 2010). Alternatively, signaling may occur via the Wnt/Calcium pathway. Interaction between the ligand and receptor leads to an increase in the levels of intracellular calcium. This results in the activation of the calcium binding proteins, Calcium calmodulin-dependent protein kinase II (CAMKII), and Protein kinase C (PKC, reviewed in Gao and Chen 2010; De 2011).

5.2 Wnt Signaling in Lens Development

Wnt signaling is involved in many processes during eye organogenesis (reviewed by Fuhrmann 2008), including the lens. In the early stages of vertebrate lens development, canonical Wnt signaling in the presumptive lens ectoderm needs to be inhibited in order for the initiation of lens formation (Smith et al. 2005; Kreslova et al. 2007; Machon et al. 2010). However, during later stages of lens development, canonical Wnt signaling becomes required for proper differentiation of the lens epithelium (Stump et al. 2003) and lens fiber cells (Chen et al. 2006). The expression patterns of various components and regulators of Wnt signaling during these processes have been well characterized and include most of the *wnt* ligands, the *fzd* receptors, as well as *dkks* and *sfrps* (Ang et al. 2004; Chen et al. 2004), the latter two being antagonists of the Wnt signaling pathway. More recently, the non-canonical Wnt/PCP pathway has emerged as a player in the organization of the lens through the elongation and orientation of lens fiber cells (Chen et al. 2008; Sugiyama et al. 2010, 2011).

5.3 Wnt Signaling in Wolffian Lens Regeneration

While much is known about the role that Wnt signaling plays during lens development, very little is known about the involvement of this signaling pathway during lens regeneration. The sole functional analysis of Wnt signaling during the process of Wolffian lens regeneration was carried out by Hayashi et al. (2006) in the newt. Using RT-PCR of iris tissue collected at various time points during regeneration, Hayashi and colleagues observed the expression of *wnt2b*, *wnt5a*, *fz2*, and *fz4*. Two of these genes, *wnt2b* and *fz4*, showed elevated expression levels in late lens regeneration, specifically in the dorsal iris (day 8 and day 12, respectively). Lens regeneration rates were dramatically reduced when cultured in the presence of FGF2 and the Wnt signaling antagonists DKK1 and SFRP1 (Fig. 3d). When WNT3A (a canonical Wnt) was added with FGF2 in culture medium, larger lenses regenerated from the dorsal iris and limited lens regeneration was even observed from the ventral iris, which is not normally capable of regeneration. The findings of this study suggest that canonical Wnt signaling is necessary for lens regeneration from the dorsal iris and is sufficient for lens regeneration to occur in ventral irides that have already initiated the early steps of lens regeneration, which are triggered by FGF2. While a screen of genes expressed during the initiation of depigmentation and proliferation of dorsal and ventral irides did not produce any ESTs representing Wnt pathway members (Maki et al.

2010), the authors point out this could have been due to the particular stage of the irides collected or may have been a caveat of the cloning methods.

5.4 Wnt Signaling in Cornea Lens Regeneration

No functional studies have yet been undertaken to examine the role of Wnt signaling during cornea lens regeneration in *Xenopus*. However, Wnt signaling has been implicated in two screens designed to characterize gene expression during this process. As mentioned above, Henry et al. (2002) created a subtracted cDNA library of genes expressed during the process of lens regeneration in *X. laevis* using control corneas and corneas in the first four days of lens regeneration. From this study, several ESTs were identified as members of the Wnt signaling pathway, including the ligand *wnt7b*, as well as the Wnt antagonists secreted frizzled-related proteins 3 and 5 (*sfrp3* and *sfrp5*; Malloch et al. 2009). These findings suggest some involvement of Wnt signaling during the process of cornea lens regeneration. The presence of these ESTs is also consistent with what is known about the expression of these genes during vertebrate lens development in mice. For instance, *sfrp3* is expressed in the lens epithelium (Stump et al. 2003; Chen et al. 2004), *wnt7b* and *sfrp5* are expressed in the lens primordia, epithelium, and fiber cells (Ang et al. 2004; Chen et al. 2004).

In a more recent microarray study using a *X. laevis* Affymetrix GeneChip, Day and Beck (2011) identified the differential expression of numerous components of the Wnt signaling pathway during cornea lens regeneration. Furthermore, qRT-PCR analysis of two of these genes, *fzd7* and *wnt7a*, revealed significant changes (up-regulation and down-regulation, respectively) in expression take place during regeneration when compared to sham operated control corneas at three days following surgery. Taken together, the presence of Wnt signaling transcripts in these two studies suggests that Wnt signaling is likely involved in cornea lens regeneration. How closely related these processes may be to those taking place during vertebrate lens formation or Wolffian lens regeneration has yet to be determined.

6 Hedgehog Signaling

6.1 The HH Signaling Pathway

Hedgehog (HH) proteins are secreted signaling molecules that are present in the form of three homologs in vertebrates: Sonic hedgehog (SHH), Indian hedgehog (IHH), and Desert hedgehog (DHH) (Ingham and McMahon 2001). Hedgehog signaling occurs through Patched (PTC), which is a 12-transmembrane receptor that mediates Smoothened activity (Fig. 3e). Smoothened is characterized as a heptahelical transmembrane protein that is a functionally divergent member of the G-protein coupled receptor (GPCR) superfamily (Goodrich et al. 1996; Johnson et al. 1996; Stone et al. 1996). Under conditions where HH is limited, Smoothened activity is normally inhibited by PTC. When HH is present, Patched-mediated repression of Smoothened is overcome and Smoothened positively regulates HH pathway activation. This results in transduction of the HH signal into the cell and attenuates the processing of GLI proteins (van den Heuvel and Ingham 1996; Ingham et al. 2011). Zinc-finger motif transcription factors GLI1, GLI2 and GLI3 are activated and translocated to the nucleus where their accumulation controls transcription of HH target genes (Sasaki et al. 1999). Cell surface proteins also mediate HH signaling, such as Hedgehog interacting protein (HHIP) and GAS1, which negatively or positively modulate the pathway, respectively (Chuang and McMahon 1999; Allen et al. 2007; Martinelli and Fan 2007).

6.2 HH Signaling in Lens Development

The HH signaling pathway significantly influences vertebrate embryonic development and is vital for proper eye morphogenesis. More specifically, SHH regulates the spatial expression of *Pax-6* in the ventral forebrain, which is instrumental in establishing the midline and proper formation of the two eyes (Ekker et al. 1995; Macdonald et al. 1995). Further, *Shh* expression in the ganglion cell layer of the developing retina is important for proper retinal organization and differentiation of photoreceptors (Perron et al. 2003). Based on the important role HH signaling plays during embryonic eye development, it would seem likely that this pathway would be involved during lens regeneration.

6.3 HH Signaling in Wolffian Lens Regeneration

One study has explored HH signaling during newt lens regeneration (Tsonis et al. 2004). In this study, expression data was presented for members of the HH pathway in embryonic lenses, and in adult and regenerating lenses. During development, newt embryonic lenses express *Shh*, *Ihh*, *Ptc1* and *Ptc2*. Yet, once the adult lens has formed, only one HH related gene (*Ptc1*) is detected in the lens epithelium. However, during the process of lens regeneration, members of the HH pathway (*Shh*, *Ihh*, *Ptc1* and *Ptc2*) are expressed again in the regenerating lens epithelium and lens fibers.

Tsonis et al. (2004) explored the necessity of the HH pathway with two known HH inhibitors. The first experiment involved the use of a synthetic cyclopamine inhibitor, KAAD, which interferes with the activation of Smoothened and prevents GLI translocation to the nucleus (Fig. 3e). In the presence of KAAD, 28.6 % of lens regenerates were of smaller size in comparison to a normal regenerating lens.

Perturbation of the HH signaling pathway was also accomplished through the introduction of mammalian cells transiently expressing HHIP (Fig. 3e). Results were similar to those seen for KAAD, and the authors reported that lenses were smaller in 33.3 % of cases when compared to normal regenerating lenses. Inhibition of Ptc1 was demonstrated through semi-quantitative PCR and corroborated the effects of KAAD and HHIP. Further, inhibition of the HH pathway during lens regeneration affected cell proliferation and lens fiber differentiation. Eyes undergoing the process of regeneration in the presence of KAAD had lower rates of cell proliferation, indicating that the HH pathway regulates cell division during this process. An assay for lens fiber differentiation also showed that those eyes mostly affected by KAAD did not synthesize β-crystallin, which suggests the HH pathway is important for proper lens fiber differentiation during regeneration.

In order to better understand the dedifferentiation process (i.e., the earliest events during lens regeneration in newts), Maki et al. (2010) examined dorsal and ventral iris expression profiles early during newt lens regeneration at eight days after lens removal. While other signaling pathway members were present in their ESTs (e.g., TGFβ and BMPs), HH pathway members were not present. However, this absence could best be explained by the stage of irides examined and the fact that pathways important for later lens cell differentiation may not be active this early in the regenerative process (Maki et al. 2010).

6.4 HH Signaling in Cornea Lens Regeneration

No studies have been undertaken to explore the presence or absence of HH signaling during the process of cornea lens regeneration in *Xenopus*. The only implication for HH signaling in *Xenopus* comes from subtracted cornea cDNA library screens during lens regeneration (Henry et al. 2002; Malloch et al. 2009). In those studies a member of the HH pathway, *Ptc1*, was found to be up-regulated in the cornea within the first four days of regeneration. While no other members of the HH family were uncovered during this screen, it is possible that this signaling pathway may participate in the process of lens regeneration in *Xenopus* and this should be explored in future studies.

7 Evidence for Involvement of Other Signaling Pathways in Lens Regeneration

As previously mentioned, a small number of EST surveys have been undertaken to characterize gene expression during the process of lens regeneration in newts (Makarev et al. 2007; Maki et al. 2010) and *Xenopus laevis* (Henry et al. 2002; Malloch et al. 2009; Day and Beck 2011). In addition to the pathways discussed above, those studies implicate additional signaling pathways in these processes of lens regeneration.

7.1 Studies of Gene Expression During Wolffian Lens Regeneration

The study by Maki et al. (2010) examined gene expression during Wolffian lens regeneration in the newt *Cynops pyrrhogaster*. A cDNA library was constructed using mRNA isolated from both dorsal and ventral iris tissues collected eight days after lens removal, when the pigmented epithelial cells were undergoing the process of de-differentiation (i.e. depigmentation and proliferation). Aside from the expression of certain elements of pathways described above (i.e., the BMP/TGF-β pathways), they did not observe significant expression of elements from other signaling pathways.

Though more limited, an earlier study by Makarev et al. (2007) also examined gene expression in dorsal versus ventral iris tissues collected eight days following lens removal in the newt *Notophthalmus viridescens*. That study used a small microarray representing 373 genes previously found to be expressed in regenerating newt limb blastemas (Atkinson et al. 2006). No evidence for specific signaling pathways was reported for Wolffian lens regeneration in that study.

A proteomic analysis was undertaken by Roddy et al. (2008) to compare protein expression changes in dorsal versus ventral iris tissues following removal of the lens in the newt *N. viridescens*. They also compared expression for selected proteins in dorsal iris tissues of the axolotl (*Ambystoma mexicanum*), which is unable to regenerate lenses. Overall, a great deal of similarity was found in the expression of numerous proteins in newt dorsal versus ventral iris tissues, suggesting that both respond similarly to initiate the process of lens regeneration, immediately following removal of the lens. There were, however, some interesting differences. For instance, Epidermal growth factor receptor (EGFR) was found to be significantly up-regulated in dorsal iris tissues of the newt when compared to ventral iris tissues, and EGFR was not found to be expressed in the dorsal iris of the axolotl. Correspondingly, a negative regulator of EGFR signaling, Suppressor of cytokine signaling 4, was found to be up-regulated in newt ventral iris tissues and in the dorsal iris of the axolotl, suggesting that EGFR signaling may play an important role in permitting lens regeneration to take place in the dorsal iris of the newt. Roddy et al. (2008) also detected significant increases in the expression of Protein-serine kinase Cα, calcium/calmodulin-dependent protein-serine kinase 2α, MAP kinase phosphatase 1 and p21 activated serine kinase 3 in dorsal iris tissue of the newt following lens removal.

7.2 Studies of Gene Expression During Cornea Lens Regeneration

In the case of cornea lens regeneration in *Xenopus,* Day and Beck (2011) carried out a microarray analysis to characterize gene expression three days following lens removal (using an Affymetrix GeneChip). As described above, that analyses implicated the involvement of the Wnt signaling pathways in lens regeneration. Evidence for other specific signaling pathways was not described. An earlier study by Malloch et al. (2009) revealed that elements of other signaling pathways are expressed following lens removal in *Xenopus*. These included a number of genes encoding proteins involved in Rho/Ras signal transduction, including: Rho-related GTP-binding protein RhoB, Rho GDP-dissociation inhibitor 2, Rho GTPase-activating protein 10, Rho-related BTB domain-containing protein 1, and Ras-related protein Rab-4B. In addition, components of the MAPK/ERK signal transduction pathway involved in transducing a number of different signal pathways (including receptor tyrosine kinases such as FGF and EGF receptors) were also expressed (e.g., MAP kinase signal-integrating kinase 2, MAPK activator 2, RAS guanyl-releasing protein and UPF0485 or putative MAPK-activating protein PM18/20/22). Also detected was a mediator of delta-notch signaling (*Mindbomb*).

8 Summary

These studies indicate that lens regeneration is a complex process that appears to involve a number of different signaling pathways that are deployed in many different contexts, including the establishment of lens regeneration competence in specific tissues, the initiation of lens regeneration, and subsequent processes related to lens cell proliferation, patterning, and differentiation. Furthermore, there appear to be some interesting differences in the deployment of specific signaling pathways in Wolffian versus cornea lens regeneration in these animals.

References

Allen BL, Tenzen T, McMahon AP (2007) The Hedgehog-binding proteins Gas1 and Cdo cooperate to positively regulate Shh signaling during mouse development. Genes Dev 21:1244–1257

Ang SJ, Stump RJ, Lovicu FJ, McAvoy JW (2004) Spatial and temporal expression of *Wnt* and *Dickkopf* genes during murine lens development. Gene Expr Patterns 4:289–295

Arresta E, Bernardini S, Gargioli C, Filoni S, Cannata SM (2005) Lens-forming competence in the epidermis of *Xenopus laevis* during development. J Exp Zool 303A:1–12

Atkinson DL, Stevenson TJ, Park EJ, Riedy MD, Milash B, Odelberg SJ (2006) Cellular electroporation induces dedifferentiation in intact newt limbs. Dev. Biol. 299:257–271

Belecky-Adams TL, Adler R, Beebe DC (2002) Bone morphogenetic protein signaling and the initiation of lens fiber cell differentiation. Development 129:3795–3802

Blobe G, Liu X, Fang SJ, How T, Lodish HF (2001) A novel mechanism for regulating transforming growth factor β (TGF-β) signaling. Functional modulation of type III TGF-b receptor expression through interaction with the PDZ domain protein. GIPC. J. Biol. Chem 276:39608–39617

Bosco L, Venturini G, Willems D (1994) First evidence of lens-transdifferentiation of larval *Xenopus laevis* induced by brain-derived acidic FGF. Rendiconti Lincei 5:261–268

Bosco L, Venturini G, Willems D (1997) In vitro lens transdifferentiation of *Xenopus laevis* outer cornea induced by Fibroblast Growth Factor (FGF). Development 124:421–428

Chamberlain C, McAvoy J (1997) Fibre differentiation and polarity in the mammalian lens: a key role for FGF. Progress in Retinal and Eye Research 16:443–478

Chen Y, Stump RJ, Lovicu FJ, McAvoy JW (2004) Expression of frizzleds and secreted frizzled-related proteins (Sfrps) during mammalian lens development. Int J Dev Biol 48:867–877

Chen Y, Stump RJ, Lovicu FJ, McAvoy JW (2006) A role for Wnt/planar cell polarity signaling during lens fiber cell differentiation? Semin Cell Dev Biol 17:712–725

Chen Y, Stump RJ, Lovicu FJ, Shimono A, McAvoy JW (2008) Wnt signaling is required for organization of the lens fiber cell cytoskeleton and development of lens three-dimensional architechture. Dev Biol 324:161–176

Chuang PT, McMahon AP (1999) Vertebrate hedgehog signaling modulated by induction of a hedgehog-binding protein. Nature 397(6720):617–621

Clevers H (2006) Wnt/beta-catenin signaling in development and disease. Cell 127:469–480

Cousins SW, McCabe MM, Danielpour D, Streilein W (1991) Identification of transforming growth factor-beta as an immunosuppressive factor in aqueous humor. Invest Ophthalmol Vis Sci 32:2201–2211

Cvekl A, Wang WL (2009) Retinoic acid signaling in mammalian eye development. Exp Eye Res 89:280–291

Dailey L, Ambrosetti D, Mansukhani A, Basilico C (2005) Mechanisms underlying differential responses to FGF signaling. Cytokine Growth Factor Rev 16:233–247

Day RC, Beck CW (2011) Transdifferentiation from cornea to lens in *Xenopus laevis* depends on BMP signaling and involves upregulation of Wnt signaling. BMC Dev Biol 11:54

De A (2011) Wnt/Ca2 + signaling pathway: a brief overview. Acta Biochim Biophys Sin 43:745–756

Del Rio-Tsonis K, Jung JC, Chiu I-M, Tsonis PA (1997) Conservation of fibroblast growth factor function in lens regeneration. Proc Natl Acad Sci 94:13701–13706

Del Rio-Tsonis K, Trombley MT, McMahon G, Tsonis PA (1998) Regulation of lens regeneration by fibroblast growth factor receptor 1. Dev Dyn 213:140–146

DeRobertis EM, Kuroda H (2004) Dorsal–ventral patterning and neural induction in *Xenopus* embryos. Annu Rev Cell Dev Biol 20:285–308

Donner A, Lachke S, Maas R (2006) Lens induction in vertebrates: variations on a conserved theme of signaling events. Semin Cell Dev Biol 17:676–685

Dorey K, Amaya E (2010) FGF signaling: diverse roles during early vertebrate embryogenesis. Development 137:3731–3742

Ekker SC, Ungar AR, Greenstein P, von Kessler DP, Porter JA, Moon RT, Beachy PA (1995) Patterning activities of vertebrate hedgehog proteins in the developing eye and brain. Curr Biol 5(8):944–955

Enwright JF, Grainger RM (2000) Altered retinoid signaling in the heads of small eye mouse embryos. Dev. Biol 221:10–22

Faber SC, Robinson ML, Makarenkova HP, Lang RA (2002) Bmp signaling is required for development of primary lens fiber cells. Development 129:3727–3737

Filoni S, Bernardini S, Cannata SM (2006) Experimental analysis of lens-forming capacity in Xenopus borealis larvae. J Exp Zool 305:538–550

Freeman G (1963) Lens regeneration from the cornea in *Xenopus laevis*. J Exp Zool 154:39–66

Fuhrmann S (2008) Wnt signaling in eye organogenesis. Organogenesis 4:60–67

Fukui L, Henry JJ (2011) FGF signaling is required for lens regeneration in *Xenopus laevis*. Biol Bull 221:137–145

Furuta Y, Hogan BLM (1998) BMP4 is essential for lens induction in the mouse embryo. Genes Dev 12:3764–3775

Gao C, Chen YG (2010) Dishevelled: the hub of Wnt signaling. Cell Signal 22:717–727

Garcia CM, Huang J, Madakashira BP, Liu Y, Rajagopal R, Dattilo L, Robinson ML, Beebe DC (2011) The function of FGF signaling in the lens placode. Dev Biol 351:176–185

Goodrich LV, Johnson RL, Milenkovic L, McMahon JA, Scott MP (1996) Conservation of the hedgehog/patched signaling pathway from flies to mice: induction of a mouse *patched* gene by Hedgehog. Genes Dev 10:301–312

Gopal-Srivastava R, Cvekl A, Piatigorsky J (1998) Involvement of retinoic acid/retinoid receptors in the regulation of murine alphaB-crystallin/small heat shock protein gene expression in the lens. J Biol Chem 273:17954–17961

Granstein RD, Staszewski R, Knisely TL, Zeira E, Nazareno R, Latina M, Albert DM (1990) Aqueous humor contains transforming growth factor- beta and a small (<3500 daltons) inhibitor of thymocyte proliferation. J Immunol 144:3021–3027

Grogg MW, Call MK, Okamoto M, Vergara MN, Tsonis PA (2005) BMP inhibition-driven regulation of six-3 underlies induction of newt lens regeneration. Nature 438:858–862

Gunhaga L (2011) The lens: a classical model of embryonic induction providing new insights into cell determination in early development. Philos Trans R Soc Lond B Biol Sci 366:1193–1203

Hayashi T, Mizuno N, Owaribe K, Kuroiwa A, Okamoto M (2002) Regulated lens regeneration from isolated pigmented epithelial cells of newt iris in culture in response to FGF2/4. Differentiation 70:101–108

Hayashi T, Mizuno N, Ueda Y, Okamoto M, Kondoh H (2004) FGF2 triggers iris-derived lens regeneration in newt eye. Mech Dev 121:519–526

Hayashi T, Mizuno N, Takada R, Takada S, Kondoh H (2006) Determinative role of Wnt signals in dorsal iris-derived lens regeneration in newt eye. Mech Dev 123:793–800

Henry JJ (2003) Cell and molecular biology of lens regeneration. Int Rev Cytol Surv Cell Biol 228:195–264

Henry JJ, Elkins ME (2001) Cornea lens transdifferentiation in *Xenopus tropicalis*. Dev Genes Evol 211:377–387

Henry JJ, Tsonis PA (2010) Molecular and cellular aspects of amphibian lens regeneration. Prog Retinal Eye Res 29:543–555

Henry JJ, Carinato ME, Schaefer JJ, Wolfe AD, Walter BE, Perry KJ, Elbl TN (2002) Characterizing gene expression during lens formation in *Xenopus laevis*: evaluating the model for embryonic lens induction. Dev Dyn 224:168–185

Henry JJ, Wever JA, Veragara MN, Fukui L (2008) Ch. 6. *Xenopus*, an ideal vertebrate system for studies of eye development and regeneration. In: Tsonis, PA (ed) Animal models for eye research. Academic Press, San Diego, CA

Hyatt GA, Schmitt EA, Marsh-Armstrong N, McCaffery P, Drager UC, Dowling JE (1996) Retinoic acid establishes ventral retinal characteristics. Development 122:195–204

Ikeda Y (1936a) Neue versuche zur analyse de Wolffschen linsregeneration. Arb Anat Inst Kais Jpn Univ Sendai 18:1–16

Ikeda Y (1936b) Beiträge zur frage der fähigkeit zur linsenregeneration bei einer Art von *Hynobius* (*Hynobius unnangso* Tago). Arb Anat Inst Kais Jpn Univ Sendai 18:17–50

Ikeda Y (1939) Zur frage der lensenpotenz der hornhaut in spätembryonalen und larvvalen stadien bei ener Art von *Hynobius* (*Hynobius unnangso* Tago). Arb Anat Inst Kais Jpn Univ Sendai 22:27–52

Ingham PW, McMahon AP (2001) Hedgehog signaling in animal development: paradigms and principles. Genes Dev 15:3059–3087

Ingham PW, Nakano Y, Seger C (2011) Mechanisms and functions of hedgehog signaling across the metazoa. Nat Rev Genet 12(6):393–406

Itoh N (2007) The Fgf families in humans, mice, and zebrafish: their evolutional processes and roles in development, metabolism, and disease. Biol Pharm Bull 30:1819–1825

Itoh N, Ornitz DM (2011) Fibroblast growth factors: from molecular evolution to roles in development, metabolism and disease. J Biochem 149:121–130

Jampel HD, Roche N, Stark WJ, Roberts AB (1990) Transforming growth factor-beta in human aqueous humor. Curr Eye Res 9:963–969

Johnson RL, Rothman AL, Xie J, Goodrich LV, Bare JW, Bonifas JM, Quinn AG, Myers RM, Cox DR, Epstein EH Jr, Scott MP (1996) Human homolog of *patched*, a candidate gene for the basal cell nevus syndrome. Science 272:1668–1671

Kastner P, Grondona JM, Mark M, Gansmuller A, LeMeur M, Decimo D, Vonesch JL, Dollé P, Chambon P (1994) Genetic analysis of RXR alpha developmental function: convergence of RXR and RAR signaling pathways in heart and eye morphogenesis. Cell 78(6):987–1003

Kreslova J, Machon O, Ruzickova J, Lachova J, Wawrousek EF, Kemler R, Krauss S, Piatigorsky J, Kozmik Z (2007) Abnormal lens morphogenesis and ectopic lens formation in the absence of beta-catenin function. Genesis 45:157–168

Kurosaka D, Nagamoto T (1994) Inhibitory effect of TGF-beta 2 in human aqueous humor on bovine lens epithelial cell proliferation. Invest Ophthalmol Vis Sci 35:3408–3412

Liu J, Hales AM, Chamberlain CG, McAvoy JW (1994) Induction of cataract-like changes in rat lens epithelial explants by transforming growth factor beta. Invest Ophthalmol Vis Sci 35:388–401

Logan CY, Nusse R (2004) The Wnt signaling pathway in development and disease. Annu Rev Cell Dev Biol 20:781–810

Lovicu FJ, Overbeek PA (1998) Overlapping effects of different members of the FGF family on lens fiber differentiation in transgenic mice. Development (Cambridge, England) 125:3365–3377

Luo G, Hofmann C, Bronckers AL, Sohocki M, Bradley A, Karsenty G (1995) BMP-7 is an inducer of nephrogenesis, and is also required for eye development and skeletal patterning. Genes Dev 9:2808–2820

Macdonald R, Barth KA, Xu Q, Holder N, Mikkola I, Wilson SW (1995) Midline signaling is required for *Pax* gene regulation and patterning of the eyes. Development 121(10):3267–3278

MacDonald BT, Tamai K, He X (2009) Wnt/beta-catenin signaling: components, mechanisms, and diseases. Dev Cell 17:9–26

Machon O, Kreslova J, Ruzickova J, Vacik T, Klimova L, Fujimura N, Lachova J, Kozmik Z (2010) Lens morphogenesis is dependent on Pax6-mediated inhibition of the canonical Wnt/beta-catenin signaling in the lens surface ectoderm. Genesis 48:86–95

Makarev E, Call M, Crogg M, Atkinson DL, Milash B, Odelberg SJ, Tsonis PA (2007) Gene expression signatures in the newt irises during lens regeneration. FEBS Lett 581:1865–1870

Maki N, Martinson J, Nishimura O, Tarui H, Meller J, Tsonis PA, Agata K (2010) Expression profiles during dedifferentiation in newt lens regeneration revealed by expressed sequence tags. Mol Vis 16:72–87

Malloch EM, Perry KJ, Fukui L, Johnson V, Wever J, Beck CW, King MW, Henry JJ (2009) Gene expression profiles of lens regeneration and development in *Xenopus laevis*. Dev Dyn 238:2340–2356

Martinelli DC, Fan CM (2007) Gas1 extends the range of Hedgehog action by facilitating its signaling. Genes Dev 21:1231–1243

Mason I (2007) Initiation to end point: the multiple roles of fibroblast growth factors in neural development. Nat Rev Neurosci 8:583–596

Massagué J, Seoane J, Wotton D (2005) Smad transcription factors. Genes Dev 19:2783–2810

McDevitt D, Brahma SK, Courtois Y, Jeanny J-C (1997) Fibroblast growth factor receptors and regeneration of the eye lens. Dev Dyn 208:220–226

Molotkov A, Deltour L, Foglio MH, Cuenca AE, Duester G (2002a) Distinct retinoid metabolic functions for alcohol dehydrogenase genes Adh1 and Adh4 in protection against vitamin A toxicity or deficiency revealed in double null mutant mice. J Biol Chem 277:13804–13811

Molotkov A, Fan X, Duester G (2002b) Excessive vitamin A toxicity in mice genetically deficient in either alcohol dehydrogenase Adh1 or Adh3. Eur J Biochem 269:2607–2612

Molotkov A, Molotkova N, Duester G (2006) Retinoic acid guides eye morphogenetic movements via paracrine signaling but is unnecessary for retinal dorsoventral patterning. Development 133:1901–1910

Moustakas A, Heldin CH (2009) The regulation of TGFbeta signal transduction. Development 136:3699–3714

Niederreither K, Dolle P (2008) Retinoic acid in development: toward an integrated view. Nat Rev Genet 9:541–553

Ornitz DM, Xu J, Colvin JS, McEwen DG, MacArthur CA, Coulier F, Gao G, Goldfarb M (1996) Receptor specificity of the fibroblast growth factor family. J Biol Chem 271:15292–15297

Perron M, Boy S, Amato MA, Viczian A, Koebernick K, Pieler T, Harris WA (2003) A novel function for hedgehog signaling in retinal pigment epithelium differentiation. Development 130:1565–1577

Perry KJ, Henry JJ (2012) Expression of pluripotency factors in the cornea epithelium of the frog *Xenopus*: evidence for the presence of epithelial stem cells (Submitted for Publication***)

Robinson ML (2006) An essential role for FGF receptor signaling in lens development. Semin Cell Dev Biol 17:726–740

Robinson ML, Ohtaka-Maruyama C, Chan CC, Jamieson S, Dickson C, Overbeek PA, Chepelinsky AB (1998) Disregulation of ocular morphogenesis by lens-specific expression of FGF-3/int-2 in transgenic mice. Dev Biol 198:13–31

Roddy M, Tsonis PA (2008) The newt as a model for eye regeneration. In: Tsonis PA, (ed). Animal models for eye research. Academic Press, San Diego, CA

Roddy M, Fox TP, McFadden JP, Nakamura K, Del Rio-Tsonis K, Tsonis PA (2008) A comparative proteomic analysis during urodele lens regeneration. Biochem Biophys Res Commun 377:275–279

Saika S, Okada Y, Miyamoto T, Ohnishi Y, Ooshima A, McAvoy JW (2001) Smad translocation and growth suppression in lens epithelial cells by endogenous TGFbeta2 during wound repair. Exp Eye Res 72:679–686

Saika S, Kono-Saika S, Ohnishi Y, Sato M, Muragaki Y, Ooshima A, Flanders KC, Yoo J, Anzano M, Liu C, Kao WW, Roberts AB (2004) Smad3 signaling is required for epithelial-mesenchymal transition of lens epithelium after injury. Am J Pathol 164:651–663

Santander C, Brandan E (2006) Betaglycan induces TGF-beta signaling in a ligand-independent manner, through activation of the p38 pathway. Cell Signal 18:1482–1491

Sasaki H, Nishizaki Y, Hui C, Nakafuku M. Kondoh H (1999) Regulation of Gli2 and Gli3 activities by an amino-terminal repression domain: implication of Gli2 and Gli3 as primary mediators of Shh signaling. Development 126:3915–3924

Schulz MW, Chamberlain CG, de Iongh RU, McAvoy JW (1993) Acidic and basic FGF in ocular media and lens: implications for lens polarity and growth patterns. Development 118:117–126

Smith AN, Miller LA, Song N, Taketo MM, Lang RA (2005) The duality of beta-catenin function: a requirement in lens morphogenesis and signaling suppression of lens fate in periocular ectoderm. Dev Biol 285:477–489

Stolen CM, Jackson MW, Griep AE (1997) Overexpression of FGF-2 modulates fiber cell differentiation and survival in the mouse lens. Development 124:4009–4017

Stone DM, Hynes M, Armanini M, Swanson TA, Gu Q, Johnson RL, Scott MP, Pennica D, Goddard A, Phillips H, Noll M, Hooper JE, de Sauvage F, Rosenthal A (1996) The tumor-suppressor gene *patched* encodes a candidate receptor for sonic hedgehog. Nature 384(6605):129–134

Stump RJ, Ang S, Chen Y, von Bahr T, Lovicu FJ, Pinson K, de Iongh RU, Yamaguchi TP, Sassoon DA, McAvoy JW (2003) A role for Wnt/beta-catenin signaling in lens epithelial differentiation. Dev Biol 259:48–61

Sugiyama Y, Stump RJ, Nguyen A, Wen L, Chen Y, Wang Y, Murdoch JN, Lovicu FJ, McAvoy JW (2010) Secreted frizzled-related protein disrupts PCP in eye lens fiber cells that have polarized primary cilia. Dev Biol 338:193–201

Sugiyama Y, Lovicu FJ, McAvoy JW (2011) Planar cell polarity in the mammalian eye lens. Organogenesis 7:191–201

Tsonis PA, Trombley MT, Rowland T, Chandraratna RAS, Del Rio-Tsonis K (2000) Role of retinoic acid in lens regeneration. Dev Dyn 219:588–593
Tsonis PA, Tsavaris M, Call MK, Chandraratna RAS, Del Rio-Tsonis K (2002) Expression and role of retinoic acid receptor alpha in lens regeneration. Dev Growth Differ 44:391–394
Tsonis PA, Vergara MN, Spence JR, Madhavan M, Kramer EL, Call MK, Santiago WG, Vallance JE, Robbins DJ, Rio-Tsonis KD (2004) A novel role of the hedgehog pathway in lens regeneration. Dev Biol 267:450–461
van den Heuvel M, Ingham PW (1996) Smoothened encodes a receptor-like serpentine protein required for hedgehog signaling. Nature 382:547–551
Wagner E, McCaffery P, Dräger UC (2000) Retinoic acid in the formation of the dorsoventral retina and its central projections. Dev Biol 222(2):460–470
Wawersik S, Purcell P, Rauchman M, Dudley AT, Robertson EJ, Maas R (1999) BMP7 acts in murine lens placode development. Dev. Biol 207:176–188
Yang EV, Wang L, Tassava RA (2005) Effects of exogenous FGF-1 treatment on regeneration of the lens and the neural retina in the Newt, Notophthalmus viridescens. J Exp Zool 303A:837–844
Yang C, Yang Y, Brennan L, Bouhassira EE, Kantorow M, Cvekl A (2010) Efficient generation of lens progenitor cells and lentoid bodies from human embryonic stem cells in chemically defined conditions. FASEB J 24:3274–3283
Zhang X, Ibrahimi OA, Olsen SK, Umemori H, Mohammadi M, Ornitz DM (2006) Receptor specificity of the fibroblast growth factor family. The complete mammalian FGF family. J Biol Chem 281:15694–15700

Part III
Musculoskeletal Repair and Regeneration

Regenerative Medicine: Challenges and Perspectives for Successful Therapies

S. Viswanathan and C. Joshi

Abstract Regenerative Medicine (RM) has the promise to revolutionize the treatment of many debilitating diseases for which the current therapies are inadequate. To realize the full potential of RM, a pragmatic approach needs to be taken by all stakeholders keeping in mind the lessons learnt from recombinant protein manufacturing, gene therapy trials, etc., to develop novel service delivery models for economic viability and regulatory processes in the absence of long-term data. In this chapter, we focus on the three main drivers of RM field and discuss the potential pitfalls and possible ways to mitigate them in order to move the field closer to clinical implementation.

Contents

S. Viswanathan (✉)
University Health Network, Princess Margaret Hospital, 610 University Avenue, Suite 5-303, Toronto, ON M5G 2M9, Canada
e-mail: Sowmya.viswanathan@uhnresearch.ca

C. Joshi
Cell Therapy Program, University Health Network, Princess Margaret Hospital, Toronto, ON, Canada

Current Topics in Microbiology and Immunology (2013) 367: 101–112
DOI: 10.1007/82_2012_290

© Springer-Verlag Berlin Heidelberg 2012
Published Online: 4 December 2012

1 Introduction

As the median age of the population shifts higher and as improvements in clinical and pharmaceutical management push the lifespan longer, debilitating degenerative diseases are gaining the spotlight as the new medical challenge that needs to be addressed in ever greater demographic. In this context, Regenerative Medicine (RM) promises to be the next leap in innovating the standard of care, and possibly offering curative solutions to many degenerative conditions.

In this chapter, we will focus on three main aspects of RM, cell-based therapies, tissue engineering and harnessing the regenerative capacity of endogenous organs, and review the recent progress and major obstacles in each.

2 RM and Cell-Based Therapies

RM and stem cells are intricately linked. Stem cells provide the nodal points for obtaining a variety of cell subtypes with specialized functions—the very functions that RM aims to restore such as insulin producing beta cell or, in case of cyto-immunotherapy, novel functions such as targeted anti-tumor effect. Starting from the discovery of hematopoietic stem cells (HSC) (Becker et al. 1963), and building on the knowledge gained from decades of research on HSC, the discoveries of other somatic tissue adult stem cell-like cells with restricted potential such as mesenchymal stromal cells (MSC) (Pittenger et al. 1999), muscle satellite cells (Seale and Rudnicki 2000), and neuronal stem cells (NSC) (Reynolds and Weiss 1992) have incrementally moved the RM field forward. However, the discovery of embryonic stem cells (ESC) (Thomson et al. 1998) has been seminal in augmenting interest in this area by several orders of magnitude. This is because ESCs afforded access to previously unavailable compartments such as cardio vascular and mesendo-dermal derivatives (such as beta cells) owing to their pluripotency. Also, this cell type in principle, removes the obstacle of obtaining stem cells in sufficient quantities to remedy a particular disease, as it would truly be used for cell-replacement therapy. However, ESCs also brought with them ethical controversies and technological limitations. The field of RM got a second enormous boost following the discovery of methods to "induce" multi or pluripotency in post-mitotic somatic cells—by generation of induced pluripotent cells (iPS) (Takahashi and Yamanaka 2006). Through iPS cells, it is now possible to obtain disease-specific and patient-specific stem cells, their differentiated progeny and model the disease progression, discover appropriate intervening actions, achieve a desirable number of target cells for administration, and even mitigate immunorejection without using embryonic tissue and thereby eliminate ethical concerns (Robinton and Daley 2012). Advances in this area are also being made with "direct" reprogramming where it may be feasible to reprogram one mature phenotype to another without transmitting to a less mature state (Davis et al. 1987;

Vierbuchen et al. 2010; Pang et al. 2011; Huang et al. 2011), although it is unclear how similar this re-programming process is to normal development.

2.1 Recent Progress in Cell-Based Therapies

Considerable effort and research investment is being made to harness the potential of cell-based therapies in almost all physiological and pathological conditions. There are over 5,000 clinical trials ongoing globally with various stem cells, and a similar number or more in the preclinical translational stage of research (www.clinicaltrials.org; (Culme-Seymour et al. 2000)). Despite this tremendous interest and progress, only a handful of cell-based therapies are commercially available including Osiris's Prochymal for pediatric Graft-versus-Host Disease (GvHD) in Canada and New Zealand (2012), Dendreon's Provenge for metastatic castrate-resistant prostrate cancer (2010), Genzyme's Carticel (1997) for articular cartilage injuries, TETEC's Novocart® (for joint cartilage), etc. In the pipeline though are a few companies employing adult stem cells in late phase clinical trials such as Aastrom's Ixmyelocel-T for Critical Limb Ischemia, Mesoblast's Mesenchymal Precursor Cells for Type II Diabetes, Recent Acute Myocardial Infarction, Heart Failure, etc.; positive outcomes in Phase II and Phase III trials are needed to enable market approval and justify the commercialization of these products. Others such as Geron's Phase I trial using hESC-derived oligodendrocytes have stopped investigations because of economic non-viability. There are a number of safety, efficacy, manufacturing, regulatory, and economic hurdles that need to be overcome to enable the successful commercialization of stem cell-based therapeutics.

Considerable effort has also been made into translating the promise of gene therapy. Over the past decade a number of clinical studies have provided proof of concept that genetically modified hematopoietic stem and progenitor cells can be used to treat metabolic disorders and monogenic diseases (Cavazzana-Calvo et al. 2010; Cartier et al. 2009). Recently, allogenic stem cell transplantation into an HIV-positive individual was shown to result in long-term cure demonstrating the feasibility and potency of genetic transfer using modified stem cells (Hutter and Zaia 2011). Adoptive immunotherapy using engineered T-cells has largely been shown to be safe in over 180 patients (Cruz et al. 2010) with compelling evidence in the case of melanoma (Rosenberg et al. 2011). T-cells engineered with chimeric antigen receptors (CARs) are also gaining favor especially in treating B-cell neoplasias (Brentjens et al. 2011; Porter et al. 2011).

Regenerative medicine approaches to understand disease progression in integration with high-throughput screening platforms has resulted in application of cell therapy tools in drug discovery programs. The ability to model complex diseases through use of stem cell-based assays and in vitro mini-tissue architectures has allowed the interrogation of disease states and their manipulation by pharmacological agents and small molecules. Chemical screens, libraries of previously purposed small molecules are all being screened in the hope of finding the next "hit" such as the anti-microbial agent, tigecycline and its ability to selectively inhibit leukemic stem cells (Skrtic et al. 2011). Another promising drug repurposing has been shown with Zeluton, an approved drug for asthma that inhibits activity of Alox5 gene product, which can also effectively block chronic myeloid leukemia (CML) stem cells that cause gleevac-resistant CML relapse. Other molecules such as inhibitors of HSP90 and hedgehog pathways are also being screened for this purpose (Chen et al. 2010). Commercially produced hESC-derived cardiomyocytes and hepatocytes have been in extensive use for toxicology screening (Jensen et al. 2009; Sartipy and Bjorquist 2011). Disease modeling through pluripotent stem cell-derived populations complements direct cell-or tissue-replacement therapeutics, and is getting attention from pharmaceutical companies like GE and GSK (Ebert et al. 2012).

As more commercial interest is generated and investments in cell-based therapeutics areas grow, the field will evolve towards more translational aspects of a therapeutic requirements such as product safety, quality assurance, scale-up and manufacturing, and reproducibility, delivery and dosage formats (Carmen et al. 2012). There are some significant challenges to overcome in this area as described below.

2.2 Challenges to Overcome for Cell-Based Therapies

2.2.1 Manufacturing

Stem cells, unlike tumor cell lines, cannot be easily cultured in large batches as their functionality and quality are highly susceptible to cell culture conditions. Cell doses per patient can vary depending on the type and application, but anywhere from 10^6 to 10^{10} cells may be needed for a standard dose. To treat 1,000 patients a year, correspondingly 10^9 to 10^{13} cells would be required. The current technologies of stem cell culture allow for production of a fraction of that number. Obtaining specialized, functional, terminally differentiated cells from stem cells in that quantity remains a serious hurdle. New technologies for scalable bioreactor culture and processing are needed to derive enough target cells continuously in an industrial setting (Zweigerdt 2009).

Manufacturing of these cells under current Good Manufacturing Practice (cGMP) conditions poses additional challenges requiring closed systems, USP-grade reagents, highly screened and quality controlled raw materials, and highly standardized manipulation steps. The traditional fill- and finish-model of bio-pharmaceutical manufacturing is not applicable to a cellular system with limited viability and varying functionality. A separation process to isolate, maintain, and purify the cell product, cell expansion and manipulation, and storage and delivery system that retains viability and functionality of the product needs to be developed and validated at appropriate scales for each therapeutic product (Amos et al. 2012).

There are a lot of lessons to learn from production of cells for recombinant proteins and vaccines including aspects of biotherapeutic protein supply chain such as standardized raw materials (cell banks), cold chain storage and distribution, and lot validation. However the relatively shorter shelf-life of the cell-therapy product poses additional challenges often requiring lot-release and validation to be performed post-shipping. Autologous cell therapy products pose additional challenges such as variable raw materials and variable end product with wide tolerances in potency as well as on-site manufacturing, validation, supply, and delivery facilities. Even if technologies are licensed and stringent SOPs are in place, local variations in cell manipulation during manufacture and delivery to patients may result in heterogeneous success rates across centers.

To implement genetically modified cells as standard-of-care, larger Phase II and III trials need to be undertaken and the challenges of manufacturing GMP-grade stem cells apply to this field as well. Additionally, there are the challenges of manufacturing, purifying, and concentrating clinical grade viral vectors, although alternatives such as nanoparticles (Hosseinkhani and Tabata 2006) and lipid-based complexes (Fenske et al. 2008) are beginning to emerge.

2.2.2 Regulation

Regulatory and marketing authorities rely heavily on historical data and animal studies for safety and efficacy results. In the case of most cell-based products, long-term safety and efficacy data are not available in animal models or in humans. Thus initiation of Phase I trials is often challenging for both regulators and sponsors in this nascent field, as the properties of the cells, their mechanism of action, their biodistribution, and long-term safety effects are all usually not fully-defined in a preclinical setting. Clinical studies are often undertaken in patients who are not in the ideal disease-progression stage for efficacious cell-therapy investigations casting doubts on the utility of such treatments when inevitably mixed results are obtained in a Phase I/II setting. Increased dialogue and interaction between the regulators and sponsors will allow this field to advance. There is promising evidence that this is already happening; the FDA has been open to dialoguing with sponsors and investigators prior to initiation of clinical trials via their pre-Investigative New Drug (IND) and even pre–pre-IND meetings. Additional guidelines specific to various tissues have been developed and put out by the FDA. Out-of-the-box thinking has been also demonstrated by reviewers at Health Canada who provided market approval for pediatric use of Prochymal to treat GvHD in the absence of complete efficacy data, and compromising by allowing Phase IV studies to collect such data.

It will be up to the scientists, regulators, industry, and medical professionals to manage risks and expectations from cell therapies without hype but also without risk-averse bias since the emerging experimental therapies are most likely to provide variably efficacy data and lack the precision that years of pharmaceutical experience has provided.

2.2.3 Economics

Cell-based therapies are a highly specialized branch of medicine with multiple steps requiring continuous integration with healthcare providers, laboratory staff, cell manufacturing facilities, and administrators. Autologous therapies are typically not suited for acute conditions; off-the shelf, allogeneic products may work more broadly; however, this requires better understanding of the long-term implications of culture-expanded cells. The normal service delivery model of biopharmaceutical manufacturing may also not apply, except for off-the-shelf allogeneic products. Cell-based therapies may require healthcare facility to house special infrastructure, for example, cGMP facilities or clean-rooms, or additional validated laboratory equipment within an operating theater to provide such therapies to patients. A different delivery model, knowledgeable personnel, infrastructure, and equipment are needed to support this process. Insurance companies and governments are not necessarily ready to assume these costs, especially when efficacy and long-term benefits are still being answered in clinical trials. To support these therapeutics and to make them economically viable, a new service delivery model system would be required (Luijten et al. 2012). Public-Private partnership models currently in use in the Netherlands may provide a blueprint for a system where cell manufacturing is carried out by private companies and the administering of the product, follow-up for safety, efficacy, and clinical oversight are performed at the publicly-funded healthcare institutions. In an environment where healthcare costs are ballooning, individual subscriptions to healthcare insurance plans may need to be enforced either through indirect taxation or by way of user fees. It may require innovative political approaches to educate the public and obtain sufficient participation from all stakeholders.

3 Tissue Engineering and RM

Tissue engineering has a number of applications ranging from traditional replacement and repair of structural tissues such as skin, bone, cartilage to engineering complex organs (liver, pancreas, kidney, heart, etc.) including engineered blood vessels (Miller et al. 2012), and bioprosthetic heart valves (porcine, bovine, cadaveric, or pulmonary-to-aortic autografts) (Mendelson and Schoen 2006) to providing research tools to understand tissue functioning. For example, using the latest advances in microfluidics technology, it is now possible to study fluid dynamics and blood cell interactions using a 3-D microvascular network on a chip (Zheng et al 2012).

Tissue engineering approaches are providing novel solutions to cell-and tissue-repair and replacement issues. For example, in vivo tissue-integration of iPSC-derived cardiomyocytes may prove difficult, however, use of emerging technologies such as 3D-angiogenic printing (Miller et al. 2012) and biomaterial scaffolds allows one to mimic the in vivo niche microenvironment and enable creation of cardiac-patches that may more easily integrate with the host tissue (Dengler et al. 2011).

Examples of commercial success of tissue engineered products include Organogenesis' Apligraf (1998) which combines a collagen matrix with fibroblasts and keratinocytes, Genzyme's Epicel (1998) which uses cultured autologous patient keratinocytes to form sheets that are stapled onto petrolatum gauze backing, Tissue Regenix's dCELL® Vascular Patch (2010), Cytograft's Lifeline, a tissue engineered blood vessel that has approval for autologous use in Germany (2009), and others.

3.1 Challenges to Tissue Engineering

3.1.1 Biomaterials

Different types of biomaterials ranging from natural, synthetic to composite, unmodified to modified chemically or physically and available in a variety of forms, injectable or non-injectable provide not just structural cues, but also microenvironmental cues to truly modulate surrounding cells and tissue (Davis et al. 2005). Indeed the use of microfabricated arrays of stem cell regulatory factors and extracellular matrix (ECM) components has been used to demonstrate the complex network of regulatory signals involved in self-renewal and differentiation of neural stem cells (Soen et al. 2006). Identifying, engineering, and optimizing specific biomaterials for appropriate end-uses remains a challenge, especially as multiple parameters from stability, biocompatibility, release of growth factors, physical support, etc., need to be considered and appropriately configured for specific tissue uses.

Biomaterials can also be designed to respond to variations in microenvironmental acidity, temperature, sheer stress, oxygenation, or enzyme levels (Stoop 2008; Rosso et al. 2005; Williams 2005) and thus provide directional and sequential release of growth factors resulting in appropriate spatio-temporal gradients. Smart biomaterials, although in their infancy, can therefore be developed along with advances in micromolding, laser photolithography, and microfluidic devices to create complex, controlled networks for drug/growth factor delivery.

3.1.2 Vascularization of Engineered Tissue

In addition to having biocompatible and smart biomaterials, it is important to have vascular networks as the transport of oxygen, nutrient, and waste is currently a major challenge in the field of tissue engineering. Strategies to induce network of blood vessels in vivo (Laschke et al. 2006; Lovett et al. 2009) or implanting pre-vascularized scaffolds will be critical to successful grafting, integration, survival, and functioning of these engineered tissue grafts, especially complex tissues such as kidney, heart, and liver. There are a number of strategies postulated including controlled and local release of angiogenic factors to promote neovascularization (Silva and Mooney 2007), but this still remains a major hurdle in designing and integrating viable scaffolds and constructs in vivo.

In addition, managing immune reaction to engineered tissue grafts requires separate strategies ranging from reduction of the graft immunogenicity and immunosuppressive regimes with mesenchymal stromal cells or donor-derived immunosuppressive antigen-presenting cells, and poses separate technical and regulatory issues that cannot be overlooked.

4 Regenerative Capacity of Endogenous Organs

The idea of recruiting resident tissue stem cell populations for effecting repair in injury or degenerative conditions is not new, but the widespread existence of such cells and the provoking concept of using bioactive molecules to recruit and engage such cells in an endogenous repair process are still nascent. Proof-of-concept has already been demonstrated by the use of erythropoietin to enhance blood cell formation, Granulocyte-Colony Stimulating Factor (G-CSF) for mobilization of hematopoietic precursors, and Bone Morphogenetic Protein (BMP)-2 which promotes osteogenesis from mesenchymal precursors. Several high-profile chemical screens have been published on several stem cell types, including HSCs, which identified soluble factors that inhibit ligand-induced signaling by aryl hydrocarbon receptor (AhR), and thus promote expansion of mobilized peripheral blood and umbilical cord blood (Boitano et al. 2010). Other small molecules that have been identified include a peroxisome proliferator-activated receptor-gamma (PPAR-g) inhibitor bipshenol A diglycidyl ether which has been shown to accelerate hematopoietic engraftment (Naveiras et al. 2009), and parathyroid hormone (PTH) which has been shown to improve mobilization and engraftment during sequential cord blood transplantation, and is currently in clinical investigations (NCT00393380 and NCT00299780).

Use of growth factors or small molecules may be particularly effective as treatment strategies for a variety of extreme psychiatric disorders including schizophrenia, extreme depression (having failed even electroconvulsive therapy), etc., as signals that are implicated in normal stem cell maintenance such as brain-derived neural factor (BDNF) may be disrupted in mood disorders, and thus serve as targets for pharmacological-based intervention (Benninghoff 2009).

Despite the promise of using small molecules or growth factors to harness the endogenous potential of tissue resident stem cells, there are concerns regarding delivery and retention of these molecules in required concentrations, potential harmful systemic side effects, and potential tumorigenic safety concerns for highly active bioactive molecules that promote endogenous stem cell recruitment and proliferation (rev in. (Miller and Kaplan 2012)).

Proposed solutions include identifying small molecules or drugs with a demonstrable safety profile that can be re-purposed to target endogenousstem cells or perhaps their niche environment, which may be particularly relevant for targeting cancers (rev in. (Wagers 2012)). This has been shown to some extent with the use of G-CSF mobilization, which can reduce amyloid plaque deposition in the hippocampus and improve cognition in murine models of Alzheimer's disease, although, the pilot study suggested safety but questionable efficacy in humans (Sanchez-Ramos et al. 2012).

5 Next Steps

Unlike research and delivery timescales in cancer or infectious disease medicine, RM has come a long way rapidly, largely owing to the successes in stem cell science. This has led to inflated expectations from a rapidly growing demographic consisting of aging population and their care providers to deliver better quality of life through RM along with substantial media hype (Eisenstein 2012). Therein lies the challenge for the scientific and clinical community to translate the research knowledge rapidly and safely enough to meet this demand. In the absence of significant gains, stem cell clinics are sprouting globally and offering cures to desperate patients and families although in most cases the cells have not been experimentally tested, protocols are not scientifically or ethically reviewed by third parties, and there is no independent monitoring of patient safety and welfare.

Since RM is most likely to offer benefits to an aging population, emerging ideas about stem cells and aging need to be kept in mind. An accumulation of myeloid biased versus lymphoid-biased HSC with increasing age and expenditure of skeletal muscle stem cells in the aging niche underscore that both niche and stem cell populations may be adversely affected by aging (Baumann 2012; Muller-Sieburg et al. 2012), and this factor could limit the utility of at least autologous stem cells in the very population that stands to benefit the most.

The growth of personalized medicine by way of whole-genome sequencing may help in prognosticating patients who might benefit the most with RM-therapeutics versus those who may experience poor outcomes because of failure of therapy or unacceptable side-effects. This may further streamline creation of ratiometric formulae for cost/benefit, and thus make the risk acceptable to all stakeholders.

Measured solutions from multiple fields including researchers, engineers and manufacturers, regulators, and the business development side are needed to develop viable and profitable business and/or services models to translate largely personalized regenerative medicine concepts to routine clinical practice. Hard lessons learned from gene therapy failures (Wilson 2009) can be applied to regenerative medicine applications including the importance of adhering to the protocols, proper training, accreditation, and documentation involving the staff and therapeutic product preparation and administration, and avoiding conflicts of interest to ensure the field moves forward in a responsible manner. With these in place, we can optimistically expect the field to make significant advances in providing novel therapeutic solutions to patients, and indeed become practice-changing over the next decade.

References

Amos PJ, Cagavi Bozkulak E, Qyang Y (2012) Methods of cell purification: a critical juncture for laboratory research and translational science. Cells Tissues Organs 195(1–2):26–40

Baumann K (2012) Stem cells: an ageing decline. Nat Rev Mol Cell Biol 13(11):681

Becker AJ, Mc CE, Till JE (1963) Cytological demonstration of the clonal nature of spleen colonies derived from transplanted mouse marrow cells. Nature 197:452–454

Benninghoff J (2009) Stem cell approaches in psychiatry–challenges and opportunities. Dialogues Clin Neurosci 11(4):397–404

Boitano AE, Wang J, Romeo R, Bouchez LC, Parker AE, Sutton SE et al (2010) Aryl hydrocarbon receptor antagonists promote the expansion of human hematopoietic stem cells. Science 329(5997):1345–1348

Brentjens RJ, Riviere I, Park JH, Davila ML, Wang X, Stefanski J et al (2011) Safety and persistence of adoptively transferred autologous CD19-targeted T cells in patients with relapsed or chemotherapy refractory B-cell leukemias. Blood 118(18):4817–4828

Carmen J, Burger SR, McCaman M, Rowley JA (2012) Developing assays to address identity, potency, purity and safety: cell characterization in cell therapy process development. Regen Med 7(1):85–100

Cartier N, Hacein-Bey-Abina S, Bartholomae CC, Veres G, Schmidt M, Kutschera I et al (2009) Hematopoietic stem cell gene therapy with a lentiviral vector in X-linked adrenoleukodystrophy. Science 326(5954):818–823

Cavazzana-Calvo M, Payen E, Negre O, Wang G, Hehir K, Fusil F et al (2010) Transfusion independence and HMGA2 activation after gene therapy of human beta-thalassaemia. Nature 467(7313):318–322

Chen Y, Peng C, Sullivan C, Li D, Li S (2010) Novel therapeutic agents against cancer stem cells of chronic myeloid leukemia. Anticancer Agents Med Chem 10(2):111–115

Cruz CR, Hanley PJ, Liu H, Torrano V, Lin YF, Arce JA et al (2010) Adverse events following infusion of T cells for adoptive immunotherapy: a 10-year experience. Cytotherapy 12(6):743–749

Culme-Seymour EJ, Davie NL, Brindley DA, Edwards-Parton S, Mason C (2012) A decade of cell therapy clinical trials (2000–2010). Regen Med 7(4):455–462

Davis RL, Weintraub H, Lassar AB (1987) Expression of a single transfected cDNA converts fibroblasts to myoblasts. Cell 51(6):987–1000

Davis ME, Hsieh PC, Grodzinsky AJ, Lee RT (2005) Custom design of the cardiac microenvironment with biomaterials. Circ Res 97(1):8–15

Dengler J, Song H, Thavandiran N, Masse S, Wood GA, Nanthakumar K et al (2011) Engineered heart tissue enables study of residual undifferentiated embryonic stem cell activity in a cardiac environment. Biotechnol Bioeng 108(3):704–719

Ebert AD, Liang P, Wu JC (2012) Induced pluripotent stem cells as a disease modeling and drug screening platform. J Cardiovasc Pharmacol 60(4):408–416

Eisenstein M (2012) Stem cells: don't believe the hype. Nature 484(7393):S5

Fenske DB, Chonn A, Cullis PR (2008) Liposomal nanomedicines: an emerging field. Toxicol Pathol 36(1):21–29

Hosseinkhani H, Tabata Y (2006) Self assembly of DNA nanoparticles with polycations for the delivery of genetic materials into cells. J Nanosci Nanotechnol 6(8):2320–2328

Huang P, He Z, Ji S, Sun H, Xiang D, Liu C et al (2011) Induction of functional hepatocyte-like cells from mouse fibroblasts by defined factors. Nature 475(7356):386–389

Hutter G, Zaia JA (2011) Allogeneic haematopoietic stem cell transplantation in patients with human immunodeficiency virus: the experiences of more than 25 years. Clin Exp Immunol 163(3):284–295

Jensen J, Hyllner J, Bjorquist P (2009) Human embryonic stem cell technologies and drug discovery. J Cell Physiol 219(3):513–519

Laschke MW, Harder Y, Amon M, Martin I, Farhadi J, Ring A et al (2006) Angiogenesis in tissue engineering: breathing life into constructed tissue substitutes. Tissue Eng 12(8):2093–2104

Lovett M, Lee K, Edwards A, Kaplan DL (2009) Vascularization strategies for tissue engineering. Tissue Eng Part B Rev 15(3):353–370

Luijten PR, van Dongen GA, Moonen CT, Storm G, Crommelin DJ (2012) Public-private partnerships in translational medicine: concepts and practical examples. J Control Release 161(2):416–421

Mendelson K, Schoen FJ (2006) Heart valve tissue engineering: concepts, approaches, progress, and challenges. Ann Biomed Eng 34(12):1799–1819

Miller FD, Kaplan DR (2012) Mobilizing endogenous stem cells for repair and regeneration: are we there yet? Cell Stem Cell 10(6):650–652

Miller JS, Stevens KR, Yang MT, Baker BM, Nguyen DH, Cohen DM et al (2012) Rapid casting of patterned vascular networks for perfusable engineered three-dimensional tissues. Nat Mater 11(9):768–774

Muller-Sieburg CE, Sieburg HB, Bernitz JM, Cattarossi G (2012) Stem cell heterogeneity: implications for aging and regenerative medicine. Blood 119(17):3900–3907

Naveiras O, Nardi V, Wenzel PL, Hauschka PV, Fahey F, Daley GQ (2009) Bone-marrow adipocytes as negative regulators of the haematopoietic microenvironment. Nature 460(7252):259–263

Pang ZP, Yang N, Vierbuchen T, Ostermeier A, Fuentes DR, Yang TQ et al (2011) Induction of human neuronal cells by defined transcription factors. Nature 476(7359):220–223

Pittenger MF, Mackay AM, Beck SC, Jaiswal RK, Douglas R, Mosca JD et al (1999) Multilineage potential of adult human mesenchymal stem cells. Science 284(5411):143–147

Porter DL, Levine BL, Kalos M, Bagg A, June CH (2011) Chimeric antigen receptor-modified T cells in chronic lymphoid leukemia. N Engl J Med 365(8):725–733

Reynolds BA, Weiss S (1992) Generation of neurons and astrocytes from isolated cells of the adult mammalian central nervous system. Science 255(5052):1707–1710

Robinton DA, Daley GQ (2012) The promise of induced pluripotent stem cells in research and therapy. Nature 481(7381):295–305

Rosenberg SA, Yang JC, Sherry RM, Kammula US, Hughes MS, Phan GQ et al (2011) Durable complete responses in heavily pretreated patients with metastatic melanoma using T-cell transfer immunotherapy. Clin Cancer Res 17(13):4550–4557

Rosso F, Marino G, Giordano A, Barbarisi M, Parmeggiani D, Barbarisi A (2005) Smart materials as scaffolds for tissue engineering. J Cell Physiol 203(3):465–470

Sanchez-Ramos J, Cimino C, Avila R, Rowe A, Chen R, Whelan G et al (2012) Pilot study of granulocyte-colony stimulating factor for treatment of Alzheimer's disease. J Alzheimers Dis 31(4):843–855

Sartipy P, Bjorquist P (2011) Concise review: human pluripotent stem cell-based models for cardiac and hepatic toxicity assessment. Stem Cells 29(5):744–748

Seale P, Rudnicki MA (2000) A new look at the origin, function, and "stem-cell" status of muscle satellite cells. Dev Biol 218(2):115–124

Silva EA, Mooney DJ (2007) Spatiotemporal control of vascular endothelial growth factor delivery from injectable hydrogels enhances angiogenesis. J Thromb Haemost 5(3):590–598

Skrtic M, Sriskanthadevan S, Jhas B, Gebbia M, Wang X, Wang Z et al (2011) Inhibition of mitochondrial translation as a therapeutic strategy for human acute myeloid leukemia. Cancer Cell 20(5):674–688

Soen Y, Mori A, Palmer TD, Brown PO (2006) Exploring the regulation of human neural precursor cell differentiation using arrays of signaling microenvironments. Mol Syst Biol 2:37
Stoop R (2008) Smart biomaterials for tissue engineering of cartilage. Injury. 39(1):S77–S87
Takahashi K, Yamanaka S (2006) Induction of pluripotent stem cells from mouse embryonic and adult fibroblast cultures by defined factors. Cell 126(4):663–676
Thomson JA, Itskovitz-Eldor J, Shapiro SS, Waknitz MA, Swiergiel JJ, Marshall VS et al (1998) Embryonic stem cell lines derived from human blastocysts. Science 282(5391):1145–1147
Vierbuchen T, Ostermeier A, Pang ZP, Kokubu Y, Sudhof TC, Wernig M (2010) Direct conversion of fibroblasts to functional neurons by defined factors. Nature 463(7284):1035–1041
Wagers AJ (2012) The stem cell niche in regenerative medicine. Cell Stem Cell 10(4):362–369
Williams D (2005) Environmentally smart polymers. Med Device Technol 16(4):9–10
Wilson JM (2009) Lessons learned from the gene therapy trial for ornithine transcarbamylase deficiency. Mol Genet Metab 96(4):151–157
Zheng Y, Chen J, Craven M, Choi NW, Totorica S, Diaz-Santana A et al (2012) In vitro microvessels for the study of angiogenesis and thrombosis. Proc Natl Acad Sci U S A 109(24):9342–9347
Zweigerdt R (2009) Large scale production of stem cells and their derivatives. Adv Biochem Eng Biotechnol 114:201–235

Employing the Biology of Successful Fracture Repair to Heal Critical Size Bone Defects

Jo Ann Cameron, Derek J. Milner, Jung Seok Lee, Jianjun Cheng, Nicholas X. Fang and Iwona M. Jasiuk

Abstract Bone has the natural ability to remodel and repair. Fractures and small noncritical size bone defects undergo regenerative healing via coordinated concurrent development of skeletal and vascular elements in a soft cartilage callus environment. Within this environment bone regeneration recapitulates many of the same cellular and molecular mechanisms that form embryonic bone. Angiogenesis is intimately involved with embryonic bone formation and with both endochondral and intramembranous bone formation in differentiated bone. During bone regeneration osteogenic cells are first associated with vascular tissue in the adjacent periosteal space or the adjacent injured marrow cavity that houses endosteal blood vessels. Critical size bone defects cannot heal without the assistance of therapeutic aids or materials designed to encourage bone regeneration. We discuss the prospects for using synthetic hydrogels in a bioengineering approach to repair critical size bone defects. Hydrogel scaffolds can be designed and fabricated to potentially trigger the same bone morphogenetic cascade that heals bone fractures and noncritical size defects naturally. Lastly, we introduce adult *Xenopus laevis* hind limb as a novel small animal model system for bone regeneration research. *Xenopus*

J. A. Cameron (✉) · D. J. Milner
Department of Cell and Developmental Biology, Institute for Genomic Biology,
University of Illinois at Urbana-Champaign, Urbana, IL 61801, USA
e-mail: joanncam@illinois.edu

J. S. Lee · J. Cheng
Department of Materials Science and Engineering, Institute for Genomic Biology,
University of Illinois at Urbana-Champaign, Urbana, IL 61801, USA

N. X. Fang
Department of Mechanical Engineering, Massachusetts Institute of Technology,
Cambridge, MA 02139-4307, USA

I. M. Jasiuk
Department of Mechanical Science and Engineering, Institute for Genomic Biology,
University of Illinois at Urbana-Champaign, Urbana, IL 61801, USA

Current Topics in Microbiology and Immunology (2013) 367: 113–132
DOI: 10.1007/82_2012_291
© Springer-Verlag Berlin Heidelberg 2012
Published Online: 14 December 2012

hind limbs have been used successfully to screen promising scaffolds designed to heal critical size bone defects.

Abbreviations

BMPs	Bone morphogenetic proteins
Cbfa1	Core binding factor 1
CDMP-1	Cartilage-derived morphogenetic protein-1
CSD	Critical size defect
CXCR-4	Receptor for SDF-1
ECM	Extracellular matrix
FDA	Food and Drug Administration
FGFs	Fibroblast growth factors
GDF-5	Growth/differentiation factor 5
GF-11	Skeletal growth factor
HA	Hydroxyapatite
HDDA	1,6 Hexanediol diacrylate
HIFα	Hypoxia-induced factor alpha
IGF	Insulin-like growth factor
IHH	Indian hedgehog
IL	Interleukin
M-CSF	Macrophage colony stimulating factor
MMP	Metalloproteinase
MRI	Magnetic resonance imaging
MSCs	Mesenchymal stem cells
OPG	Osteoprotegerin
PDGF	Platelet-derived growth factor
PTHrP	Parathyroid hormone related peptide
RANKL	Receptor activator of nuclear factor kappa-B ligand
SDF-1	Stromal cell-derived factor-1
SHH	Sonic hedgehog
TCP	Tricalcium phosphate
TGFβ	Transforming growth factor beta
TNF-α	Tumor necrosis factor-alpha
VEGF	Vascular endothelial growth factor

Contents

1 Successful Bone Regeneration Resembles Bone Development

Bone is a tissue that can repair by regeneration very effectively. This ability is enhanced by the postnatal characteristic mechanism of bone maintenance, whereby processes of resorption and formation balance continuous remodeling. Although there are distinct aspects of both processes, successful regeneration such as fracture healing occurs by a cellular cascade that resembles skeletal development and regeneration. The regenerative cascade requires coordinated cellular events of cell migration, cell differentiation, and proliferation of multiple cell types (Willie et al. 2010; Mehta et al. 2012). Understanding bone morphogenesis and the regenerative process of fracture repair form the basis for creating therapeutic tools designed on principles of bone tissue engineering.

1.1 Endochondral Bone Development

Skeletal precursor cells aggregate, proliferate, and form a mesenchymal condensation that becomes a temporary cartilaginous template for the future skeletal element (Hall and Miyake 2000). Embryonic long bone formation begins when skeletal precursor mesenchymal cells (MCs) form aggregations associated with capillaries (Fig. 1a). Sonic hedgehog (SHH) and bone morphogenetic proteins (BMPs) are important mediators of the correct location and patterning of undifferentiated skeletal mesenchyme in limb bud. Sox9 is expressed in cells undergoing aggregation before they condense and form cartilage. Sox9 represses Runx2 and β-catenin and regulates collagen type II expression. Core binding factor 1 (Cbfa1) is a transcription factor expressed in condensation as well (Ferguson et al. 1999).

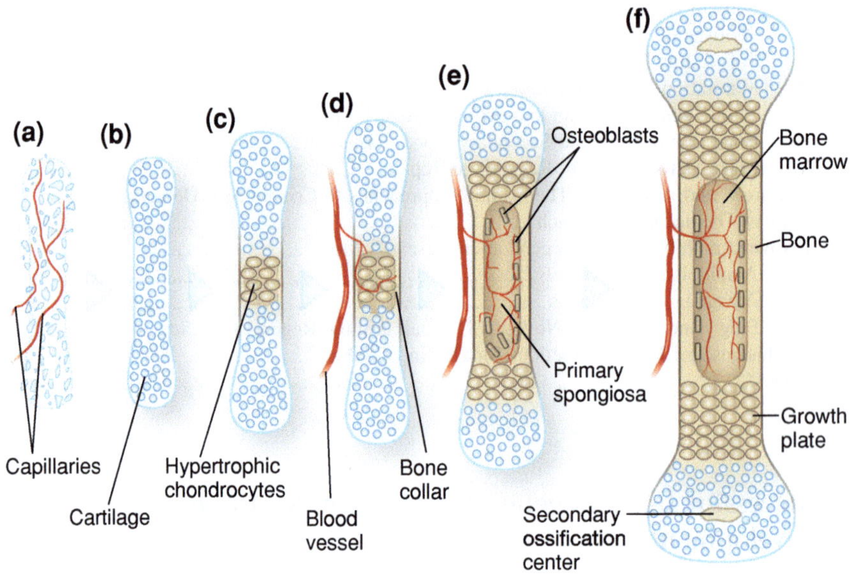

Fig. 1 Schematic of endochondral bone formation. (**a**) Mesenchymal cells aggregate near capillaries and (**b**) differentiate into chondrocytes forming an avascular cartilage model of the future bone. (**c**) At the center of the condensation the chondrocytes cease proliferating and become hypertrophic. (**d**) Perichondral cells adjacent to the hypertrophic chondrocytes differentiate into osteoblasts forming a bone collar. The hypertrophic cartilage regulates formation of mineralized matrix and release of angiogenic factors to attract blood vessels. Eventually hypertrophic chondrocytes undergo apoptosis. (**e**) The coordination of osteoblasts and vascular invasion forms the primary spongiosa. At each end of the diaphysis chondrocytes continue to proliferate with concomitant vascularization resulting in a coordinated process that lengthens the bone. Osteoblasts in the bone collar will form cortical bone, while osteoprogenitor cells in the primary spongiosa will eventually form trabecular bone. (**f**) Secondary ossification centers develop through cycles of chondrocyte hypertrophy, vascular invasion, and osteoblast activity. Columns of proliferating chondrocytes form the growth plate beneath the secondary ossification center. Finally, expansion of stromal and hematopoietic stem cells starts to take place in the marrow space. This figure is based on Kanczler and Oreffo (2008) with permission

MCs in the condensation differentiate into a cartilage model of the future bone (Fig. 1b). Proliferating cells shift from laying down a mesenchymal matrix of collagen types III and I to collagen types II, IX, XI and aggrecan characteristic of the cartilage condensation. Parathyroid hormone-related peptide (PTHrP) stimulates cartilage cell proliferation and represses cartilage differentiation. Cells in the center of the condensation produce Indian hedgehog (IHH), which couples chondrocyte maturation and osteoblast differentiation (Hartmann 2009; Karsenty et al. 2009).

In the central interior of the cartilage model maturing chondrocytes exit the cell cycle, become hypertrophic, the extracellular matrix (ECM) becomes calcified, and the hypertrophic cells eventually undergo apoptosis (Fig. 1c–e). The mature calcified ECM and type X collagen favors vascular endothelial growth factor

(VEGF) dependent vascular invasion from the adjacent blood vessels. Developing bone marrow, the primary spongiosa, is trabecular bone that forms mainly from osteoprogenitors associated with the invading blood vessels. The primary center of ossification is established near the periphery of the cartilage model midway along the length. The bone collar differentiates from external perichondrium cells that express Runx2, the master gene for osteoblast differentiation. Osteoprogenitors resident in the developing bone collar and those associated with invading blood vessels differentiate into osteoblasts and lay down bone matrix. As the collar becomes more vascularized it establishes the first cortical bone and the periosteum of the developing bone (Ferguson et al. 1999; Hartmann 2009; Karsenty et al. 2009). For a comprehensive review of the genetic and molecular control of bone formation see Hartman (2009) and Karsenty et al. (2009).

1.2 Long Bone Regenerative Repair

Bone can heal successfully without forming a fibrous scar. In response to injury a complex series of regeneration promoting cascades is triggered within the resulting fragments of bone. If the bone fragments are positioned optimally, a specific pattern of cellular and molecular events that recapitulates bone development at certain stages during the process occurs across the gap. This secondary or indirect fracture healing typically results in successful regenerative bone repair or healing by formation of endochondral and intramembranous bone. This type of fracture healing occurs under conditions that permit micro-motion and weight bearing (for reviews see Wraighte and Scammell 2007; Jahagirdar and Scammell 2009; Marsell and Einhorn 2011; Stocum 2012).

1.3 Morphology of Fracture Healing Region

Within the fracture healing region of the adjacent bone fragments there are four main spatial zones that contribute specific cellular and molecular components to the regeneration site: the medullary canal, the area between the cortices, the cambium layer of the periosteum, and the surrounding soft tissues. The medullary canal and the inter-cortical areas create the soft callus and go on to create bone by endochondral ossification. The subperiosteal region and immediately surrounding soft tissues create the hard callus and create bone by intramembranous ossification (reviewed by Phillips 2005). Intramembranous ossification is direct bone formation from osteoprogenitor and undifferentiated MCs that reside in the periosteum. Endochondral ossification requires recruitment, proliferation, and differentiation of undifferentiated MCs into cartilage, which later becomes calcified and eventually replaced by bone.

The phases of fracture repair have been described as four or five overlapping progressive temporal phases. The defined steps in bone regeneration consist of overlapping phases of regenerative cellular activity (Kolar et al. 2010; Dwek 2010; Schindeler et al. 2008). These stages of fracture repair are characterized by histological changes within the fracture-healing region and the later stages resemble embryonic bone development at the cellular and molecular level. Many of the same initial developmental cellular activities are present during long bone regeneration and fracture repair (Ferguson et al. 1999; Dwek 2010).

1.4 Inflammation and Hematoma Formation

The inflammatory response begins as soon as the bone break occurs. Complete fractures cause disruption of the bone tissue along with blood vessels and nerves that serve the bone. More extensive breaks can involve tearing of the periosteum and injury to adjacent soft tissues such as skeletal muscle, tendons, as well as blood vessels and nerves that serve the soft tissues. Disrupted blood vessels lead to ischemic necrosis of the affected bone ends.

Upon injury, peripheral blood and intramedullary blood and bone marrow cells fill the fracture-healing site (Fig. 2a). A fibrin clot forms and this initiates the healing cascade with the establishment of the hematoma. The fibrin clot will support the establishment of granulation tissue that serves as a template for a cartilage callus to form between and around the adjacent fracture fragments. Within the fibrin clot platelets degranulate releasing transforming growth factor beta (TGFβ) and platelet-derived growth factor (PDGF). TGFβ stimulates undifferentiated mesenchymal stem cells (MSCs) and PDGF encourages MSC and osteoblast proliferation as well as macrophage chemotaxis (reviewed by Nikolau and Tsiridis 2007; Tsiridis et al. 2007).

The hematoma forms in an environment that is ischemic, hypoxic, low pH, and high potassium and lactate concentrations. These initial conditions are not favorable for cells but some immune cells survive and there are subpopulations of cells that change over time and some support and some impair bone healing (Kolar et al. 2010, 2011; Schmidt-Bleek et al. 2011; Willie et al. 2010). A cascade of cellular and molecular events is triggered by low O_2 concentration in the injury site. Hypoxia-induced factor-α (HIFα), VEGF, and BMPs concentrations are upregulated and endothelial and osteoprogenitor cells will respond to these signals by initiating vascular in growth and initial osteogenesis (Towler 2007, 2008, 2011; Bianco 2011a, b).

The hematoma consists of coagulated circulating blood cells, immune cells, and MCs from the adjacent marrow cavities and periosteum of the bone. The initial blood cells that form the hematoma release cytokines and growth factors that fight infection, degrade the necrotic tissue at the fracture site, form a fibrin clot, stimulate neoangiogenesis and bone formation. Tumor necrosis factor-α (TNF-α) expressed by macrophages and inflammatory cells induces secondary

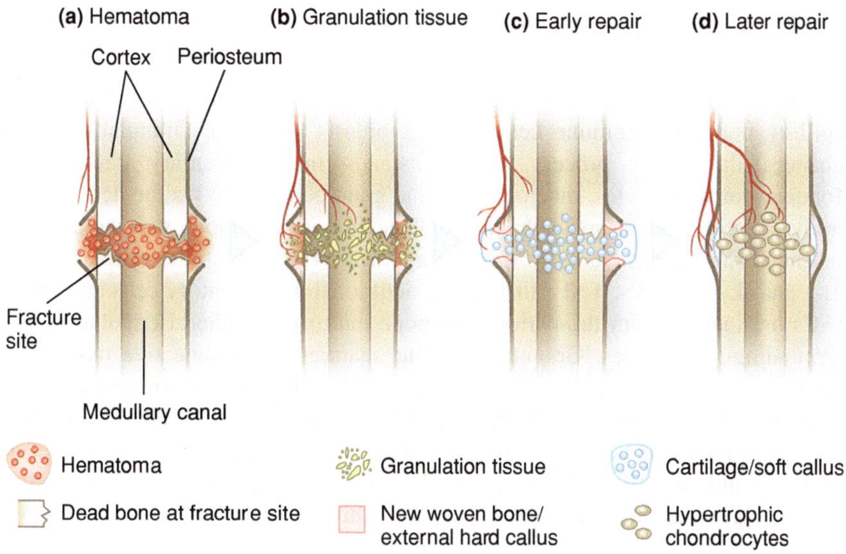

Fig. 2 Schematics of long bone fracture repair. The phases of long bone fracture repair are sequential and overlapping. Each phase may vary in length depending upon the specific bone that has been broken, the location of the break, the extent of injury, and the health status of the patient. (**a**) A hematoma forms in response to injury. Blood cells from peripheral vasculature, bone marrow (medullary canal), bone cortex, periosteum, and the adjacent surrounding soft tissues flow into the fracture site and become entrapped in a fibrin clot. (**b**) Granulation tissue forms in the fracture gap with capillary ingrowth, and inflammatory cells and MSCs migrating to the site. (**c**) A cartilaginous soft callus forms during the early repair phase. The cells that form the soft callus are osteochondral progenitors. The external hard callus is new woven bone that forms from progenitor cells originating from the cambium layer of the periosteum. (**d**) Later in the repair process, cartilage cells mature into hypertrophic chondrocytes that withdraw from the cell cycle, form mineralized cartilage matrix, and eventually undergo apoptosis. Blood vessels grow into the site and new woven bone is established across the fracture gap interior to the external hard callus. Eventually an extensive hard callus of new woven bone joins the fracture segments. The hard callus ultimately remodels to form the lamellar cortical bone and the mature trabecular bone characteristic of the bone marrow.

inflammatory signals, is chemotactic to cells, and induces osteogenesis in MSCs. TNFR1 and TNFR2 are receptors that may be specific to injury and bone regeneration (Marsell and Einhorn 2011). Coagulation activates immune cells in the hematoma as it develops and neovascularization occurs. The hematoma also has fibroblast growth factors (FGFs) released from macrophages, MSCs, chondrocytes, and osteoblasts. These are mitogenic for MSCs, chondrocytes and osteoblasts. Insulin-like growth factor (IGF) released from bone matrix, and secreted by osteoblasts and chondrocytes promotes further proliferation and differentiation of osteoprogenitor cells (Marsell and Einhorn 2011).

1.5 Formation of Granulation Tissue

The hematoma becomes transformed into granulation tissue due to increased capillary ingrowth, immune cells, and fibroblasts (Fig. 2b). Granulation tissue enhances the repair process and withstands interfragmentary deformation. Mononuclear phagocytes, macrophages, and giant cells arrive via the new vessels, remove necrotic bone, and help to build the fibrocartilaginous soft callus. Also present is macrophage colony stimulating factor (M-CSF), interleukins1 and 6 (IL-1 and IL-6), BMPs, TNFα that recruit additional inflammatory cells and attract MSCs originating from the periosteum, bone marrow, peripheral circulation, and surrounding soft tissues. Granulation tissue matures into connective tissue containing collagen types I, II, and III initially, and collagen type III gradually predominates as the granulation tissue matures (Shindeler et al. 2008). Granulation tissue is formed and the wound bed is stabilized by the periosteum (Ozaki et al. 2000). Granulation tissue forms a template and it is replaced by fibrous tissue, fibrocartilage, and then cartilage of the soft callus.

1.6 Periosteum

A bony collar forms in the subperiosteal region adjacent to distal and proximal ends of the fracture by intramembranous bone formation. This healing process is promoted by motion within the fracture site and inhibited by internal rigid fixation. The cambium layer is composed of MCs, differentiated osteoprogenitor cells, osteoblasts, and fibroblasts. A rich peripheral vascular and sympathetic neural network is also present. In addition there are endothelial pericyte cells known to have osteoblastic potential. Upon injury the cambium layer of periosteum adjacent to the fracture site thickens, undergoing a proliferative response called the periosteal reaction. The periosteal reaction produces a mass of cartilage about the fracture site that eventually contributes to the formation of an external callus of woven bone or hard callus as well as the internal soft callus. Macrophages and inflammatory cells and MCs in periosteum secrete pro-inflammatory cytokines. IL-1, IL-6, and TNF-α and these cells induce secondary inflammatory signals. Undifferentiated MSCs release BMPs that induce angiogenesis, chemotaxis, mitogenesis, and cell differentiation into osteoblasts and chondroblasts (Malizos and Papatheodorou 2005; Dwek 2010).

Osteoprogenitor cells already in the subperiosteum are ready to begin intramembranous ossification. Necessary MSCs are recruited, proliferated, and differentiated into osteogenic cells. MSCs come from surrounding soft tissues, bone marrow, and circulating MSCs to the injury site. BMP7 may be important in recruitment (Bais et al. 2009). Stromal cell-derived factor-1 (SDF-1) and the receptor CXCR-4 a G-protein-coupled receptor are important for homing MSCs to fracture site periosteum at the edges of the fracture. SDF-1 recruits CXCR-4 expressing MSCs. FGF is expressed in cells of the expanded cambial layer and it is associated with increase of fibroblast-like MCs that increases callus and bone formation (Marsell and Einhorn 2011). The external hard callus of new woven bone helps stabilize the fracture site during early repair stages through later repair stages (Fig. 2c, d).

In addition to the soft callus of the fracture site, the periosteum also participates in a process resembling endochondral ossification that forms a hard callus by recapitulating fetal skeletogenesis (Dwek 2010). The hyaline cartilage ECM becomes mineralized; osteogenic cells associated with capillaries lay down bone matrix forming a bony collar at the periphery of the callus and regions of woven bone in the core of the hard callus.

1.7 Soft Callus Formation

An internal cartilaginous soft callus forms within the fracture site (Fig. 2c). The sources of proliferating chondrocytes within the soft callus are multiple. Mesenchymal cells (MCs) within the cambium layer of periosteum, endosteum, bone marrow, and adjacent soft tissues differentiate into chondrocytes to produce the semi-rigid, avascular soft callus. Endochondral formation occurs within the granulation tissue between fracture ends and external to periosteal sites. Granulation tissue is replaced by fibrous tissue, fibrocartilage, and then hyaline cartilage. The early soft callus ECM consists of type II collagen and proteoglycan core biomarkers that are gradually replaced with type X collagen as the soft callus matures. Adhesion molecule osteonectin is present as early cartilage forms (Gerstenfeld et al. 2003, 2006).

There are several growth factors that are associated with the establishment and eventual maturation of the soft callus. TGF-β, BMPs, fibroblast growth factor-1 (FGF1), and insulin-like growth factor-II (IGF-II) influence chemotaxis, proliferation, and differentiation of progenitor cells into chondrocytes or osteoblasts. TGF-β2 and TGF-β3, and growth and differentiation factor-5 (GDF-5) are also involved with chondrogenesis and eventual endochondral ossification. The healing cascade may be initiated during the inflammation phase (Kolar et al. 2010). TGF-β2 and TFG-β3 peak during chondrogenesis. TGFs attract MSCs, pre-osteoblasts, chondrocytes, and osteoblasts and they act during chondrogenesis and endochondral bone formation. TGFs might induce synthesis of BMPs.

BMPs regulate growth, differentiation, and apoptosis of osteoblasts and chondroblasts. They induce the developmental cascade for chondro-osteogenesis. BMP2 is especially crucial for initiation of the healing cascade and possibly callus formation. BMP2 may play an important role in inducing osteoblast differentiation; most BMPs can stimulate osteogenesis in mature osteoblasts. BMPs may stimulate synthesis and secretion of IGF and VEGF, or they may directly activate endothelial cells to undergo angiogenesis. BMP7 has been especially effective in stimulating osteogenesis and is already used clinically. GDF-5 and cartilage-derived morphogenetic protein-1 (CDMP-1) induce endochondral bone growth. Endothelial cells, osteoblasts, and chondrocytes produce IGFs. IGF-11 (skeletal growth factor) acts later during endochondral bone formation. TNF-α initiates chondrocyte apoptosis (Gerstenfeld et al. 2003; 2006).

1.8 Maturation of the Soft Callus

Fracture callus chondrocytes proliferate and mature into hypertrophic chondrocytes (Fig. 2d). Mineralization of soft callus ECM proceeds from the fragment ends toward the center of the fracture site. Chondrocyte mitochondria accumulate calcium phosphate granules that are transported through the cytoplasm and released in the ECM to become seeds for growth of apatite microcrystals. A cascade involving M-CF, RANKL OPG, and TNF-α accomplishes this. Mineralization and resorption of the cartilaginous callus is a recapitulation of embryological bone development. Eventually the hypertrophic chondrocytes undergo apoptosis. Hypertrophic chondrocytes secrete VEGF, a factor that is central to neovascularization during endochondral bone formation (Carlevaro et al. 2000). In addition, VEGF is a key player in promoting both angiogenesis and osteogenesis during fracture repair (Street et al. 2002; Keramis et al. 2008). VEGF acts synergistically with BMP4 to improve bone regeneration (Peng et al. 2002; Li et al. 2009; Feng et al. 2011).

1.9 Hard Callus Formation and Primary Bone Formation

This is an active period of osteogenesis and the formation of mineralized bone matrix proceeds from the periphery toward the center of the fracture site. Primary bone formation at the fracture site resembles endochondral ossification. Matrix metalloproteinases 9 and 13 (MMP-9, MMP-13) are responsible for ECM degradation during the soft callus remodeling. These are expressed in osteoclasts and osteoblasts and mature hypertrophic chondrocytes. In addition, MMPs are expressed in vascular endothelial cells and perivascular cells associated with angiogenesis during endochondral bone repair. Resorption of mineralized cartilage takes place and a cascade involving M-CF, RANKL OPG, and TNF-α accomplishes this. M-CSF, RANKL, and OPG help to recruit bone cells and osteoclasts to form woven bone. TNF-α initiates chondrocyte apoptosis. Eventually the calcified cartilage is replaced with woven bone and the fracture site becomes more solid and mechanically stable. There is no more cell proliferation (Shindeler et al. 2008).

The external hard callus derived from the periosteum and internal soft callus together are called the bridging callus. This structure increases the strength and stiffness within the fracture gap and allows formation of lamellar or secondary bone. At the end of the repair phase the injured bone has regained enough strength and rigidity to allow low impact exercise.

1.10 Bone Remodeling and Secondary Bone Formation

Woven bone from either intramembranous or endochondral ossification is replaced by lamellar bone, which will form either cortical or trabecular bone depending on the location. This process can continue for years after successful union of the fracture segments. The original external bony hard callus and the internal hard callus are gradually remodeled to integrate with the adjacent bone segments. If the callus is formed across a relatively small size gap, lamellar bone first forms perpendicular to the adjacent bone segment then it is remodeled to the correct orientation. If the callus is formed across a large gap woven bone is formed first then transformed into lamellar bone with the correct orientation to the adjacent bone segments (Shindeler et al. 2008).

There are several key features of bone formation that are associated with bone regeneration as well. First, bone formation is always intimately associated with neoangiogenesis or vasculogenesis. Another example of the intimate association of bone formation with blood vessels is in ectopically induced bone in tissues or organs that would not normally be associated with bone (Ripamonti et al. 2006, 2010). Second, within both the developing bone and the regenerating fracture, osteogenic cells associated with the external periosteal region form a bony collar that becomes the first cortical bone in the developing bone and repairs the cortical bone defect during fracture repair. Third, a cartilage model in the developing bone or a cartilaginous soft callus in fracture repair characterizes endochondral bone formation. These cartilaginous tissues are eventually replaced by woven bone that matures mainly into trabecular bone found in the bone marrow (Ferguson et al. 1999; Kanzcler and Oreffo 2008).

2 Critical Size Defect does not Regenerate/Repair

Critical size defect (CSD) injuries in bone by definition are those that do not undergo regenerative healing. A CSD can vary depending on the species, the particular bone, and the location of the defect within the bone. Segmental long bone CSDs can be caused by several injuries such as high-energy trauma, infections and cancerous bone tumors that must be surgically removed, as well as revision surgery (surgery to correct a failed implant or results from a previous surgery). Preexisting patient risk factors such as immune compromise and osteoporosis can affect the surgical outcome resulting in delayed bone healing, cartilaginous nonunions, or infection (Willie et al. 2010).

An important study to examine the biological processes resulting in a lack of healing demonstrates that except for impaired chondrogenesis, no differences in callus tissue distribution could be observed at 2 weeks postoperation between a successful bone healing and a nonunion rat model. Differences become apparent only at 4 weeks postoperation (Kolar et al. 2010). In a second study of delayed bone healing employing sheep tibial osteotomy model, all the stages of the regenerative response take place except that the hematoma is prolonged with a different spatial distribution of new bone, and there is delayed or prolonged endochondral bone formation present compared to the successful bone regeneration (Kolar et al. 2010, 2011).

3 Traditional and Alternative Therapeutic Approaches to Critical Size Defects

Of the traditional therapeutic methods currently available to treat CSDs, amputation is the treatment of last resort or is the emergency treatment to save the patient's life in limb crush injuries resulting from earthquakes or battlefield explosions. In cases where treatment options are available, the choice is complex and depends on the cause, size, and location of the defect and the experience of the surgeon. Ideally, treatment options should provide three essential characteristics: osteoconductivity, osteoinductivity, and osteogenic potential. Osteoconductive materials promote in-growth of local capillaries and osteoprogenitor cells from the patient to the implant. Osteoinductive materials stimulate the osteoprogenitor cells to form bone directly, while osteogenic material contains cells that can differentiate into osteoblasts and form new bone (reviewed by Willie et al. 2010; Mehta et al. 2012).

Autogenic cortical vascularized bone graft continues to be the gold standard for bone healing and restoration. However, problems can develop at the graft site due to stress fracture, bone resorption, and ultimately nonintegration of the graft. Adverse effects can occur at the site where the bone is harvested, including hematoma formation, bone fracture, infection, and nerve injury resulting in persistent pain. Autografts are restricted in availability and often result in donor site morbidity (Willie et al. 2010; De Long et al. 2007). Allogenic bone grafts from human cadaver sources will eliminate donor site complications, but these bone grafts have higher complications due to reduced revascularization and remodeling, increased bacterial and viral infection, and immune rejection (Willie et al. 2010; De Long et al. 2007).

Alternative therapeutic approaches employ bone graft substitutes that incorporate osteoconductive extracellular matrices, osteoinductive proteins, and often, osteogenic cells. BMP2 and BMP7 have been FDA approved but treatments require supraphysiological dosages to get an effect and outcomes are inconsistent. Platelet-rich plasma treatment is safe and feasible but there is no clinical evidence of benefit (reviewed by Willie et al. 2010).

Grafts of synthetic bone substitute calcium phosphate ceramics such as hydroxyapatite (HA) or tricalcium phosphate (TCP) with the addition of features that enhance the osteogenic and mechanical performance have been used in selected suitable cases. Synthetic bone substitutes must incorporate materials with structural properties that encourage bioactivity, osteoconductivity, osteoinductivity, or osteogenesis, as well as the necessary mechanical properties or compressive strength (reviewed by Hannink and Arts 2011). A major concern with engineered tissue implants in general and bone implants in particular are that they often do not become vascularized sufficiently to keep the implant alive and become integrated into the host. Furthermore there is often a lack of sufficient vascularity when new bone grows on the bone graft substitute scaffold. Scaffolds that do not fully integrate into the bone fragment can cause delayed healing, future fractures, nonunions, and cartilaginous nonbone unions.

There are currently two approaches to solving the problem of providing sufficient vascularity to bioengineered implants. The first approach is to prevascularize the scaffold prior to implantation. This can be accomplished in vitro by culturing vascular endothelial cells on the scaffold. Alternatively the scaffold can be prevascularized by implanting the prospective scaffold implant in a region of the body that will promote vascularization from the patient and subsequently placing the vascularized implant in the bone defect. The second approach is to apply biomolecules known to attract blood vessel sprouting to the scaffold prior to implantation (Nomi et al. 2002; Rouwkema et al. 2008; Kanczler and Oreffo 2008; Novosel et al. 2011).

4 A Novel Therapeutic Approach

Knowledge of the cellular and molecular mechanisms that lead to successful fracture repair will be vital for establishing effective bioengineered therapies. Postnatal bone formation recapitulates, in part, the same cellular mechanisms that establish embryonic bone. Angiogenesis is intimately associated with both endochondral and intramembranous bone formation. During bone regeneration osteogenic cells are associated with vascular tissue invading from the adjacent periosteal space or the adjacent injured marrow cavity. Through the design of bioengineered scaffold material that supports these cellular mechanisms, novel therapeutic approaches can be fabricated that enhance regenerative repair of bone. Willie et al. (2010) have suggested mechanical signals necessary to initiate successful regeneration: ECM that provides growth factor release and a surface to support inflammation, cell migration and differentiation, establishment of vasculature and soft cartilage callus formation.

There are several materials that could be potentiallyused successfully in a soft tissue therapeutic approach to bone regeneration. One such material is synthetic hydrogel (Fig. 3). Recently hydrogels have been used widely as a three-dimensional (3D) soft scaffold in tissue engineering, since they can be designed and fabricated to be excellent physicochemical mimetics of natural ECMs (Tibbitt and Anseth 2009; Lutolf and Hubbell 2005; DeForest and Anseth 2012). The hydrogels are water-swollen polymer networks and the molecular architecture can result in tissue-like viscoelastic material, diffusive transport, and interstitial flow

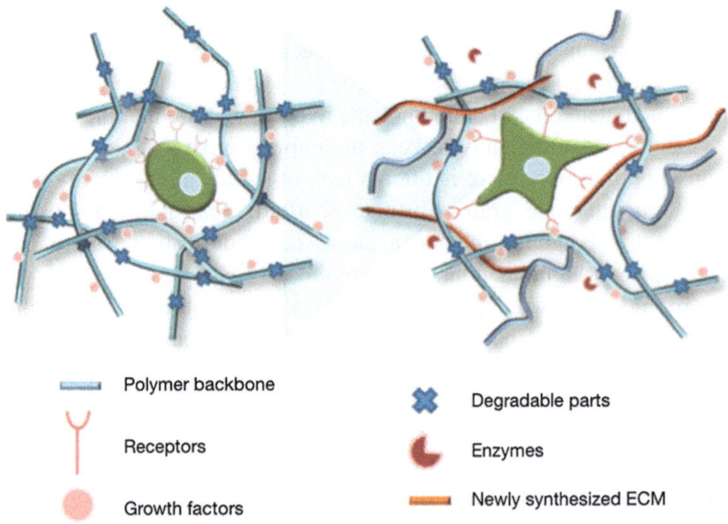

Fig. 3 Depiction of synthetic hydrogels fabricated from polymers conjugated with growth factors that facilitate binding of these factors to cell surface receptors. Hydrogel scaffolds can direct cell behavior through signaling cascades. The gradual degradation of the polymer backbone by either hydrolysis or enzymatic reactions will allow cells to create their own microenvironment and drive regenerative events

characteristics. Despite their distinctive features the hydrogels must meet several criteria to be used for bone tissue engineering. They have to be biologically compatible to minimize adverse inflammatory responses. Biodegradable hydrogels are typically desired such that they degrade, over time, to yield space for new ECM formation. Moreover, it is important to control spatial–temporal presentation of growth factors for the hydrogels to be angiogenic and osteoinductive scaffolds.

A novel approach for enhancing fracture repair and potentially improve regeneration repair of CSDs is to make a scaffold biomimetic of cartilage soft callus stage fracture repair. Various types of hydrogels can be formed from synthetic or naturally derived polymer materials in situ (Chung and Park 2009). Photo-crosslinked hydrogels can be prepared by using vinyl-conjugated monomers in the presence of a light source and initiator. Thermosensitive amphiphilic block copolymers form hydrogels in concentrated aqueous solutions by micelle formation and packing in response to change in temperature. Stereo-complexation or hydrogen bonding interaction between polymer backbones enables gelation. Introducing ionizable groups to copolymers can form pH sensitive hydrogels. Peptide oligomers designed to go through self-assembly are able to form hydrogels. VEGF and BMPs can be entrapped within hydrogel scaffolds. They can also be physically or chemically immobilized on the hydrogels for control of the growth factor release in a predefined manner.

5 A Novel Amphibian Bone Regeneration Research Model

The cellular and molecular mechanisms of long bone development, remodeling, and regeneration in amphibians are highly conserved evolutionarily, and these mechanisms are typical of other vertebrate bone studied (Pritchard and Ruzicka 1950; Hall 2003; Hall and Miyake 2000; Hutchison et al. 2007; Miura et al. 2008; Slack et al. 2008; Song et al. 2010). There has been a renewed interest in *Xenopus laevis* tadpoles and post-metamorphosed adult frogs as model systems for regenerative medicine due to the wide range of micromanipulative surgical procedures and transgenic methods available for studying differential gene expression and gradual loss of function as these animals transition from regeneration capable of larvae to regeneration-incapable adults (Slack et al. 2008; Feng et al. 2011). As described above for mammals, many of the same genes and cellular mechanisms are expressed during amphibian limb embryonic development, epimorphic regeneration of the vertebrate limb, and during long bone fracture repair. Transcription factors SOX9, RUNX2 and OSTERIX and GDF5, and later transcription factor Cbfa-1 and PTHrP and collagen type II are actively expressed in all vertebrate long bone development and regeneration examined so far (Miura et al. 2008; Hutchison et al. 2007).

5.1 Amphibian CSD Models

Amphibians, just as mammals, fail to regenerate CSDs (Goss 1969; Hutchison et al. 2007; Satoh et al. 2010; Feng et al. 2011). In 2007, Hutchinson reported that urodele long bone would not regenerate when the defect was greater than a certain size. Recent studies of CSDs in amphibians have demonstrated that neither larval urodeles (salamanders) nor adult anurans (frogs) can regenerate bone across a CSD without therapeutic aid. CSDs in the cartilaginous radius of juvenile axolotls underwent repair when BMP2 soaked beads were applied to the defect site (Satoh et al. 2010). CSDs in adult *Xenopus laevis* principal tarsus bone healed when 1,6 hexanediol diacrylate (HDDA) scaffolds soaked with BMP4 and VEGF were implanted in the defect (Feng et al. 2011).

5.2 Advantage Amphibians

A potentially fruitful approach to bone regeneration would be to chemically induce the regeneration of damaged tissues or whole complex structures in situ (Stocum 2012). This would involve activation of resident stem cells. Urodele amphibians can naturally reprogram dedifferentiated cells to an earlier mesenchymatous state. Identification of natural molecules that encourage regeneration in situ would

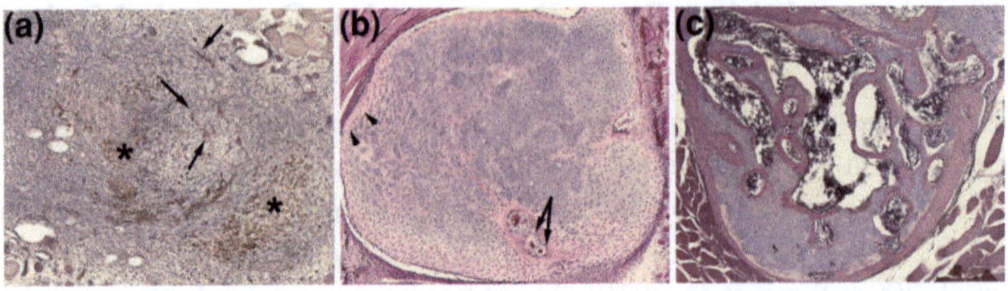

Fig. 4 Histological stages of *Xenopus* CSD bone regeneration. An HDDA scaffold with VEGF/BMP4 was applied to the CSD. (**a**) Hematoma (*asterisks*) with granulation tissue (*arrows*) is present at 2 weeks. (**b**) Cartilage rod starting to ossify (*arrows*) at 8 weeks. (**c**) Extensive woven bone ossification with a lamellar bone outer shell is present at 24 weeks

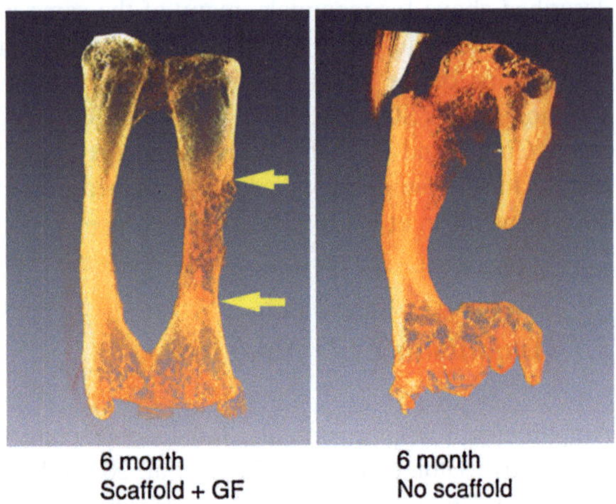

Fig. 5 MicroCT image of *Xenopus* hind limb CSD regeneration. An HDDA scaffold with VEGF/BMP4 was applied to the CSD. After 6 months of regeneration microCT image revealed that extensive woven bone was present in the growth factor treated CSD. *Arrows* indicate the original CSD boundary

permit a more effective search of resources such as combinatorial chemical libraries for synthetic small molecules that trigger the cascades activating dedifferentiation and transdifferentiation (Song et al. 2010). Since both frogs and salamanders can repair fractures like mammals and do not regenerate CSDs, both groups of amphibians could be used as inexpensive and effective screens to examine therapeutic effectiveness in enhancing bone regeneration.

5.3 Frog Hind Limb as a Novel CSD Small Animal Regeneration Screen

A study of CSD in the adult *Xenopus laevis* hind limb has demonstrated that successful regenerative repair can be triggered with a single application of VEGF/BMP4 after 6 months (Feng et al. 2011). Additional studies demonstrate that the progressive stages of *Xenopus* CSD healing resemble those of mammalian fracture

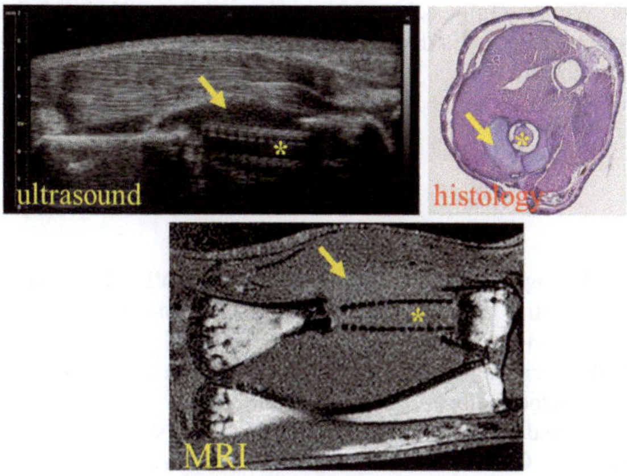

Fig. 6 Noninvasive monitoring of *Xenopus* hind limb CSD regeneration with ultrasound and MRI. An HDDA scaffold with growth factors was applied to the CSD. After 8 weeks of regeneration a cartilage callus (*arrow*) and the HDDA scaffold (*asterisk*) are visible in both ultrasound and MRI images. Histological analysis of the same case at the end of the regeneration period confirmed the presence and location of the scaffold (*asterisk*) and a cartilage callus (*arrow*)

repair (Figs. 4, 5). The progress and potential efficacy of bioengineered therapies designed and fabricated to improve regenerative repair in *Xenopus* CSDs can be monitored and evaluated by noninvasive imaging such as ultrasound and MRI (Fig. 6).

6 Conclusions and Future Directions for Healing CSDs

Fractures undergo endogenous successful regenerative healing that recapitulates many of the same cellular mechanisms and molecular cascades found in embryonic endochondral bone development. Advances in the knowledge of bone development and bone regeneration will permit more effective treatment of bone injures. In addition, vascularization is intimately involved with the initiation of every type of bone formation. Supplying regenerating bone with adequate sustainable vascularization continues to be a concern with all current therapies. Bioengineered therapies that induce and support the regenerative morphogenetic cascade and vascularization offer a potentially valuable approach to healing CSDs. One promising material is synthetic hydrogel, which can be designed and fabricated into scaffolds to trigger and extend the successful morphogenesis of fracture repair to heal CSDs. Theoretically, 3D hydrogels could be designed to support the establishment and function of any regeneration stage: hematoma, granulation tissue, soft callus, or hard callus. Adult *Xenopus laevis* hind limb is a novel small animal model system for bone regeneration research that offers several advantages. *Xenopus* hind limbs have been used successfully to screen promising scaffolds designed to heal critical size bone defects.

Acknowledgments We gratefully acknowledge support of the National Science Foundation (IJ, NF, JC). A special thank you is due to Heidi Richter of Precision Graphics, Champaign IL, for creating the illustrations for Figs. 1 and 2.

References

Bais MV, Wigner N, Young M, Toholka R, Graves DT, Morgan EF, Gerstenfeld LS, Einhorn TA (2009) BMP2 is essential for post natal osteogenesis but not for recruitment of osteogenic stem cells. Bone 45:254–266

Bianco P (2011a) Minireview: the stem cell next door: skeletal and hematopoietic stem cell "niches" in bone. Endocrinology 152:2957–2962

Bianco P (2011b) Bone and hematopoietic niche: a tale of two stem cells. Blood 117:5281–5288

Carlevaro MF, Cermelli S, Cancedda R, Cancedda FD (2000) Vascular endothelial growth factor (VEGF) in cartilage neovascularization and chondrocyte differentiation: auto-paracrine role during endochondral bone formation. J Cell Sci 113:59–69

Chung HJ, Park TG (2009) Self-assembled and nanostructured hydrogels for drug delivery and tissue engineering. Nano Today 4:429–437

DeForest CA, Anseth KS (2012) Advances in bioactive hydrogels to probe and direct cell fate. Annu Rev Chem Biomol Eng 3:421–444

Delong WG, Einhorn TA, Koval K, McKee M, Smith W, Sanders R, Watson T (2007) Bone grafts and bone graft substitutes in orthopaedic trauma surgery: a critical analysis. J Bone Joint Surg Am 89:649–658

Dwek JR (2010) The periosteum: what is it, where is it, and what mimics it in its absence? Skeletal Radiol 39:319–323

Feng L, Milner D, Chunguang X, Nye HLD, Redwood P, Cameron JA, Stocum DL, Fang N, Jasiuk I (2011) *Xenopus laevis* as a novel model to study long bone critical size defect repair by growth factor mediated regeneration. Tissue Eng Part A 17:691–701

Ferguson C, Alpern E, Miclau T, Helms JA (1999) Does adult fracture repair recapitulate embryonic skeletal formation? Mech Dev 87:57–66

Gerstenfeld LC, Cullinane DM, Barnes GL (2003) Fracture healing as a post-natal developmental process: molecular, spatial, and temporal aspects of its regulation. J Cell Biochem 88:873–884

Gerstenfeld LC, Alkhiary YM, Krall EA, Nicholls FH, Stapleton SN, Fitch JL, Bauer M, Kayal R, Graves DT, Jepsen KJ, Einhorn TA (2006) Three-dimensional reconstruction of fracture callus morphogenesis. J Histochem Cytochem 54:1215–1228

Goss RJ (1969) The genesis of form. In: Principles of regeneration. Academic Press, Inc., NY, p 166–170

Hall BK (2003) Developmental and cellular origins of the amphibian skeleton. In: Heatwole H, Davies M (eds) Amphibian biology, vol 5., OsteologySurrey Beatty and Sons, Chipping Norton, pp 1551–1597

Hall BK, Miyake T (2000) All for one and one for all: condensations and the initiation of skeletal development. BioEssays 22:138–147

Hannink G, Chris Arts JJ (2011) Bioresorbability, porosity and mechanical strength of bone substitutes: what is optimal for bone regeneration? Inj-Int J Care Inj 42:522–525

Hartmann C (2009) Transcriptional networks controlling skeletal development. Curr Opin Genet Dev 19:437–443

Hutchison C, Pilote M, Roy S (2007) The axolotl limb: a model for bone development, regeneration and fracture healing. Bone 40:45–56

Jahagirdar R, Scammell BE (2009) Principles of fracture healing and disorders of bone union. Surgery (Oxford) 27:63–69

Kanczler JM, Oreffo ROC (2008) Osteogenesis and angiogenesis: the potential for engineering bone. Eur Cells Mater 15:100–114

Karsenty G, Kronenberg H, Settember C (2009) Genetic control of bone formation. Ann Rev Cell Dev Biol 25:629–648

Keramis NC, Callori GM, Nikolaou VS, Schemitsch EH, Giannoudis PV (2008) Fracture vascularity and bone healing: a systematic review of the role of VEGF. Inj-Int J Care Inj 39S2:S45–S57

Kolar P, Schmidt-Bleek K, Schell H, Gaber T, Toben D, Schmidmaier G, Perka C, Buttgereit F, Duda GN (2010) The early fracture hematoma and its potential role in fracture healing. Tissue Eng Part B 16:427–433

Kolar P, Gaber T, Perka C, Duda GN, Buttgereit F (2011) Human early fracture hematoma is characterized by inflammation and hypoxia. Clin Orthop Relat Res 469:3118–3126

Li G, Corsi-Payne K, Zheng B, Usas A, Peng H, Huard J (2009) The dose of growth factors influences the synergistic effect of vascular endothelial growth factor on bone morphogenetic protein 4-induced ectopic bone formation. Tissue Eng Part A 15:2123–2133

Lutolf MP, Hubbell JA (2005) Synthetic biomaterials as instructive extracellular microenvironments for morphogenesis in tissue engineering. Nat Biotech 23:47–55

Malizos KN, Papatheodorou LK (2005) The healing potential of the periosteum: molecular aspects. Inj-Int J Care Inj 36S:S13–S19

Marsell R, Einhorn TA (2011) The biology of fracture healing. Inj-Int J Care Inj 42:551–555

Mehta M, Schmidt-Bleek K, Duda GN, Mooney DJ (2012) Biomaterial delivery of morphogens to mimic the natural healing cascade in bone. Adv Drug Deliv Rev. doi:10.1016/j.addr.2012.05.006

Miura S, Hanaoka K, Togashi S (2008) Skeletogenesis in *Xenopus tropicalis*: characteristic bone development in an anuran amphibian. Bone 43:901–909

Nikolaou E, Tsiridis E (2007) Minisymposium: fracture healing (i) Pathways and signaling molecules. Curr Orthop 21:249–257

Nomi M, Atala A, DeCoppi P, Soker S (2002) Principals of neovascularization for tissue engineering. Mol Aspects Med 23:463–483

Novosel EC, Kleinhans C, Kluger PJ (2011) Vascularization is the key challenge in tissue engineering. Adv Drug Deliv Rev 63:300–311

Ozaki A, Tsunoda M, Kinoshita S, Saura R (2000) Role of fracture hematoma and periosteum during fracture healing in rats: interaction of fracture hematoma and the periosteum in the initial step of the healing process. J Orthop Sci 5:64–70

Peng H, Wright V, Usas A, Gearhart B, Shen H-C, Cummins J, Huard J (2002) Synergistic enhancement of bone formation and healing by stem cell-expressed VEGF and bone morphogenetic protein-4. J Clin Invest 110:751–759

Phillips AM (2005) Overview of the fracture healing cascade. Inj-Int J Care Inj 36S:S5–S7

Pritchard JJ, Ruzicka AJ (1950) Comparison of fracture repair in the frog, lizard, and rat. J Anat 84:236–261

Ripamonti U (2010) Soluble and insoluble signals sculpt osteogenesis in angiogenesis. World J Biol Chem 1:109–132

Ripamonti U, Feretti C, Heliotis M (2006) Soluble and insoluble signals and the induction of bone formation: molecular therapeutics recapitulating development. J Anat 209:447–468

Rouwkema J, Rivron NC, van Blitterswijk C (2008) Vascularization in tissue engineering. Trends Biotech 26:434–441

Satoh A, Cummings GMC, Bryant SV, Gardiner DM (2010) Neurotrophic regulation of fibroblast dedifferentiation during limb skeletal regeneration in the axolotl (*Ambystoma mexicanum*). Dev Biol 337:444–457

Schmidt-Bleek K, Schell H, Schulz N, Hoff P, Perka C, Buttgereit F, Volk H-D, Lienau J, Duda GN (2011) Inflammatory phase of bone healing initiates the regenerative healing cascade. Cell Tissue Res. doi:10.1007/s00441-011-1205-7

Shindeler A, McDonald MM, Bokko P, Little DG (2008) Bone remodeling during fracture repair: The cellular picture. Semin Cell Dev Biol 19:459–466

Slack JM, Lin G, Chen Y (2008) The *Xenopus tadpole*: a new model for regeneration research. Cell Mol Life Sci 65:54–63

Song F, Li B, Stocum DL (2010) Amphibians as research models for regenerative medicine. Organogenesis 6:141–150

Stocum DL (2012) Regenerative therapies for musculoskeletal tissues. In: Regenerative biology and medicine, 2nd Edn. Elsevier/Academic Press, San Diego

Street J, Bao M, deGuzman L, Bunting S, Peale FV, Ferrara N, Steinmetz H, Hoeffel J, Cleland JL, Daugherty A, van Bruggen N, Redmond HP, Carano RA, Filvaroff EH (2002) Vascular endothelial growth factor stimulates bone repair by promoting angiogenesis and bone turnover. PNAS 99:9656–9661

Tibbitt MW, Anseth KS (2009) Hydrogels as extracellular matrix mimics for 3D cell culture. Biotech Bioeng 103:655–663

Towler DA (2007) Vascular biology and bone formation: hints from HIF. J Clin Invest 17:1477–1480

Towler DA (2008) The osteogenic-angiogenic interface: novel insights into the biology of bone formation and fracture repair. Curr Osteoporos Rep 6:67–71

Towler DA (2011) Skeletal anabolism, PTH, and the bone-vascular axis. J Bone Miner Res 26:2579–2582

Tsiridis E, Upadhyay N, Giannoudis P (2007) Molecular aspects of fracture healing: which are the important molecules? Inj-Int J Care inj 38S1:S11–S25

Willie BM, Petersen A, Schmidt-Bleek K, Cipitria A, Mehta M, Strube P, Lienau J, Wildemann B, Fratzl P, Duda G (2010) Designing biomimetic scaffolds for bone regeneration: why aim for a copy of mature tissue properties if nature uses a different approach? Soft Matter 6:4976–4987

Wraighte PJ, Scammell BE (2007) Principles of fracture healing. Found Years 3:243–251

Muscle Repair and Regeneration: Stem Cells, Scaffolds, and the Contributions of Skeletal Muscle to Amphibian Limb Regeneration

Derek J. Milner and Jo Ann Cameron

Abstract Skeletal muscle possesses a robust innate capability for repair of tissue damage. Natural repair of muscle damage is a stepwise process that requires the coordinated activity of a number of cell types, including infiltrating macrophages, resident myogenic and non-myogenic stem cells, and connective tissue fibroblasts. Despite the proficiency of this intrinsic repair capability, severe injuries that result in significant loss of muscle tissue overwhelm the innate repair process and require intervention if muscle function is to be restored. Recent advances in stem cell biology, regenerative medicine, and materials science have led to attempts at developing tissue engineering-based methods for repairing severe muscle defects. Muscle tissue also plays a role in the ability of tailed amphibians to regenerate amputated limbs through epimorphic regeneration. Muscle contributes adult stem cells to the amphibian regeneration blastema, but it can also contribute blastemal cells through the dedifferentiation of multinucleate myofibers into mononuclear precursors. This fascinating plasticity and its contributions to limb regeneration have prompted researchers to investigate the potential for mammalian muscle to undergo dedifferentiation. Several works have shown that mammalian myotubes can be fragmented into mononuclear cells and induced to re-enter the cell cycle, but mature myofibers are resistant to fragmentation. However, recent works suggest that there may be a path to inducing fragmentation of mature myofibers into proliferative multipotent cells with the potential for use in muscle tissue engineering and regenerative therapies.

D. J. Milner (✉) · J. A. Cameron
Department of Cell and Developmental Biology, Institute for Genomic Biology,
University of Illinois at Urbana-Champaign, Urbana, IL 61801, USA
e-mail: d-milner@life.uiuc.edu

Current Topics in Microbiology and Immunology (2013) 367: 133–159

DOI: 10.1007/82_2012_292
© Springer-Verlag Berlin Heidelberg 2012
Published Online: 8 December 2012

Abbreviations/Terms

ABCG2	ATP-Binding cassette sub-family G member 2
ARF	Alternate reading frame, cyclin-dependent kinase inhibitor 2A
BrdU	Bromodeoxyuridine
CD34	Cluster of differentiation molecule 34
CD133	Cluster of differentiation molecule 133
CXCR4	Chemokine receptor type 4
ECM	Extracellular matrix
EDL	Extensor digitorum longus
GFP	Green fluorescent protein
IGF-1	Insulin-like growth factor 1
ink4a	Cyclin-dependent kinase inhibitor 2A
mdx	Mouse strain, X-linked muscular dystrophy
Msx1	Muscle segment homeobox 1
MSC	Mesenchymal Stem Cell
p16	Cyclin-dependent kinase inhibitor 2A
p21	Cyclin-dependent kinase inhibitor 1A
p27	Cyclin-dependent kinase inhibitor 1B
p57	Cyclin-dependent kinase inhibitor 1C
Pax7	Paired-box protein 7
pRB	Retinoblastoma protein
PW1	Pw1/Peg3 Paternally expressed gene 3
Sca-1	Stem cell antigen 1
SIS	Small intestine submucosa
SP	Side population
Tcf4	Transcription factor 4
TGF-β1	Transforming growth factor beta 1
VML	Volumetric muscle loss

Contents

1 Introduction

As a consequence of both their function and their location, skeletal muscles can frequently become damaged. Whether the result of traumatic injury, or the consequence of a genetic disorder, severe damage and degeneration of skeletal muscle can have a profoundly deleterious effect on an individual's mobility, functional independence, and overall quality of life.

The basic unit of skeletal muscle tissue is the myofiber, a syncytial, tubular cell that can contain thousands of nuclei and run several centimeters in length. Myofibers are formed during embryonic development by the fusion of committed precursors (myoblasts) to form immature myotubes. As development proceeds, myotubes grow by both hypertrophy and hyperplasia from further incorporation of myoblasts. As they grow, the actin-myosin based contractile apparatus is assembled, synapse-like connections (called neuromuscular junctions) are made with innervating motor neurons, and connections are made at the fiber ends (myotendinous junctions) with fibrous connective tissue that ultimately form tendons or fascia that link muscles to bones.

The mature myofiber is a highly architecturally organized cell specialized for the generation of contractile force. This architectural organization is extended to the tissue and organ level, as individual muscles consist of bundles of myofibers arranged and packed for the efficient generation and transmission of contractile force (Fig. 1). Individual myofibers are intimately surrounded by a basal lamina consisting of collagen IV and laminin, which is in turn surrounded by a fibrous sheath called the *endomysium*. Bundles of sheathed fibers are organized into fascicles, which are surrounded by a connective tissue *perimysium*, and fascicles are bundled to form the complete muscle, which is surrounded by an *epimysium*. The mysia are also collectively called muscle interstitial tissue, or interstitium. The interstitium is highly vascularized and contains tracts of innervating motor neurons.

Similar to bone, skeletal muscle has a robust and elaborate natural mechanism to repair damage and restore this complex tissue structural architecture and its resultant function. In this review, we will provide an overview of the natural repair process of skeletal muscle and discuss how tissue engineering holds promise to extend our capacity to repair muscles that are damaged beyond what the endogenous process can handle. We will also discuss the contributions skeletal muscle makes to the fascinating process of epimorphic limb regeneration in urodele amphibians, and explore how insights from amphibians could potentially lead to the ability to dedifferentiate mammalian mature myofibers in order to provide a source of precursors for mammalian limb regeneration and muscle tissue engineering.

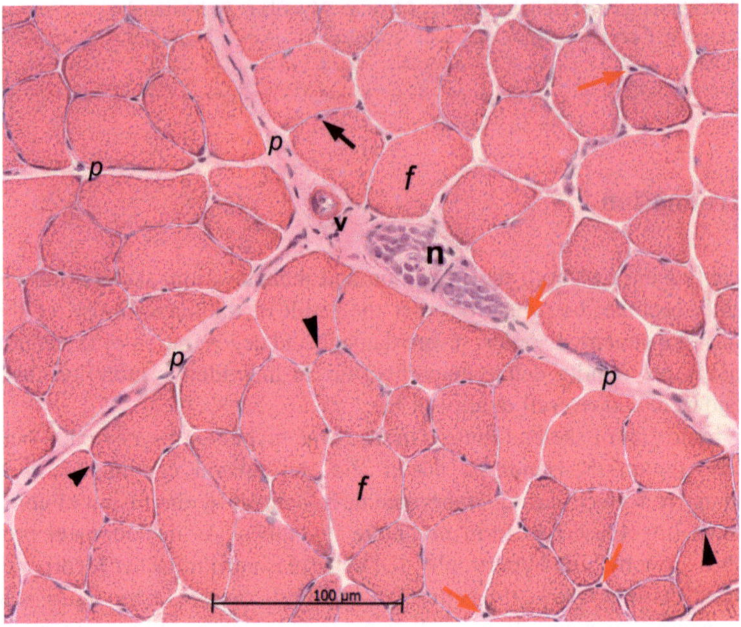

Fig. 1 Mammalian skeletal muscle in cross section. Murine tibialis anterior muscle in cross section showing portions of four fascicles surrounded by perimysium (p). A vessel (v) and nerve bundle (n) can be observed in the perimysium. Individual myofibers (f) display peripherally localized myonuclei (*arrowheads*) and are surrounded by endomysium. Both endomysium and perimysium contain numerous interstitial cells (*red arrows*), which can be any number of different cell types, ranging from fibroblasts or capillary endothelial cells, to one of several different cells with myogenic potential. A presumptive satellite cell (*black arrow*) sitting in a furrow on the myofiber surface and under the basal lamina. Definitive identification as a satellite cell requires staining for one of any number of satellite cell markers, such as Pax7 or CXCR4. Bar: 100 μm

2 Repair and Regeneration of Mammalian Skeletal Muscle

2.1 Endogenous Skeletal Muscle Repair: An Overview

The natural skeletal muscle repair process works extremely well for the repair of muscle damage resulting from tears, strains, toxin damage, and relatively small lacerations, and is functional through all stages of life. This repair process has been investigated and reviewed extensively in a number of publications (see Jarvinen et al. 2005; Huard et al. 2002 for review) so only an overview of the process will be presented here. Endogenous skeletal muscle repair follows a basic pattern, which can be roughly divided into three stages. The first stage is the *demolition stage*, characterized by sealing of damaged myofiber ends, inflammation, and the necrosis and clearing of damaged myofiber segments. The next stage is the *repair stage*, characterized by the activation, proliferation, and differentiation of satellite cells to replace myofiber segments or form new muscle fibers. Importantly, fibrous scar tissue formation also develops concomitant with muscle fiber formation during this stage. Finally, the *remodeling stage* takes place, where remodeling of scar tissue and repaired muscle fibers occurs, and muscle function is restored. In most muscle injuries, the repair stage and remodeling stage are often overlapping (Jarvinen et al. 2005).

Following the initial injury to the muscle, myofibers are ruptured or severed, resulting in the exposure of the sarcoplasm to the extracellular environment. A combination of exposure to Ca^{2+} in the extracellular fluid and massive release of Ca^{2+} from damaged sarcoplasmic reticulum leads to the activation of calcium-dependent proteases. The *demolition stage* begins with this elevated Ca^{2+} and protease activity initiating the rapid hypercontraction and disintegration of the myofibrils in the damaged area, resulting in a dense contraction band of myofibrillar and cytoskeletal proteins forming in order to wall off the damaged portions of the fiber, sealing the undamaged sarcolemma and salvaging the viable, undamaged myofiber segments. The severed ends of the muscle pull back from one another, forming an open *central zone*. During the initial injury, capillaries are ruptured, giving blood-borne cells access to the injury site. A hematoma forms in the central zone and an inflammatory reaction begins with the activation of the complement cascade, which induces chemotactic recruitment of neutrophils and macrophages. These cells carry out the disposal of necrotic myofiber fragments and other debris by phagocytosis. Beyond merely removing the trash, neutrophils and macrophages release cytokines that amplify (and, in later stages of repair, subdue) the inflammatory response, recruit muscle satellite cells to the injury site, and protect satellite cells from apoptosis in the early repair environment (Chazaud et al. 2003; Tidball and Villalta 2010). Thus, circulating cells brought in by hematoma formation and the inflammatory reaction during demolition help to set the stage for successful muscle repair, much like they do during the initial stages of bone fracture repair (Kolar et al. 2010; Schmidt-Bleek et al. 2011).

As the demolition phase progresses and macrophages proceed with the removal of necrotic myofiber fragments, the basal lamina sheathing the necrotic myofiber segments are left unmolested, yielding empty tubes of lamina, usually extending 1–3 mm from the hematoma-filled central zone. The region of hollowed-out lamina generated by the clearing action of macrophages is termed the *regenerative zone*. Meanwhile, the hematoma clots, forming a lattice of fibrin and fibronectin. As macrophages begin clearing the clot, fibroblasts begin to invade the hematoma along the clot matrix and start forming a nascent, collagenous extracellular matrix (ECM). The end result of the demolition stage is a central zone filled with resorbing clot and invading fibroblasts, and flanking regenerative zones of hollow cylindrical basal lamina, which will serve as a scaffold for myotube formation.

Once the demolition stage has progressed sufficiently, the *repair stage* begins. The repair stage is actually a combination of two simultaneously supportive and competitive processes: the formation of new muscle tissue, and the formation of a connective tissue scar. A balanced progression of both of these processes is a prerequisite for optimal recovery of muscle function (Jarvinen et al. 2005; Serrano et al. 2011; Murphy et al. 2011; Mann et al. 2011). During the repair stage, activated satellite cells proliferate and migrate through the hollow lamina scaffolding and differentiate, becoming committed myoblasts that fuse with other myoblasts or with existing myofiber ends to reform lost myofiber segments. Concurrent with the repair of muscle by the satellite cells, fibroblasts form a fibrotic scar in the central zone, which cements the link between the central zone and the flanking regenerative zones and preserves the transduction of force along the muscle. This early scar formation is important, as it allows for physical use of the muscle during the healing process, which in turn promotes better capillary ingrowth and better subsequent muscle formation (Jarvinen et al. 2005).

Gradually, as the tubes of basal lamina become completely filled with regenerating muscle, additional myoblasts extend myotube formation from the regenerative zones and penetrate the scar tissue of the central zone, eventually linking with myotubes forming from the other side of the injured muscle. The *remodeling stage* is a continuation of the repair process in which the scar tissue shrinks and new myofibers formed during the repair stage subsequently mature and form attachments to the surrounding ECM. However, muscle repair is not perfect, and this remodeling process can often be associated with marked reorganization of the muscle tissue including the formation of forked fibers, satellite myofibers, or orphan myofibers that form outside the original basal lamina scaffolding (Jarvinen et al. 2005).

2.2 Macrophages: Key Players in Muscle Repair

The involvement of macrophages in the muscle repair process has been evident for many years (Robertson et al. 1993; St. Pierre and Tidball 1994; McClennan 1996), but work over the past decade has shown how these cells do much more than simply take out the trash. Proper macrophage activity is critical for successful muscle repair, as depletion of macrophages or suppression of their activity severely compromises skeletal muscle repair (Summan et al. 2006; Segawa et al. 2008).

Two distinct macrophage types are involved in the repair of injured muscle tissue. First, classically activated pro-inflammatory M1 phenotype macrophages infiltrate the muscle from both the epimysium and the circulatory system, and phagocytose the necrotic muscle tissue. These macrophages also promote the proliferation of satellite cells, protect them from pro-apoptotic signals, and enhance their migratory ability (Chazaud et al. 2003; Sonnet et al. 2006). The phagocytosis of necrotic muscle remnants induces these cells to gradually switch their phenotype to the tissue remodeling, M2 phenotype (Summan et al. 2006; Arnold et al. 2007). This macrophage phenotype promotes myotube formation. Consistent with this observation, M2 macrophages secrete IGF-1 (Summan et al. 2006; Pelosi et al. 2007), which acts as a trophic factor for muscle, promoting muscle growth (Musaro et al. 2001). M2 macrophages also secrete TGF-β1 (Arnold et al. 2007), which promotes fibroblast growth and migration. It is possible that in chronic muscle damage, or in dystrophic pathologies where muscle is constantly undergoing rounds of degeneration and regeneration, sustained M2 macrophage activity and secretion of TGF-β1 contributes to the gradual replacement of muscle with fibrotic scar tissue (Vidal et al. 2008).

2.3 *Stem Cells and Skeletal Muscle Repair: Satellite Cells and the Cells of the Interstitium*

The main driver of endogenous skeletal muscle repair is the satellite cell. Satellite cells are resident mononuclear stem cells that are intimately associated with the myofibers, sitting tightly in grooves on the myofiber surface and underneath the myofiber basal lamina. This location puts them in the perfect spot to receive and respond to signals coming from their partner myofibers, as well as blood-borne signals delivered through capillaries and signals from cells of the endomysial interstitium. Despite demonstrations of contributions of other muscle resident cells and circulating stem cells to muscle regeneration, satellite cells have been shown time and time again to be the dominant contributor to muscle repair (Sherwood et al. 2004; Sacco et al. 2008; Lepper et al. 2011; Murphy et al. 2011). Satellite cells can be identified by their tissue localization and their expression of a panel of molecular markers including Pax7, CXCR4, β1-integrin, and CD34 (Sherwood et al. 2004). Recent works have identified heterogeneity in the satellite cell population. For example, Olwin et al. have identified a progenitor present in skeletal muscle that shares characteristics of both muscle interstitial side population (SP) cells and satellite cells. These cells express ABCG2 and Sca1 like SP cells, but also express satellite cell markers Pax7 and Syndecan 3/4. Like satellite cells, these cells are found intimately associated with myofibers, are highly myogenic, and generate canonical satellite cell progeny in vivo, suggesting they may serve as a satellite cell progenitor population (Tanaka et al. 2009).

As a consequence of their profound importance in muscle repair, there has been an intense focus on the study of satellite cells over the past 20 years. During this time, numerous advances have been made in our understanding of the embryonic origins of satellite cells, how satellite cells function during muscle repair, how satellite cell populations are maintained through self-renewal, how satellite cell function is regulated by a multitude of growth factors and interactions with other cells, and how their function is disrupted with aging and in disease states. Space constraints for this review limits our discussion of these fascinating cells, so readers are directed to a number of recent, outstanding reviews discussing satellite cell biology (Zammit et al. 2006; Ten Broek et al. 2010; Scharner and Zammit 2011; Yablonka-Reuveni 2011).

While it is now basically agreed that the satellite cell is the major cellular contributor to skeletal muscle repair, other cells residing in skeletal muscle are capable of contributing to the muscle repair process. Once thought to be a simple connective tissue consisting of fibroblasts and capillaries, the interstitium is proving to be a complicated mixture of cells that contribute to the muscle repair process, either through direct myofiber formation, co-fusion with other myogenic cells, support of satellite cell function, or a mixture of these functions. Numerous studies have identified vascular pericytes (Dellavalle et al. 2007) SP cells (Asakura et al. 2002), muscle-derived stem cells (Qu-Petersen et al. 2002) $PW1^+$ interstitial cells (Mitchell et al. 2010), and mesoangioblasts (Cossu and Bianco 2003) as resident muscle interstitial cells that have myogenic potential and participate in the muscle repair process.

Adding a further layer of complexity, there are additional nonmyogenic cell populations residing in the skeletal muscle interstitia which influence and support satellite cell activity though secretion of paracrine factors. Joe et al. (2010) have recently identified a mesenchymal stem cell-like cell residing in the interstitium. In response to injury, these cells proliferate and secrete trophic factors that support and enhance satellite cell activity. Unlike other interstitial cells, however, these stem cells serve as progenitors for fibroblasts and adipocytes. In situations where muscle repair fails or is deficient, these cells are theorized to contribute to fat and fibrous tissue accumulation in place of muscle (Rodeheffer 2010). It is interesting to speculate that sustained activity of M2 macrophages and their TGF-β1 secretion in these situations might also influence these fibroadipogenic progenitors to undergo fibroblastic differentiation. Not to be outdone, connective tissue fibroblasts also contribute to facilitation of satellite cell activity and muscle repair. Fibroblasts expressing the transcription factor Tcf4 have been shown to influence satellite cell behavior in regenerating muscle. Genetic ablation of $Tcf4^+$ fibroblasts resulted in premature satellite cell differentiation and production of smaller myofibers during repair of muscle damage, while ablation of satellite cells resulted in increased muscle fibrosis and $Tcf4^+$ fibroblast density (Murphy et al. 2011). Thus, it appears that satellite cells and fibroblasts of the interstitum can engage in reciprocal interactions to ensure a proper balance between new muscle formation and connective tissue deposition.

The interstitial tissue of skeletal muscle is rapidly becoming a crowded space. With each passing month it seems that a new study is released detailing the isolation of a new cell type found in the mysia that contributes to skeletal muscle regeneration. Further work will be required to precisely sort out the function and interactions of each of these cell types, and how they contribute to not only muscle repair, but also normal muscle function and homeostasis. Understanding what goes on in the muscle interstitium, as well as furthering our understanding of macrophage function in muscle, are important for another reason: better understanding of the muscle repair process and the functional choreography of the cellular players will help us to better imitate and replicate it through tissue engineering, as some injuries are too devastating to be repaired by endogenous repair processes.

2.4 *When Nature Is not Enough: Tissue Engineering Approaches to Skeletal Muscle Repair*

While the intrinsic muscle repair system has a robust capacity to repair muscle damage, there are limits to what it can accomplish. Satellite cell-based repair works extremely well for small, common injuries such as strains, tears, or lacerations. Severe traumatic injuries resulting in significant muscle tissue loss and gaps in muscle tissue are another matter. Such injuries are termed volumetric muscle loss (VML) injuries (Grogan and Hsu 2011). The endogenous muscle repair process is of little help in VML, as not only are muscle fibers and their supporting matrix lost, but satellite cells and myogenic cells of the interstitium are obliterated as well. Treatment options for VML are currently limited to limb bracing, free muscle transfers, and amputation. However, advances in materials science and stem cell technology, along with our increasing understanding of the endogenous muscle repair process, are now making tissue engineering approaches to treating VML more feasible.

Tissue engineering is a multidisciplinary field that applies the principles of life sciences and engineering to the development of implantable materials that can restore lost tissues or organs, or maintain and improve structure and function of degenerating tissues and organs (Langer and Vacanti 1993). Typical tissue engineering approaches utilize biocompatible scaffolds seeded with stem cells, growth factors, or combinations of both, which are implanted into a target to carry out repair or reformation of lost tissue. Scaffolds are three-dimensional (3D) constructs engineered to support tissue formation. They are usually designed to support cell attachment and migration, retain or deliver growth factors or other bioactive compounds, and display mechanical properties to support tissue formation and function. Scaffolds can be fabricated from biocompatible synthetic materials, or they can be produced using purified matrix proteins or complex multicomponent ECM preparations. ECM preparations have become an area of intense focus in tissue engineering (Badylak 2007; Lu et al. 2011). Numerous factors must be taken into account when designing a scaffold, but two parameters are particularly important: The first is pore size and extent of porosity of the scaffold, as this will affect the diffusion of nutrients and migration of cells throughout the scaffold, and influence the formation of a vascular network within the scaffold and the integration with vasculature of surrounding tissue (Novosel et al. 2011). Second, biodegradability is also a key, as the ideal scaffold should gradually break down and be resorbed while tissue is forming, eliminating the necessity for surgical removal of the scaffold material after completion of its function. Once designed and fabricated, scaffolds can be seeded with stem cells and growth factors and allowed to begin tissue formation in vitro before eventual implantation, or they can be implanted immediately after seeding.

Progress in engineering of skeletal muscle tissue has lagged compared to other tissues such as bone. This is due primarily to the complexity of skeletal muscle tissue. Skeletal muscle is a highly organized tissue, with bundles of muscle fibers running in parallel, surrounded by layers of connective tissue and penetrated by dense capillary networks, so scaffold design must imitate this organization, or be able to promote its formation. Furthermore, the hard, synthetic materials used in bone tissue engineering are typically easier to shape and fabricate into structures for implantation compared to the softer materials required for muscle tissue engineering scaffolds. Nevertheless, skeletal muscle tissue engineering has begun to make significant steps forward in recent years. Early attempts at skeletal muscle tissue engineering focused on use of synthetic polymer patches seeded with cells and implanted into muscle (Levenberg et al. 2005). While some muscle tissue engineering work continues to explore use of synthetic materials (Thorrez et al. 2008; Fernandes et al. 2012), recent works have focused more on utilizing biomatrix-based scaffolding for muscle tissue engineering. Some investigations have used purified proteins to form gels or fibrous meshes which are seeded with myogenic cells and subsequently implanted into surgically generated defects (Huang et al. 2005; Thorrez et al. 2008; Page et al. 2011). Page et al. were able to fabricate bundles of fibrin microthreads and seed them with human myogenic cells. When implanted into large defects in the tibialis anterior muscle of nude mice, these fibrin patches promoted healing and extensive ingrowth of new muscle tissue, with significantly reduced fibrosis when compared to control injuries (Page et al. 2011).

Other works have tested the use of complex ECM preparations as patches to implant in muscle defects to promote muscle repair. A number of different matrices have been used, including acellularized skeletal muscle matrix (Merritt et al. 2010), porcine bladder matrix (Machingal et al. 2011), and porcine small intestine submucosa (SIS) (Valentin et al. 2010; Turner et al. 2010). Using a porcine bladder matrix preparation seeded with rat myoblasts and cultured in a bioreactor for 1 week, Machingal et al. (2011) were able to partially restore function in mouse EDL muscle which had approximately 50 % of the muscle mass resected. Similarly, work using acellularized skeletal muscle matrix implanted into a rat gastrocnemius muscle defect and subsequently injected with mesenchymal stem cells (MSCs) isolated from bone marrow was able to partially restore muscle tissue and function (Merritt et al. 2010). Work from the Badylak group has demonstrated the ability of SIS-ECM patch implants to restore muscle histology and function in rat and canine models (Valentin et al. 2010, Turner et al. 2010). Implantation of a SIS-ECM patch into a rat abdominal wall defect resulted in significant muscle regeneration, along with restoration of nerves and blood vessels in densities approaching those seen in unaltered muscle tissue, and significant restoration of contractile properties (Valentin et al. 2010). SIS-ECM implants were able to restore both muscle and tendon tissue in a canine model with a distal portion of the gastrocnemius muscle and proximal portion of the Achilles tendon removed. At 6 months post implantation, muscle histology, fiber type distribution, and vascular and nerve densities in the regenerated region resembled those seen in unaltered muscle (Turner et al. 2010). Impressively, these SIS-ECM patches were able to accomplish regeneration without prior seeding with myogenic cells before implantation.

Another method for producing scaffolds for tissue engineering takes the ECM preparation approach to the extreme. As whole organs are structurally very complex and it is exceedingly difficult to fabricate scaffolds approximating organ structure, let alone tissue architecture, researchers have opted to attempt acellularization of whole organs, followed by repopulation with cultured cells and subsequent reformation of the organ in vitro. This approach was used recently by Taylor et al. to regrow rat hearts. Hearts produced by this method displayed proper cardiac histology, displayed endogenous pacemaker activity, and could pump fluid, albeit with a low ejection fraction (Ott et al. 2008). While hearts produced in this study were not strong enough to implant and test for functionality in vivo, the results nevertheless suggest that generating organs in vitro from acellularized templates could eventually be used for implantation. Such an approach would be especially attractive for VML injuries, as very large muscle defects, or even defects ablating adjacent muscle groups of a limb could be potentially treated by muscle-group implants. Perniconi et al. (2011) have recently attempted this strategy using acellularized tibialis anterior implants in mice. This group demonstrated that acellularized whole muscles retain the biochemical and architectural features of the original skeletal muscle, including the vascular bed, mysia, and muscle basal lamina. Scaffolds could also be stored for weeks either at 4 °C or at 37 °C in tissue cultures. When implanted into mice, the scaffolds were invaded with myogenic cells that subsequently formed muscle (Fig. 2), although there was a significant infiltration of macrophages and immune cells as well (Perniconi et al. 2011). When implants were coupled with immunosuppressive therapy, myogenesis in the implants was significantly enhanced (Perniconi et al. 2011). It will be of interest to see how whole muscle acellularized scaffolds perform if they are preloaded or cultured with myogenic and vasculogenic cells before implantation into muscle defects.

In addition to scaffold development, utilization of myogenic cells seeded into scaffolds is another important parameter of muscle tissue engineering. While some works simply implant scaffolds that rely on infiltration of nearby resident myogenic cells or circulating stem cells with mesenchymal properties (Valentin et al. 2010, Turner et al. 2010; Perniconi et al. 2011), many others use myogenic cells loaded and cultured in the scaffolds before implantation (Levenberg et al. 2005; Huang et al. 2005; Thorrez et al. 2008; Machingal et al. 2011; Page et al. 2011). Most works have utilized myogenic cells isolated from skeletal muscle directly, but other works have utilized MSCs isolated from bone marrow, as these cells exhibit myogenic potential in addition to osteogenic and adipogenic potential (Gimble et al. 2008). From a therapeutic standpoint, MSCs represent an attractive cell type for utilization in tissue engineering, as they can be isolated from a number of tissues, particularly bone marrow and adipose tissue (Zuk et al. 2001; Gimble and Guilak 2003), which would negate the need to isolate and expand myogenic cells from muscle in patients that have already suffered extreme muscle trauma (Fig. 3). Another cell type that has come under scrutiny as a potential source of myogenic cells for tissue engineering is the $CD133^{+}$ subpopulation of bone-marrow derived MSCs (Torrente et al. 2007). These cells can also be found

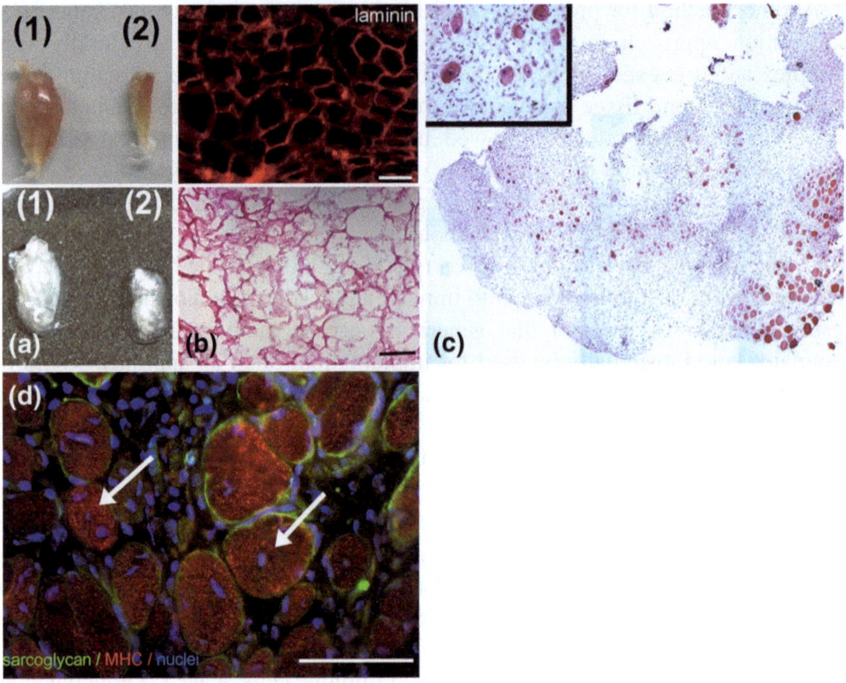

Fig. 2 Skeletal muscle tissue engineering: the organ scaffold approach. **a** Intact *tibialis anterior* (1) and *extensor digitorum longus* (2) muscles immediately after excision (*top*) and after subsequent decellularization in 0.2 % sodium dodecyl sulfate for 48 h to produce acellular, whole muscle scaffolding (*bottom*). **b** Acellular muscle scaffolding retains muscle tissue architecture even after storage for 2 weeks, as exhibited by immunostaining laminin (*top*) and H&E staining (*bottom*). Scale bars: 50 μm. **c** When implanted into excised muscle of host animals, the scaffold integrates with host tissue. At 1 week postimplantation, low-magnification histological sections from the midbelly of the implant show extensive infiltration of cells. 20X magnification inset image demonstrates the formation of nascent myofibers with eosinophilic cytoplasm and central nuceli. **d** Sections from scaffolds at 2 weeks post-\implantation immunostained for skeletal muscle markers α-sarcoglycan (*green*) and fast sarcomeric myosin heavy chain (*red*) further confirm the formation of skeletal muscle in these scaffolds. Scale bar: 50 mm. Figure reproduced from Perniconi et al. 2011, with permission

residing in skeletal muscle, and in at least one muscle tissue engineering study were found to migrate into and populate an implanted acellular scaffold that directed significant muscle repair (Turner et al. 2010). In addition to their multiple differentiation potentials, MSCs also possess another attractive ability for tissue engineering work: they are strongly immunosuppressive and anti-inflammatory (Bartholomew et al. 2002; Caplan 2007). Thus, their inclusion in muscle tissue engineering constructs may potentially provide a dual benefit of producing new muscle and protecting an implant from rejection.

Finally, chronic muscle damage that develops as a resulting phenotype of genetic muscle defects, such as Duchenne muscular dystrophy or limb-girdle

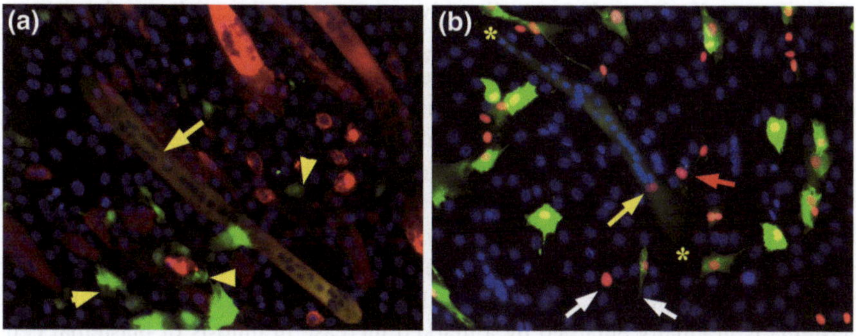

Fig. 3 Myogenic potential of mesenchymal stem cells. **a** Cultures of mesenchymal stem cells (MSCs) isolated from adipose tissue of GFP$^+$ transgenic pigs display myogenic potential when co-cultured with murine myoblasts. Immunostaining of co-cultures for sarcomeric myosin heavy chain (*red*) and GFP (*green*) demonstrates the presence of GFP$^+$ myotubes (*arrow*), indicating participation of MSCs in myogenesis. *Arrowheads* mark GFP$^+$ MSCs. **b** Participation of porcine MSCs in co-culture myotube formation is confirmed by immunostaining with an antibody specific for porcine nuclear Lamin C (*red*). A GFP$^+$ myotube (demarcated by *) clearly contains porcine nuclei (*yellow arrow*). Another MSC (*red arrow*) can be seen presumably fusing with the myotube. *White arrows* indicate non-fused GFP$^+$ MSCs

muscular dystrophies, eventually exhausts and overwhelms the intrinsic muscle repair system, leading to gradual fatty and fibrotic replacement of muscle tissue and declining muscle function. Genetic muscle damage disorders require systemic therapies, as they can affect large groups of muscles or all muscles of the body, making tissue engineering, scaffold implant-based approaches likely untenable in most cases. Most work to develop therapies for these diseases has focused on gene therapies, myogenic cell transfer therapies, or drug therapies designed to enhance expression of proteins that fill in for or modify the activity of the protein encoded by the defective gene (Pichavant et al. 2011). Recently, protein therapy has emerged as a promising method for potential treatment of some genetic muscle disorders, as a number of studies have shown that intramuscular or intravenous injections of several different proteins can ameliorate dystrophic pathology in *mdx* mice (Rooney et al. 2009; Amenta et al. 2011; Weisleder et al. 2012).

3 Skeletal Muscle in Amphibian Limb Regeneration

Urodele amphibians such as newts and salamanders have the fascinating ability to regenerate complex body structures that are lost to amputation through a process called epimorphic regeneration (for review, see Stocum and Cameron 2011; Nacu and Tanaka 2011). These animals are able to regenerate lost digits, limbs, tails and jaws, and can also regenerate parts of organs, including ventricular tissue of the heart and lenses. This powerful regenerative ability has been known to biology and studied for more than two centuries (Stocum and Cameron 2011). While a number of recent works have dramatically advanced our understanding of how this regenerative process occurs (Kumar et al. 2007; Rao et al. 2009; Kragl et al. 2009; Kumar et al. 2010; Jhamb et al. 2011), the precise mechanism(s) these animals use to fully regenerate lost structures remains elusive.

Limb and tail regeneration in urodeles are initiated by the formation of a blastema that self-organizes and gradually reforms into the ablated structures. After amputation, proteolytic degradation of tissue extracellular matrices liberates differentiated cells from their tissues. The liberated cells dedifferentiate and migrate into the space under the wound epidermis to form an avascular, early bud blastema. Blastema cells morphologically resemble MSCs. Once formed, the blastema is gradually enlarged by a marked increase in mitosis of blastemal cells and concomitant vascularization. Marker studies indicate that blastema cells derived from each tissue redifferentiate into the same tissue (Kragl et al. 2009), although some cells derived from the dermis and muscle differentiate into cartilage as well (Cameron and Hinterberger 1984; Morrison et al. 2006, 2010).

The elucidation and characterization of the source of blastemal cells is an area of ongoing study. Both tissue dedifferentiation and stem cell activation can provide cells to populate the blastema (Stocum and Cameron 2011; Nacu and Tanaka 2011; Jopling et al. 2011). As a major component of limb and tail tissue, skeletal muscle contributes heavily to the process of populating the regeneration blastema.

3.1 Dedifferentiation and Cellularization of Myofibers in Epimorphic Limb Regeneration

In urodele limb and tail regeneration, myofiber dedifferentiation contributes to the cellular population of the blastema. Early histological studies (Thornton 1938), followed by electron microscopic studies and radioactive tracer studies (Hay 1959; Hay and Fischman 1961) provided strong evidence that damaged myofibers in the stump of an amputated limb dedifferentiate, resulting in mononuclear cells budding off from the damaged fibers to contribute to the population of the regeneration blastema.

Modern studies have further verified these observations. Kumar et al. (2000) implanted retrovirally labeled myofibers in larval salamander limb stumps, and were able to observe fragmentation and accumulation of labeled mononuclear cells in the resulting blastema. Subsequent studies demonstrated that the ability of salamander myofibers to dedifferentiate was dependent on expression of Msx1, a homeobox-containing transcriptional repressor that is also a marker for regeneration competence. Confirming the importance of this protein to the process of myofiber dedifferentiation, uptake of Msx1 antisense morpholinos inhibited myofiber fragmentation (Kumar et al. 2004).

By microinjecting tail myofibers of larval salamanders with fluorescently labeled dextran, Echiverri et al. (2001) were able to observe fragmentation of myofibers and migration of the resulting mononuclear cells. Production of transgenic axolotl larvae expressing green fluorescent protein in targeted tissues has demonstrated that skeletal muscle fibers fragment and contribute to the regeneration blastema (Kragl et al. 2009). This study also verified that dedifferentiation of other tissues contributes to the blastema, and that cells derived from dedifferentiated tissues retain memory of their origin, reforming tissues they originated from in redeveloped limb structures (Kragl et al. 2009). Providing more support to the principle that dedifferentiation of tissue provides precursors used for subsequent regeneration, zebrafish cardiac regeneration has been recently shown to be driven by dedifferentiation and proliferation of cardiomyocytes (Jopling et al. 2010).

3.2 Satellite Cells Contribute to Epimorphic Limb Regeneration

In virtually all studies demonstrating myofiber dedifferentiation as a source of blastemal cells, the organisms studied have been larval urodeles. The myofibers of larval urodeles can readily dedifferentiate and fragment into mononuclear precursors, but fibers from adult salamanders appear to be refractory to at least some aspects of this process (Kumar and Brockes 2007). Like mammalian species, urodele amphibians possess satellite cells in a similar location in mature muscle, nestled closely to the myofibers, although separated from the myofiber by their own basal lamina (Cameron et al. 1986; Morrison et al. 2006). In at least one species of adult urodele, satellite cells seem to be the main source of muscle tissue contribution to the blastema. Morrison et al. (2006) were able to isolate satellite cells from the newt *Notophthalmus viridescens* and demonstrate that they express similar markers to mammalian satellite cells, including Pax7 and M-cadherin. Isolated satellite cells were multipotent, demonstrating the ability to undergo adipogenesis and osteogenesis in culture, in addition to displaying the ability to differentiate into myotubes. Importantly, when newt satellite cell cultures were labeled with bromodeoxyuridine (BrdU) and injected into fresh blastemas, BrdU-labeled nuclei were found in regenerating cartilage, muscle, and dermis (Morrison et al. 2006). Further work by the same group demonstrated that satellite cell populations are renewed during multiple rounds of limb regeneration, and confirmed the contribution of satellite cells to the formation of both muscle and cartilage in regenerating limbs by GFP^+ lineage tracing (Morrison et al. 2010).

It is interesting to note that an earlier work using adult newts reported that animals subjected to several rounds of repeated limb amputation and regeneration eventually began to demonstrate defects in limb regeneration in later rounds (Dearlove and Dresden 1976). It is tempting to speculate that this may be due to gradual exhaustion of the satellite cell pool or of other tissue stem cells, similar to how muscle repair in dystrophic animals gradually fails, eventually leading to fatty and fibrous replacement of muscle tissue. While these recent works strongly indicate that satellite cells are crucial for adult urodele blastema formation, more work will be necessary to determine if myofiber dedifferentiation also can occur in adult salamanders, and if so, what the relative contribution to the blastema is compared to the contributions of satellite cells.

Unlike urodeles, anuran amphibians are unable to regenerate lost limbs as adults. However, tadpoles retain the ability to regenerate limbs and tails, though this ability gradually diminishes as development proceeds and metamorphosis approaches (Stocum and Cameron 2011). In these amphibians, dedifferentiation of muscle fibers does not appear to play a significant role in the formation of the regeneration blastema. Garagioli and Slack (2004) found no evidence of myofiber dedifferentiation in *Xenopus* tadpole tail regeneration. More recent work by Cavaco-Rodrigues et al. (2012) has also demonstrated that myofiber dedifferentiation does not contribute to cells of the blastema in either zebrafish tail regeneration, or tadpole limb regeneration. This group was able to observe partial dedifferentiation of muscle in the tadpole tail. Myofibers near the amputation plane became thinner and displayed loss of myofibrils. Analysis of gene expression in these tail amputates confirmed a significant decrease in muscle transcripts, including myosin heavy chain and alpha actin, 3 days postamputation. A similar decrease in muscle protein levels is also observed by mass spectroscopy in the initial days after amputation in adult *Ambystoma mexicanum* (Rao et al. 2009). Despite the partial dedifferentiation, lineage-tracing experiments undertaken in the same work were unable to detect myofiber fragmentation into mononuclear cells in amputated tails (Cavaco-Rodrigues et al. 2012).

Satellite cells, however, do contribute to anuran appendage regeneration. Satellite cells clearly take part in *Xenopus* tail regeneration (Chen et al. 2006). Moreover, treatments that influence satellite cell activity improve the regeneration of muscle in regeneration deficient *Xenopus* limbs (Satoh et al. 2005), and treatment of amputated hind limbs in regeneration deficient *Xenopus* froglets by injection of gelatin microspheres loaded with sonic hedgehog protein results in the migration of satellite cells into the blastemas and improvement in the regenerative outcome (Nye et al. 2012 in preparation).

4 Dedifferentiation and Cellularization of Mature Myofibers in Mammals: Potential or Pipe Dream?

Unlike some amphibian myofibers, mammalian myofibers do not display an ability to dedifferentiate and fragment into mononuclear precursors capable of proliferation and differentiation. However, with the observation that salamander myofibers can dedifferentiate and contribute to limb regeneration, and the need for a source of muscle precursors to populate tissue engineering constructs designed to treat large muscle defects, recent interest has been directed at understanding the limits on mammalian skeletal muscle plasticity and determining if they can be overcome.

While researchers were demonstrating the ability of urodele myofibers to dedifferentiate and fragment into proliferative mononuclear cells that can take part in blastema formation and limb regeneration, works were also carried out that demonstrated mammalian muscle cultures could be dedifferentiated in vitro. Investigations into muscle plasticity have generally focused on understanding prohibition of myonuclear entry into the cell cycle, or induction of fragmentation of muscle syncytia into a mononuclear cellulate. Several studies have demonstrated the ability of C_2C_{12} myotube nuclei to reenter the cell cycle by inhibition of pRB activity (Gu et al. 1993). Similarly, the nuclei of newt myofibers undergoing dedifferentiation also reenter the cell cycle and begin DNA synthesis through the inactivation of pRB by phosphorylation (Tanaka et al. 1997). Additionally, several works demonstrated the ability of C_2C_{12} myotubes to undergo fragmentation into mononuclear cells after treatment with different factors, including newt blastema extract (McGann et al. 2001) and components of mammalian serum activated by thrombin (Lööf et al. 2007), or after inducing ectopic expression of transcription factors (Fig. 4), such as Msx1 (Odelberg et al. 2000) and Twist (Hijantoniou et al. 2008).

Concurrently, with the advent of high-throughput screening technologies that enable researchers to screen chemical libraries for compounds that have novel effects of interest, molecules were discovered that had effects on myotube structure. One of these, a trisubstituted purine called myoseverin (Rosania et al. 2000), was demonstrated to inhibit myogenic differentiation and fragment myotubes into mononuclear fragments (Perez et al. 2002) (Fig. 4). Another group using a microtubule-interacting molecule with structural similarity to myoseverin was also able to achieve cellularization of myotubes (Duckmanton et al. 2005), but the resulting mononuclear cells were nonproliferative. Myoseverin inhibits microtubule assembly (Rosania et al. 2000), although subsequent studies have suggested the myotube fragmenting activity works through disassembly of the contractile apparatus independently of its effect on microtubules (Ng et al. 2008).

Despite the ability of both cultured urodele muscle nuclei and C_2C_{12} myotube nuclei to reenter the cell cycle through pRB inactivation, myotubes formed from cultures of primary mammalian myoblasts do not do so when pRB activity is reduced or eliminated (Huh et al. 2004; Camarda et al. 2004). This suggested that mammalian muscle employs an additional factor or factors to ensure myofiber nuclei do not reenter the cell cycle. Based on observations that urodele amphibians do not possess p16/ARF/ink4a, a known cell cycle suppressor, Pajcini et al. hypothesized that this protein serves as an additional block to keep differentiated mammalian myotube nuclei from undergoing cell cycle entry. By simultaneously suppressing pRB and ARF expression, they were able to induce committed myocytes (mononuclear myoblasts that have exited the cell cycle and begun to express Myogenin) to reenter the cell cycle. After suppression, these cells could proliferate, and later redifferentiate into muscle (both in vitro and in vivo) when suppression of pRB and ARF was removed (Pajcini et al. 2010). Additionally, this group also investigated the effect of simultaneous pRB and ARF suppression in myotubes derived from primary satellite cell cultures. As expected, myotube nuclei were able to reenter the cell cycle, as demonstrated by BrdU labeling.

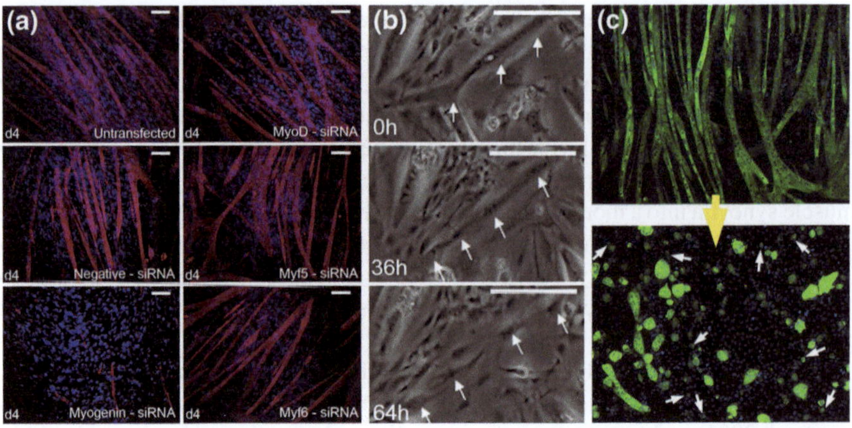

Fig. 4 Plasticity of differentiated muscle in vitro. Manipulation of gene expression (**a**, **b**) or treatments with small molecules(**c**) can induce fragmentation of differentiated myotubes into mononuclear cells. **a** Myotube cultures stained for sarcomeric myosin heavy chain expression after treatments with siRNAs designed to knock down myogenic transcription factor expression. While suppression of MyoD, Myf5, and MRF4 (Myf6) show no effect, suppression of Myogenin expression dramatically reduces sarcomeric myosin heavy chain expression. Scale bars: 50 μm. **b** Suppression of Myogenin expression in myotubes also results in fragmentation of myotubes into mononuclear cells. Time-lapse photomicroscopy of a myotube treated with Myogenin siRNA demonstrates the gradual fragmentation of the myotube into mononuclear cells over a period of 3 days. Scale bars: 50 μm. **c** Fragmentation of myotubes by treatment with myoseverin. Four-day-old myotube cultures stained for sarcomeric myosin heavy chain (*top panel*). After 24 h incubation in 20 μM myoseverin (*bottom panel*), myotubes fragment into smaller myotubes and mononuclear cells (*white arrows*). Note however, sarcomeric myosin heavy chain expression is maintained. (**a**, **b**) reproduced with permission from Mastroyiannopoulos et al. 2012

Myotubes also dedifferentiated, losing their tubular morphology and dramatically decreasing expression of muscle proteins. However, the myotubes did not fragment as with myoseverin treatment. Instead they collapsed into amorphous syncytial structures with clustered nuclei. Moreover, even though myotube nuclei could reenter the cell cycle, they could not complete it, as the nuclei did not undergo mitosis. Further experiments confirmed that the myotube nuclei were blocked at the G2/M transition (Pajcini et al. 2010). Thus, mammalian primary myotubes must express additional factors that prevent completion of the cell cycle even after forced entry by suppression of pRB and ARF.

True dedifferentiation of myotubes requires the breakdown of the myotube syncytia into mononuclear cells, reestablishment of proliferative ability by nuclear reentry into the cell cycle with complete progression through the cycle without blocks, and loss of myotube gene expression. Moreover, the dedifferentiated cells should have the ability to undergo myogenic differentiation again. In the previous mammalian myotube work, syncytia could be cellularized, but the resulting cells had little proliferative ability and could not redifferentiate (Duckmanton et al. 2005; Jung and Williams 2011), or myotube nuclei could be induced to reenter the cell cycle, but myotubes would not cellularize (Pajcini et al. 2010). Jung and

Williams recently demonstrated that a combined chemical and genetic approach can yield myotube dedifferentiation into proliferative multipotent cells. Using C_2C_{12} myotubes, they fragmented the tubes to mononuclear cells with myoseverin, and then used siRNA to suppress p21 expression. The suppression of p21 expression induced reentry into the cell cycle and proliferation of the cellulate. This proliferative cellulate retained myogenic differentiation capacity, and with additional treatment with the small molecule reversine (Chen et al. 2003) the cellulate gained the ability to undergo adipogenesis or osteogenesis in inductive media. Suppression of p21 expression was extremely important, as treatment of the cellulate with reversine alone did not allow redifferentiation into myotubes unless p21 expression was also downregulated. The effect was specific for p21, as suppression of other cyclin-dependent kinase inhibitors, such as p27 or p57 did not induce cycle entry and proliferation (Jung and Williams 2011). Proliferation of cells produced by myoseverin fragmentation could also be induced by small molecules isolated from additional chemical library screens (Kim et al. 2012). It is of interest to note that similar to p16/ARF/ink4a, amphibians apparently lack a strict homolog of p21 as well (Heber-Katz et al. this volume).

Through treatment with compounds and suppression of regulator expression, it is evident that mammalian skeletal muscle cells, once thought to be irreversibly differentiated, have much more potential for plasticity than previously imagined. However, all mammalian work exploring muscle plasticity has focused on dedifferentiating myotubes—a differentiated, but still relatively immature cell. Mature myofibers found in adult muscle tissue represent the pinnacle of muscle differentiation, and have a much more intricate and complicated cellular architecture compared to myotubes. Myofibers have a fully developed and implemented contractile apparatus and specialized junctions with innervating neurons and with the ECM at their ends. These characteristics, arising from further differentiation and maturation from the myotube state, make them potentially a much harder target to dedifferentiate and fragment (Stocum and Cameron 2011). In our laboratory we have attempted to fragment cultured mature myofibers using myoseverin (Fig. 5). While we can easily fragment C_2C_{12} myotubes and myotubes formed from primary satellite cells (see Fig. 4), mature myofibers are resistant. Isolated myofibers from the flexor digitorum brevis muscle of adult male mice were exposed to myoseverin for up to 1 week in culture, but exhibited no fragmentation or change in structure. When isolated myofibers are attached to a matrigel-coated substrate and maintained in long-term culture, they undergo a structural dedifferentiation to a more myotube-like state. Myotendinous and neuromuscular junctions disappear, striations are lost, and myonuclei reposition from their normal peripheral localization into clusters or chains. Even after losing their architecturally mature phenotype, these more myotube-like cells still remain resistant to cellular fragmentation by myoseverin treatment (Milner and Cameron, unpublished results). Thus, dedifferentiation and cellularization of this hardened target will require much more than treatment with a single factor. Still, discoveries of molecular interventions that can dedifferentiate myotubes continue to be reported

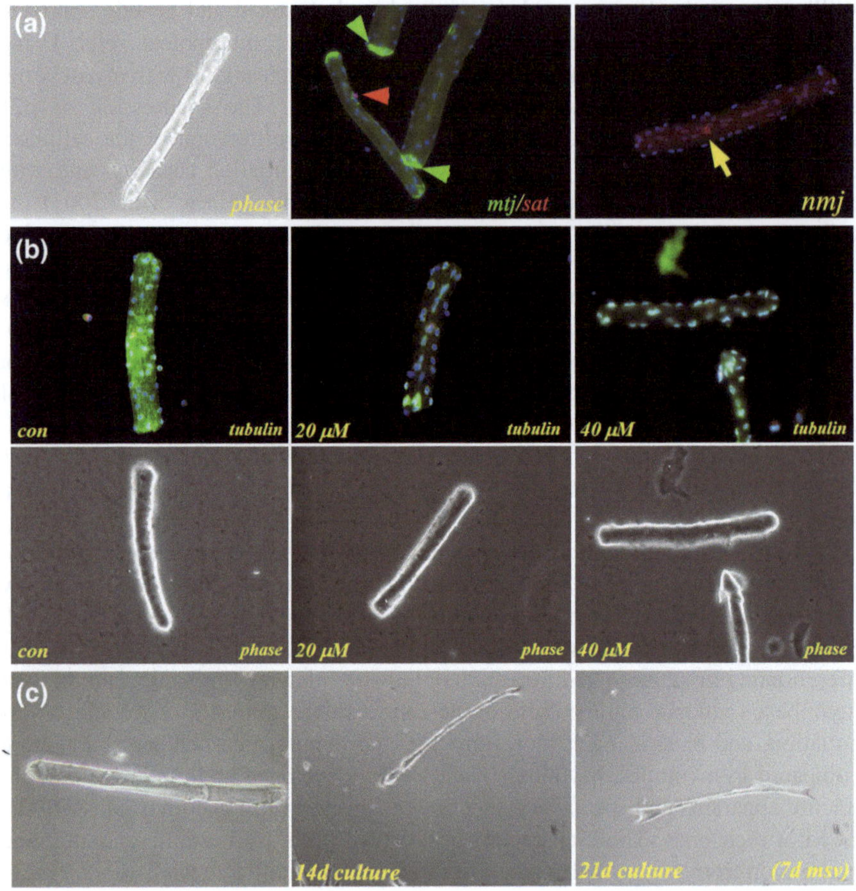

Fig. 5 Mature myofibers resist treatments that fragment myotubes. **a** Myofibers can be isolated from the *flexor digitorum brevis* muscle of the mouse and maintained in long-term culture. Myofibers display maintenance of myotendinous junction structure (*green arrowheads*), associated satellite cells (*red arrowhead*), and neuromuscular junctions (*yellow arrow*) at 4 days postisolation. **b** Treatment of myofibers with myoseverin does not result in myofiber fragmentation as with myotubes. Myofibers cultured for 4 days and subsequently treated for an additional 72 h with either 20 or 40 μM myoseverin retain their differentiated phenotype. Myoseverin microtubule assembly-inhibiting activity is confirmed by the loss of microtubule network in treated myofibers. **c** Myofibers cultured for 2 weeks gradually lose their mature morphology and take up a more myotube-like appearance. Despite the change in morphological phenotype, these cells are resistant to myoseverin activity, even after treatment with myoseverin for 1 week (7d msv)

(Meech et al. 2010; Mastroyiannopoulous et al. 2012), so perhaps the ability to dedifferentiate and fragment mature myofibers from mammals will only depend on eventually finding the right combination of factors to manipulate.

5 Summary

It has long been known that skeletal muscle possesses a robust innate ability to repair damage. Over the past two decades, our understanding of this process has been dramatically advanced by research into the cell and molecular biology of the satellite cell, and its role in the repair of skeletal muscle. More recent works detailing the importance of macrophages in skeletal muscle repair and uncovering the rich diversity of cells populating the interstitial tissues in muscle will not only open more avenues for potential therapies to enhance muscle repair, but also generate more questions for basic research to ponder. Why are there so many different myogenic cells in the interstitium, and what is their relative contribution to muscle repair? What do these cells contribute to normal muscle tissue function? Can we control and utilize them for tissue engineering? These are questions which will hopefully be addressed in the coming years.

Tissue engineering of skeletal muscle has lagged behind bone tissue engineering, as the combination of relatively soft matrix material coupled with the high level of tissue organization has been difficult for engineering and materials science to mimic. Synthetic biodegradable materials may yet be useful for muscle tissue engineering, as work with synthetic biodegradable polymers and hydrogels progresses, and technological advancements in fabrication methods such as 3D cell printing, bioplotting, and stereomicrolithography continue to move forward. For the time being, however, biomatrix-based approaches to muscle tissue engineering seem to hold the most promise. Although it is still early, a number of impressive recent works using ECM preparations as implantable muscle scaffolds have demonstrated promising results in healing sizable muscle defects. Potentially, whole muscles or whole groups of muscles from cadavers could be acellularized to make structurally competent scaffolds for implantation into large gaps in an individual muscle, or implantation into a severe defect which has ablated portions of multiple muscles.

Development of scaffolding for muscle tissue engineering is one part of the problem. Another is a source of cells to populate the scaffold and generate muscle, vasculature, and cells of the interstitium. MSCs from bone and fat hold promise, and myogenic cells isolated from muscle are another source. Tailed amphibians suggest the possibility of an additional source: dedifferentiated myofibers. Even in large muscle defects, enough muscle tissue may remain in the stumps such that if it could be dedifferentiated to proliferative myogenic cells, it would provide a robust source of cells in situ to migrate into an implanted scaffold and help integrate the scaffold with the endogenous tissue. By at least one criterion (telomere length) mammalian adult myofiber nuclei remain relatively young as animals age (O'Connor et al. 2009), so if they could be coaxed back into the cell cycle (Miyoshi et al. 2012), they would presumably proliferate robustly. Several years ago, such an idea would at best seem far-fetched. Yet, larval tailed amphibians possess the ability to dedifferentiate myofibers into proliferative cells, and through numerous recent works, it is more and more apparent that mammalian myotubes can be dedifferentiated and fragmented into proliferative cells with the potential to redifferentiate into muscle, or even other cell types.

Still, a myotube, in a culture dish or in a limb, is *not* a myofiber. All of the works that have been completed demonstrating mammalian skeletal muscle plasticity have been done using myotubes, not mature myofibers, which are what would be encountered in the muscle stumps of a VML injury. Much more work will need to be done to determine how a myofiber can be dedifferentiated and fragmented into proliferative progeny. Thanks to advancements in DNA sequencing and mass spectroscopic technology, transcriptomic profiling through RNA sequencing, and proteomic analysis via mass spectroscopy provide avenues to globally compare the transcriptomes and proteomes of myotubes and myofibers. Such comparisons will help us to understand not only how a myotube matures to become a myofiber, but also to pick out potential molecules and pathways to manipulate in order to go backwards. Coupled with recent findings on amphibian and mammalian muscle plasticity, such an approach may yet turn a pipe dream into a possibility.

References

Amenta AR, Yilmaz A, Bogdanovich S, McKechnie BA, Abedi M, Khurana TS, Fallon JR (2011) Biglycan recruits utrophin to the sarcolemma and counters dystrophic pathology in mdx mice. Proc Nat Acad Sci USA 108:762–767

Arnold L, Henry A, Poron F, Baba-Amer Y, van Rooijen N, Plonquet A, Gherardi RK, Chazaud B (2007) Inflammatory monocytes recruited after skeletal muscle injury switch into antiinflammatory macrophages to support myogenesis. J Exp Med 204:1057–1069

Asakura A, Seale P, Girgis-Gabardo A, Rudnicki MA (2002) Myogenic specification of side population cells in skeletal muscle. J Cell Biol 159:123–134

Badylak S (2007) The extracellular matrix as a biologic scaffold material. Biomaterials 28:3587–3593

Bartholomew A, Sturgeon C, Siatskas M, Ferrer K, McIntosh Patil S, Hardy W, Devine S, Ucker D, Deans R, Moseley A, Hoffman R (2002) Mesenchymal stem cells suppress lymphocyte proliferation in vitro and prolong skin graft survival in vivo. Exp Hematol 30:42–48

Camarda G, Siepi F, Pajalunga D, Bernardini C, Rossi R, Montecucco A, Meccia E, Crescenzi C (2004) A pRB-independent mechanism preserves the postmitotic state in terminally-differentiated skeletal muscle cells. J Cell Biol 167:417–423

Cameron JA, Hinterberger TJ (1984) Regional differences in the distribution of myogenic and chondrogenic cells in axolotl limb blastemas. J Exp Zool 232:269–275

Cameron JA, Hilgers AR, Hinterberger TJ (1986) Evidence that reserve cells are a source of regenerated adult newt muscle in vitro. Nature 321:607–610

Caplan A (2007) Adult mesenchymal stem cells for tissue engineering versus regenerative medicine. J Cell Physiol 213:341–347

Cavaco-Rodrigues AM, Christen B, Marti M, Belmonte JCI (2012) Skeletal muscle regeneration in *Xenopus* tadpoles and zebrafish larvae. BMC Dev Biol 12:9

Chazaud B, Sonnet C, Lafuste P, Bassez G, Rimaniol AC, Poron F, Authier FJ, Dreyfus PA, Gherardi RK (2003) Satellite cells attract monocytes and use macrophages as a support to escape apoptosis and enhance muscle growth. J Cell Biol 163:1133–1143

Chen S, Zhang Q, Wu X, Schultz PG (2003) Dedifferentiation of lineage-committed cells by a small molecule. J Am Chem Soc 126:410–411
Chen Y, Lin G, Slack JMW (2006) Control of muscle regeneration in the *Xenopus* tadpole tail by Pax7. Development 133:2303–2313
Cossu G, Bianco P (2003) Mesoangioblasts-vascular progenitors for extravascular mesodermal tissues. Curr Opin Genet Dev 13:537–542
Dearlove GE, Dreseden MH (1976) Regenerative abnormalities in *Notophthalmus viridescens* induced by repeated amputations. J Exp Zool 196:251–262
Dellavalle A, Sampaolesi M, Tonlorenzi R, Tagliafico E, Sacchetti B, Perani L, Innocenzi A, Galvez BG, Messina G, Morosetti R, Li S, Belicchi M, Peretti G, Chamberlain JS, Wright WE, Torrente Y, Ferrari S, Bianco P, Cossu G (2007) Pericytes of human skeletal muscle are myogenic precursors distinct from satellite cells. Nat Cell Biol 9:255–267
Duckmanton A, Kumar A, Chang YT, Brockes JP (2005) A single-cell analysis of myogenic dedifferentiation induced by small molecules. Chem Biol 12:1117–11126
Echeverri K, Clarke JD, Tanaka EM (2001) *In vivo* imaging indicates muscle fiber dedifferentiation is a major contributor to the regenerating tail blastema. Dev Biol 236:151–164
Fernandes S, Kuklok S, McGonigle J, Reinecke H, Murry CE (2012) Synthetic matrices to serve as niches for muscle cell transplantation. Cell Tissue Org 195:8–59
Garagioli C, Slack JMW (2004) Cell lineage tracing during *Xenopus* tail regeneration. Development 131:2669–2679
Gimble JM, Guilak F (2003) Adipose-derived adult stem cells: isolation, characterization, and differentiation potential. Cytotherapy 5:362–369
Gimble JM, Guilak F, Nuttall ME, Sathishkumar S, Vidal M, Bunnell BA (2008) *In vitro* differentiation potential of mesenchymal stem cells. Transfus Med Hemother 35:228–238
Grogan BF, Hsu JR (2011) Volumetric muscle loss. J Am Acad Orthop Surg 19:S35–S37
Gu W, Schneider JW, Condorelli G, Kaushal S, Mahdavi V, Nadal-Ginard B (1993) Interaction of myogenic factors and the retinoblastoma protein mediates muscle cell commitment and differentiation. Cell 72:309–324
Hay ED (1959) Electron microscopic observations of muscle dedifferentiation in regenerating *Ambystoma* limbs. Dev Biol 3:26–59
Hay ED, Fischman DA (1961) Origin of the blastema in the regenerating newt, *Triturus viridescens*. An autoradiographic study using tritiated thymidine to follow cell proliferation and migration. Dev Biol 3:26–59
Hijantoniou E, Anayasa M, Nicolaou P, Bantounas I, Saito M, Iseki S, Uney JB, Phylactou LA (2008) Twist induces reversal of myotube formation. Differentiation 76:182–192
Huang YC, Dennis RG, Larkin L, Barr K (2005) Rapid formation of functional muscle in vitro using fibrin gels. J Appl Physiol 98:706–716
Huard J, Li Y, Fu FH (2002) Muscle injuries and repair: current trends in research. J Bone Joint Surg Am 84:822–832
Huh MS, Parker MH, Scime A, Parks R, Rudnicki MA (2004) Rb is required for progression through myogenic differentiation but not maintenance of terminal differentiation. J Cell Biol 166:865–876
Jarvinen TAH, Jarvinen TLN, Kaariainen M, Kalimo H, Jarvinen M (2005) Muscle injuries: biology and treatment. Am J Sports Med 33:745–764
Jhamb D, Rao N, Milner DJ, Song F, Cameron JA, Stocum DL, Palakal M (2011) Network based transcription factor analysis of regenerating axolotl limbs. BMC Bioinform 12:80–90
Joe AW, Yi L, Natarajan A, LeGrand F, So L, Wang J, Rudnicki MA, Rossi FM (2010) Muscle injury activates resident fibro/adipogenic progenitors that facilitate myogenesis. Nat Cell Biol 12:153–163
Jopling C, Sleep E, Raya M, Martí M, Raya A, Izpisua-Belmonte JC (2010) Zebrafish heart regeneration occurs by cardiomyocyte dedifferentiation and proliferation. Nature 464:606–609
Jopling C, Boue S, Izpisua-Belmonti JC (2011) Dedifferentiation, transdifferentiation and reprogramming: three different routes to regeneration. Nat Rev Cell Mol Biol 12:79–89

Jung DW, Willams DR (2011) Novel chemically defined approach to produce multipotent cells from terminally differentiated tissue syncitia. ACS Chem Biol 6:553–562

Kim WH, Jung DW, Kim J, Im SH, Hwang SY, Williams DR (2012) Small molecules that recapitulate the early steps of urodele amphibian limb regeneration and confer multipotency. ACS Chem Biol 7:732–743

Kolar P, Schmidt-Bleek K, Schell H, Gaber T, Toben D, Schmidmaier G, Perka C, Buttgereit F, Duda GN (2010) The early fracture hematoma and its potential role in fracture healing. Tissue Eng Part B 16:427–434

Kragl M, Knapp D, Nacu E, Khattak S, Maden M, Epperlein HH, Tanaka EM (2009) Cells keep a memory of their tissue origin during axolotl limb regeneration. Nature 460:60–65

Kumar A, Velloso C, Imokawa Y, Brockes JP (2000) Plasticity of retrovirus labeled myotubes in the newt regeneration blastema. Dev Biol 218:125–136

Kumar A, Velloso C, Imokawa Y, Brockes JP (2004) The regenerative plasticity of isolated urodele myofibers and its dependence on MSX1. PLoS Biol 2:218–227

Kumar A, Brockes JP (2007) Preparation of cultured myofibers from larval salamander limbs for cellular plasticity studies. Nat Protoc 2:939–947

Kumar A, Godwin JW, Gates PB, Garza-Garcia AA, Brockes JP (2007) Molecular basis for the nerve dependence of limb regeneration in an adult vertebrate. Science 318:772–777

Kumar A, Nevill G, Brockes JP, Forge A (2010) A comparative study of gland cells implicated in the nerve dependence of salamander limb regeneration. J Anat 217:16–25

Langer R, Vacanti JP (1993) Tissue engineering. Science 260:920–926

Lepper C, Partridge TA, Fan CM (2011) An absolute requirement for pax7-positive satellite cells in acute injury-induced skeletal muscle regeneration. Development 138:3639–3646

Levenberg S, Rouwkema J, Macdonald M, Garfein ES, Kohane DS, Darland DC, Marini R, van Blitterswijk CA, Mulligan RC, D'Amore PA, Langer R (2005) Engineering vascularized skeletal muscle tissue. Nat Biotechnol 23:879–884

Lööf S, Straube WL, Drechsel D, Tanaka EM, Simon A (2007) Plasticity of mammalian myotubes upon stimulation with a thrombin-activated serum factor. Cell Cycle 6:1096–1101

Lu H, Hoshiba T, Kawazoe N, Chen G (2011) Autologous extracellular matrix scaffolds for tissue engineering. Biomaterials 32:2489–2499

Machingal MA, Corona BT, Walters TJ, Kesireddy V, Koval CN, Dannahower A, Zhao W, Yoo JJ, Christ GJ (2011) A tissue-engineered muscle repair construct for functional restoration of an irrecoverable muscle injury in a murine model. Tissue Eng Part A 17:2291–2303

Mann CJ, Prediguero E, Kharraz Y, Aguilar S, Pessina P, Serrano AL, Munoz-Cavones P (2011) Aberrant repair and fibrosis development in skeletal muscle. Skel Muscle 1:21

Mastroyiannopoulos NP, Nicolaou P, Anayasa M, Uney JB, Phylactou LA (2012) Down-regulation of myogenin can reverse terminal muscle cell differentiation. PLoS One 7:e29896

McLennan IS (1996) Degenerating and regenerating skeletal muscles contain several subpopulations of macrophages with distinct spatial and temporal distribution. J Anat 188:17–28

McGann M, Odelberg SJ, Keating MT (2001) Mammalian myotube dedifferentiation induced by newt regeneration extract. Proc Nat Acad Sci USA 98:13699–13704

Meech R, Gomez M, Woolley C, Barro M, Hulin JA, Walcott EC, Delgado J, Makarenkova HP (2010) The homeobox transcription factor Barx2 regulates plasticity of young primary myofibers. PLoS One 5:11612–11621

Merritt EK, Cannon MV, Hammers DW, Le LN, Gokhale R, Sarathy A, Song TJ, Tierney MT, Suggs LJ, Walters TJ, Farrar RP (2010) Repair of traumatic skeletal muscle injury with bone-marrow-derived mesenchymal stem cells seeded on extracellular matrix. Tissue Eng Part A 16:2871–2882

Mitchell KJ, Pannerec A, Cadot B, Parlakian A, Besson V, Gomes ER, Marazzi G, Sassoon DA (2010) Identification and characterization of a non-satellite cell muscle resident progenitor during postnatal development. Nat Cell Biol 12:257–266

Morrison JI, Loof S, He P, Simon A (2006) Salamander limb regeneration involves the activation of a multipotent skeletal muscle satellite cell population. J Cell Biol 172:433–440

Morrison JI, Borg P, Simon A (2010) Plasticity and recovery of skeletal muscle satellite cells during limb regeneration. FASEB J 24:750–756

Murphy MM, Lawson JA, Mathew SJ, Hutcheson DA, Kardon G (2011) Satellite cells, connective tissue fibroblasts and their interactions are crucial for muscle regeneration. Development 138:3625–3637

Musaro A, McCullagh K, Paul A, Houghton L, Dobrowolny G, Molinaro M, Barton ER, Sweeney HL, Rosenthal N (2001) Localized IGF-1 transgene expression sustains hypertrophy and regeneration in senescent skeletal muscle. Nat Genet 27:195–200

Miyoshi T, Nakano SI, Nakamura K, Yamanouchi K, Nishihara M (2012) *In vivo* electroporation induces cell cycle reentry of myonuclei in rat skeletal muscle. J Vet Med Sci (in press)

Nacu E, Tanaka EM (2011) Limb regeneration: a new development? Ann Rev Cell Dev Biol 27:409–440

Nye HLD, Milner DJ, Miller P, Miller P, Redwood P, Li B, Stocum DL, Cameron JA (2012) Enhancement of blastema formation and limb regeneration in late-stage *Xenopus laevis* tadpoles. *In preparation*

Ng DC, Gebski BL, Grounds MD, Bogoyevitch MA (2008) Myoseverin disrupts sarcomeric organization in myocytes: an effect independent of microtubule assembly inhibition. Cell Motil Cytoskeleton 65:40–58

Novosel EC, Kleinhans C, Kluger PJ (2011) Vascularization is the key challenge in tissue engineering. Adv Drug Del Rev 63:300–311

O'Connor MS, Carlson ME, Conboy IM (2009) Differentiation rather than aging of muscle stem cells abolishes their telomerase activity. Biotechnol Prog 25:1130–1137

Odelberg SJ, Kollhoff A, Keating MT (2000) Dedifferentiation of mammalian myotubes induced by msx1. Cell 103:1099–1109

Ott HC, Matthiesen TS, Goh SK, Black LD, Kren SM, Netoff TI, Taylor DA (2008) Perfusion-decellularized matrix: using nature's platform to engineer a bioartificial heart. Nat Med 14:213–221

Page RL, Malcuit C, Vilner L, Vojtic I, Shaw S, Hedblom E, Hu J, Pins GD, Rolle MW, Dominko T (2011) Restoration of skeletal muscle defects with adult human cells delivered on fibrin microthreads. Tissue Eng Part A 17:2629–2640

Pajcini KV, Corbel SY, Sage J, Pomerantz JH, Blau HM (2010) Transient inactivation of Rb and ARF yields regenerative cells from postmitotic mammalian muscle. Cell Stem Cell 7:198–213

Pelosi L, Giacinti C, Nardis C, Borsellino G, Rizzzuto E, Nicoletti C, Wannenes F, Battisini L, Rosenthal N, Molinaro M, Musaro A (2007) Local expression of IGF-1 accelerates muscle regeneration by rapidly modulating inflammatory cytokines and chemokines. FASEB J 21:1393–1402

Perez OD, Chang YT, Rosania G, Sutherlin D, Schultz PG (2002) Inhibition and reversal of myogenic differentiation by purine-based microtubule assembly inhibitors. Chem Biol 9:475–483

Perniconi B, Costa A, Aulino P, Todori L, Adamo S, Coletti D (2011) The pro-myogenic environment provided by whole organ scale acellular scaffolds from skeletal muscle. Biomaterials 32:7870–7882

Pichavant C, Aartsma-Rus A, Clemens PR, Davies KE, Dickson G, Takeda S, Wilton SD, Wolff JA, Woodell CI, Xiao X, Tremblay JP (2011) Current status of pharmaceutical and genetic therapeutic approaches to treat DMD. Mol Ther 19:830–840

Qu-Petersen Z, Deasy B, Jankowski R, Ikezawa M, Cummins J, Pruchnic R, Mytinger J, Cao B, Gates C, Wering A, Huard J (2002) Identification of a novel population of muscle stem cells in mice: potential for muscle regeneration. J Cell Biol 157:851–864

Rao N, Jhamb D, Milner DJ, Li B, Song F, Wang M, Voss SR, Palakal M, King MW, Saranjami B, Nye HL, Cameron JA, Stocum DL (2009) Proteomic analysis of blastema formation in regenerating axolotl limbs. BMC Biol 7:83–98

Robertson TA, Maley MA, Grounds MD, Papadimitriou JM (1993) The role of macrophages in skeletal muscle regeneration with particular reference to chemotaxis. Exp Cell Res 207:321–331

Rodeheffer MS (2010) Tipping the scale: muscle versus fat. Nat Cell Biol 12:102–104
Rooney JE, Gurpur PB, Burkin DJ (2009) Laminin-111 protein therapy prevents muscle disease in the mdx mouse model for Duchenne muscular dystrophy. Proc Nat Acad Sci USA 106:7991–7996
Rosania GR, Chang YT, Perez O, Sutherlin D, Dong H, Lockhart DJ, Schultz PG (2000) Myoseverin, a microtubule binding molecule with novel cellular effects. Nat Biotechnol 18:304–308
Sacco A, Doyonnas R, Kraft P, Vitorovic S, Blau HM (2008) Self-renewal and expansion of single transplanted muscle stem cells. Nature 456:502–506
Satoh A, Ide H, Tamura K (2005) Muscle formation in regenerating *Xenopus* froglet limb. Dev Dyn 233:337–346
Scharner J, Zammit PS (2011) The muscle satellite cell at 50: the formative years. Skel Musc 1:28
Schmidt-Bleek K, Schell H, Schulz N, Hoff P, Perka C, Buttgereit F, Volk HD, Lienau J, Duda GN (2011) Inflammatory phase of bone healing initiates the regenerative healing cascade. Cell Tissue Res 11:345–349
Segawa M, Fukada S, Yamamoto Y, Yahagi H, Kanematsu M, Sato M, Ito T, Uezemi A, Hayashi S, Miyagoe-Suzuki Y, Takeda S, Tsujikawa K, Yamamoto H (2008) Suppression of macrophage functions impairs skeletal muscle regeneration with severe fibrosis. Exp Cell Res 314:3232–3244
Serrano AL, Mann CJ, Vidal B, Ardite E, Prediguero E, Munoz-Canoves P (2011) Curr Top Dev Biol 96:167–201
Sherwood RI, Christiansen JL, Conboy IM, Conboy MJ, Rando TA, Weissman IL, Wagers AJ (2004) Isolation of adult mouse myogenic progenitors: functional heterogeneity of cells within and engrafting skeletal muscle. Cell 119:543–554
Sonnet C, Lafuse P, Arnold L, Brigitte M, Poron F, Authier F, Chretien F, Gherardi RK, Chazaud B (2006) Human macrophages rescue myoblasts and myotubes from apoptosis through a set of adhesion molecular systems. J Cell Sci 119:2497–2507
St. Pierre BA, Tidball JG (1994) Differential response of macrophage subpopulations to soleus muscle reloading after rat hind limb suspension. J Appl Physiol 77:290–297
Stocum DL, Cameron JA (2011) Looking proximally and distally: 100 years of limb regeneration and beyond. Dev Dyn 240:943–968
Summan M, Warren GL, Mercer RR, Chapman R, Hulderman T, Van Rooijen N, Simeonova PP (2006) Macrophages and skeletal muscle regeneration: a clodronate-containing liposome depletion study. Am J Physiol Regul Integr Comp Physiol 290:1488–1495
Tanaka EM, Gann AA, Gates PB, Brockes JP (1997) Newt myotubes reenter the cell cycle by phosphorylation of the retinoblastoma protein. J Cell Biol 136:155–165
Tanaka KK, Hall JK, Troy AA, Cornelison DDW, Majka SM, Olwin BB (2009) Syndecan-4 expressing muscle progenitor cells in the SP engraft as satellite cells during muscle regeneration. Cell Stem Cell 4:217–225
Ten Broek RW, Grefte S, von den Hoff JW (2010) Regulatory factors and cell populations involved in skeletal muscle regeneration. J Cell Physiol 224:7–16
Thornton CS (1938) The histogenesis of muscle in the regenerating forelimb of larval *Ambystoma punctatum*. J Morph 62:17–46
Thorrez L, Shansky J, Wang L, Fast L, van den Driessche T, Chuah M, Mooney D, Vandenburgh H (2008) Growth, differentiation, transplantation and survival of human skeletal myofibers on biodegradable scaffolds. Biomaterials 29:75–84
Tidball JG, Villalta SA (2010) Regulatory interactions between muscle and the immune system during muscle regeneration. Am J Physiol Regul Integr Comp Physiol 298:R1173–R1187
Torrente Y, Belicchi M, Marchesi C, Dantona G, Cogiamanian F, Pisati P, Gavina M, Giordano R, Tonlorenzi R, Fagiolari G, Lamperti C, Porretti L, Lopa R, Sampaolesi M, Vicentini L, Grimoldi N, Tiberio F, Songa V, Baratta P, Prelle A, Forzenigo L, Guglieri M, Pansarasa O, Rinaldi C, Mouly V, Butler-Browne GS, Comi GP, Biondetti P, Moggio M, Gaini SM, Stocchetti N, Priori A, D'Angelo MG, Turconi A, Bottinelli R, Cossu G, Rebulla P, Bresolin

N (2007) Autologous transplantation of muscle-derived CD133 + stem cells in Duchenne muscle patients. Cell Transpl 16:563–577

Turner NJ, Yates AJ, Weber DJ, Qureshi IR, Stolz DB, Gilbert TW, Badylak SF (2010) Xenogenic extracellular matrix as an inductive scaffold for regeneration of a functioning musculotendinous junction. Tissue Eng Part A 16:3309–3317

Valentin JE, Turner NJ, Gilbert TW, Badylak SF (2010) Functional skeletal muscle formation with a biologic scaffold. Biomaterials 31:7475–7484

Vidal B, Serrano AL, Tjwa M, Suelves M, Ardite E, De Mori R, Baeza-Raja B, de Martinez Lagran M, Lafuste P, Ruiz-Bonilla V, Jardi M, Gherardi R, Christov C, Dierssen M, Carmeliet P, Degen JL, Dewerchin M, Munoz-Canoves P (2008) Fibrinogen drives dystrophic muscle fibrosis via a TGF beta/alternative macrophage activation pathway. Genes Dev 22:1747–1752

Weisleder N, Takizawa N, Lin P, Wang X, Cao C, Zhang Y, Tan T, Ferrante C, Zhu H, Chen PJ, Yan R, Sterling M, Zhao Z, Hwang M, Takeshima M, Cai C, Cheng H, Takeshima H, Xiao RP, Ma J (2012) Recombinant MG53 protein modulates therapeutic cell membrane repair in treatment of muscular dystrophy. Sci Transl Med 4:139ra85

Yablonka-Reuveni Z (2011) The skeletal muscle satellite cell: still young and fascinating at 50. J Histochem Cytochem 59(12):1041–1059

Zammit PS, Partridge TA, Yablonka-Reuveni Z (2006) The Skeletal muscle satellite cell: the stem cell that came in from the cold. J Histochem Cytochem 54:1177–1191

Zuk PA, Zhu M, Mizuno H, Huang J, Futrell JW, Katz AJ, Benhaim P, Lorenz HP, Hedrick MH (2001) Multilineage cells from human adipose tissue: implications for cell-based therapies. Tissue Eng Part A 7:211–228

Part IV
Neural Regeneration

Neural Regeneration

Melissa M. Steward, Akshayalakshmi Sridhar and Jason S. Meyer

Abstract Regeneration of the nervous system requires either the repair or replacement of nerve cells that have been damaged by injury or disease. While lower organisms possess extensive capacity for neural regeneration, evolutionarily higher organisms including humans are limited in their ability to regenerate nerve cells, posing significant issues for the treatment of injury and disease of the nervous system. This chapter focuses on current approaches for neural regeneration, with a discussion of traditional methods to enhance neural regeneration as well as emerging concepts within the field such as stem cells and cellular reprogramming. Stem cells are defined by their ability to self-renew as well as their ability to differentiate into multiple cell types, and hence can serve as a source for cell replacement of damaged neurons. Traditionally, adult stem cells isolated from the hippocampus and subventricular zone have served as a source of neural stem cells for replacement purposes. With the advancement of pluripotent stem cells, including human embryonic stem cells (hESCs) and human induced pluripotent stem cells (iPSCs), new and exciting approaches for neural cell replacement are being developed. Furthermore, with increased understanding of the human genome and epigenetics, scientists have been successful in the direct genetic reprogramming of somatic cells to a neuronal fate, bypassing the intermediary pluripotent stage. Such breakthroughs have accelerated the timing of production of mature neuronal cell types from a patient-specific somatic

M. M. Steward · A. Sridhar · J. S. Meyer
Department of Biology, Indiana University Purdue University, Indianapolis, IN, USA

J. S. Meyer
Indiana University Center for Regenerative Biology and Medicine, Indianapolis, IN, USA

J. S. Meyer
Department of Medical and Molecular Genetics, Indiana University, Indianapolis, IN, USA

J. S. Meyer (✉)
Stark Neurosciences Research Institute, Indiana University, Indianapolis, IN, USA
e-mail: meyerjas@iupui.edu

Current Topics in Microbiology and Immunology (2013) 367: 163–191
DOI: 10.1007/82_2012_302
© Springer-Verlag Berlin Heidelberg 2013
Published Online: 6 January 2013

cell source such as skin fibroblasts or mononuclear blood cells. While extensive hurdles remain to the translational application of such stem cell and reprogramming strategies, these approaches have revolutionized the field of regenerative biology and have provided innovative approaches for the potential regeneration of the nervous system.

Abbreviations

6-OHDA	6-Hydroxydopamine
ALS	Amyotrophic lateral sclerosis
AMD	Age-related macular degeneration
BAM	Brn-2, Ascl1 and Mytl1
BDNF	Brain-derived neurotrophic factor
bHLH	Basic helix loop helix
CNTF	Ciliary neurotrophic factor
DA	Dopaminergic
EGF	Epidermal growth factor
ESCs	Embryonic stem cells
FAD	Familial Alzheimer's disease
FALS	Familial ALS
FGF2	Fibroblast growth factor 2
FGF8	Fibroblast growth factor 8
GDNF	Glial-derived neurotrophic factor
hESCs	Human embryonic stem cells
hPSCs	Human pluripotent stem cells
iDA	Induced dopaminergic
IGF-1	Insulin-like growth factor
iMN	Induced motor neuron
iN	Induced neuronal
iNPCs	Induced neural progenitor cells
iPSCs	Induced pluripotent stem cells
L-DOPA	L-3, 4-dihydroxyphenylalanine
MEF	Mouse embryonic fibroblasts
miRNA	Microribonucleic acid
mRNA	Messenger ribonucleic acid
NCAMs	Neural cell adhesion molecules
NPCs	Neural progenitor cells
PB	Piggyback
PD	Parkinson's disease
RCS	Royal College of Surgeon's
RPE	Retinal pigmented epithelium
SCNT	Somatic cell nuclear transfer
SHH	Sonic hedgehog
SMA	Spinal muscular atrophy
SMN1/SMN2	Survival of motor neuron gene 1/2

SOD1	Superoxide dismutase
TGF	Transforming growth factor
TH	Tyrosine hydroxylase

Contents

1 Introduction

The nervous system is a complex arrangement of nerve cells that coordinates the actions of an animal, including locomotion, sensory perception, and various aspects of homeostasis. Information is transmitted through the nervous system through a series of specific cell types, largely through the actions of neurons, which are responsible for the propagation of information to various areas of the body. Glial cells comprise the other major cell types of the nervous system, including oligodendrocytes, Schwann cells, and astrocytes. Oligodendrocytes and Schwann cells serve the primary function of ensheathing the neuronal axons, allowing for efficient and rapid conduction of neural impulses through the central nervous system and peripheral nervous system, respectively. Astrocytes serve numerous supportive roles in the nervous system, including the maintenance of ion homeostasis, provision of trophic support to neurons, recycling of neurotransmitter, and repair of the nervous system in response to injury.

Damage to the nervous system may result from physical trauma or through neurodegenerative mechanisms related to disease but regardless of the underlying cause of nervous system damage, the result is typically an inability of nerve cells to transmit neural impulses to specific regions of the nervous system. In order for functionality to be restored, one of three types of nervous system repair must occur. One such type of repair would be the regrowth of neuronal axons that have been damaged, while the remaining components of the neurons, including the cell body, remain spared from damage. Alternate approaches would include the restoration of damaged nerve cells, as well as the generation of new neurons to replace those that have been lost. While these three mechanisms of nervous system repair represent potential strategies that would theoretically overcome all types of damage and/or degeneration, successful implementation of these strategies is typically accomplished in only certain regions of the nervous system, as well as within the nervous system of certain animals, with evolutionarily lower animals possessing greater regenerative capacities than higher organisms.

In those higher organisms that possess limited nervous system regeneration, such repair of the nervous systems is often successfully achieved in the peripheral nervous system, with limited capabilities in the central nervous system. The basis for this discrepancy does not appear to lie within the neurons themselves, but rather the glial cells found in the peripheral nervous system as compared to those of the central nervous system (Huebner and Strittmatter 2009). The environment of the peripheral nervous system differs from the central nervous system largely due to the presence of Schwann cells providing myelin insulation to neuronal axons, whereas central nervous system neurons are ensheathed by oligodendrocytes.

Schwann cells provide a conducive environment for regeneration in the peripheral nervous system not only for peripheral neurons, but also those central nervous system neurons whose axons extend into the peripheral nervous system. This conducive environment provided by Schwann cells is due to multiple factors. First, Schwann cells of the peripheral nervous system are known to produce and secrete numerous extracellular matrix proteins, particularly laminin and fibronectin, which serve to provide a path for the regeneration of damaged axons (Bailey et al. 1993). Additionally, while these extracellular matrix proteins highlight the path for axonal regeneration, these axons require ancillary factors to further aid in pathfinding in order to effectively regenerate across an area of injury. To overcome this problem, Schwann cells also up-regulate the expression of various cell adhesion proteins in response to injury, including neural cell adhesion molecules (NCAMs) as well as members of the cadherin family, particularly N-cadherin (Thornton et al. 2005). These cell adhesion proteins expressed by Schwann cells serve as binding partners for corresponding adhesion proteins expressed by the axonal growth cone, thereby allowing these growth cones to navigate through the site of injury and eventually form proper synaptic connections with target cells. A third regenerative mechanism provided by Schwann cells lies in their ability to produce and secrete various neurotrophic factors in response to injury, including brain-derived neurotrophic factor (BDNF), ciliary neurotrophic factor (CNTF), and glial-derived neurotrophic factor (GDNF) (Frostick et al. 1998; Terenghi 1999). The availability of these neurotrophic factors likely serves to initiate a regeneration program in damaged neurons, leading to the formation of a growth cone and the initiation of pathfinding in this growth cone. These neurotrophic factors may also serve to provide a target signal for these growth cones to ensure that they find their appropriate target.

In the central nervous system, however, regenerative ability is greatly reduced when compared to the peripheral nervous system. However, this difference is largely due to the environment within the central nervous system itself, including the glial cells (Fawcett 2006; Fawcett and Asher 1999; Yiu and He 2006). Of primary importance is the fact that in the central nervous system, neuronal axons are myelinated by oligodendrocytes rather than Schwann cells as in the peripheral nervous system. Numerous differences exist between these two cell types that may account for their differential effects on regeneration of the peripheral and central nervous systems (Fawcett 2006; Yiu and He 2006). First, following injury in the peripheral nervous system, Schwann cells can become proliferative again and thereby replace other Schwann cells that had been lost in response to injury. Oligodendrocytes of the central nervous system, however, do not proliferate in response to injury and thus, those oligodendrocytes that are lost are not replaced. Because of this, a path for axonal regeneration is not established within the central nervous system and axonal growth cones are incapable of proper pathfinding. Furthermore, the remaining oligodendrocytes following injury do not demonstrate similar growth-promoting activities compared to Schwann cells and thus, the extracellular environment within the central nervous system following injury is much less conducive to axonal regrowth.

Beyond the differences between oligodendrocytes and Schwann cells, other factors within the central nervous system also act as a barrier to neural regeneration. Another important factor is the extent of cell death that occurs in the central nervous system in response to injury (Horner and Gage 2000; Springer 2002). Following trauma to a region of the nervous system, it is often not only the directly damaged neurons that undergo apoptosis, but also those neurons in the surrounding vicinity. Thus, the area affected by injury increases greatly beyond those cells that were directly impacted by trauma and often increases the severity of the injury. This increased area of injury also has profound implications for axonal regeneration, as not only more axons might need to be regenerated, but this regeneration will likely need to occur over a greater distance. Furthermore, following apoptotic events due to central nervous system injury, it will also likely be necessary to eventually replace neurons that have been lost in this process.

An additional barrier to central nervous system regeneration comes from the many glial cells that remain in the injured region. In response to injury, glial cells of the central nervous system, particularly those that would otherwise exist in a quiescent state, can begin to proliferate and form a glial scar that essentially serves as a physical barrier to axonal regeneration (Bahr and Bonhoeffer 1994; Horner and Gage 2002). These glial cells also produce and secrete a number of signaling factors, including members of the transforming growth factor (TGF) family, that largely serve to promote cell death. Other signaling factors produced by these glial cells may serve in a neuroprotective role to help rescue damaged neurons but nonetheless, growth-promoting signals provided by these glial cells appear largely absent.

2 Traditional Approaches to Enhance Central Nervous System Regeneration

Due to the generalized inability of the central nervous system to regenerate in response to injury or disease, great efforts have been taken to experimentally enhance the regenerative abilities of central nervous system neurons. Given that a much greater degree of neuronal regeneration is seen within the peripheral nervous system, some of the first efforts to enhance central nervous system regeneration involved the grafting of peripheral nerves into the central nervous system. A great example of this came from the studies of So and Aguyao (1985) in which regeneration of the optic nerve was demonstrated to be possible through the use of peripheral nerve grafts. Following axotomy of the optic nerve, regeneration of these axons is typically not achieved due to an inhospitable environment within the central nervous system. However, if a portion of the sciatic nerve was placed alongside the site of axotomy, with the other end of the sciatic nerve apposed to the superior colliculus, these optic nerve axons were capable of significant regeneration through this peripheral nerve graft toward the superior colliculus. Such experiments provided the proof-of-principle that the inability of central nervous system axons to regenerate was not likely due to an inherent feature of these neurons, but rather due to environmental factors within the central nervous system. Thus, great efforts have been focused upon modulation of the environment within the central nervous system in the hopes that these approaches could potentially yield enhanced axonal regeneration.

In addition to the release of factors mentioned previously, astrocytes in the central nervous system are also known to produce chondroitin sulfate proteoglycans in response to injury. The expression of these chondroitin sulfate proteoglycans persists for up to several months following injury, and has been demonstrated to be a significant component underlying the inability of central nervous system axons to regenerate (Laabs et al. 2005). Based on in vitro experiments with reactive astrocytes, the up-regulated expression of chondroitin sulfate proteoglycans was largely responsible for the failure of dorsal root ganglia neurons to extend axonal projections (Smith-Thomas et al. 1995). Furthermore, in vivo experiments involving the transplantation of dorsal root ganglia neurons demonstrate that they are capable of extending axons into the intact central nervous system, but fail to do so through those injured areas rich in chondroitin sulfate proteoglycans (Davies et al. 1997, 1999; Silver and Miller 2004).

Similarly, certain proteins found within the nervous system have also been demonstrated to serve as impediments to axonal regrowth. Perhaps, the most commonly studied class of such proteins are known as the Nogo proteins, with a particular focus upon Nogo-A (Grandpre and Strittmatter 2001; Zorner and Schwab 2010). The Nogo proteins, as their name suggests, serve as an inhibitor of axonal elongation following trauma, and are most commonly found associated with myelination within the central nervous system. Studies have demonstrated that methods to circumvent Nogo interactions with elongating axons lead to enhanced neurite outgrowth and potentially extension of axons into, and sometimes through, a trauma-related glial scar. Previous studies have demonstrated that in mice lacking Nogo-A, the cortex and spinal cord showed up-regulated expression of growth-related genes (Montani et al. 2009). Furthermore, the use of Nogo-A-specific antibodies has been demonstrated to lead to enhanced neurite sprouting, including an up-regulation of growth-related proteins (Craveiro et al. 2008; Zorner and Schwab 2010). More recently, such an approach utilizing anti-Nogo-A antibodies has been utilized in Phase I clinical trials for spinal cord injury patients, with the eventual goal of enhancing neural regeneration due to trauma by removing the inhibition due to Nogo protein expression (Hawryluk et al. 2008; Zorner and Schwab 2010).

3 Stem Cells for Neural Regeneration

While a variety of strategies have been developed to enhance neuroregeneration in response to trauma, complete cell replacement will likely be necessary under circumstances in which extensive numbers of cells have been lost, such as following extensive injury or in response to degenerative diseases of the nervous system. In the hopes of regenerating tissue through cell replacement, many efforts have focused upon the use of stem cells as a source for "replacement" cells. As a general rule, stem cells possess two defining characteristics: (1) they are capable of self-renewal and (2) they can give rise to more differentiated progeny. It is these characteristics that have made stem cells a desirable vehicle for neural regeneration, as they would ideally be available in large numbers and could be induced to differentiate into the cell type of choice.

Early efforts focused mainly upon the use of "adult", or "tissue-derived" stem cells, meaning those stem cells found within particular tissues of the body that could give rise to multiple cell types of that lineage. Within the adult nervous system, populations of stem cells are generally found within two main regions. In the hippocampus, a continually proliferative population of stem cells is found within the subgranular zone of the dentate gyrus. These stem cells are speculated to play an important role in the process of learning and memory, particularly in the acquisition of new memories. Neural stem cells are also found within the subventricular zone, particularly in rodents, where from new neural progenitor cells (NPCs) migrate to the olfactory bulb through the rostral migratory stream. During development, however, neural stem cells play a vital role in the formation of the nervous system.

Owing to their ability to self-renew as well as give rise to more differentiated progeny, neural stem cells have been the subject of considerable research in the hopes that they can be utilized to replace neurons and/or glia that have been lost due to degenerative processes. Experimentally, neural stem cells have been harvested from various sources, including the fetal brain, adult hippocampus, spinal cord, and neural retina, and can typically be induced to proliferate in vitro in response to signaling by fibroblast growth factor 2 (FGF2) and epidermal growth factor (EGF). Upon removal of these proliferation-inducing mitogens, neural stem cells further differentiate into the major cell types of the nervous system, including neurons, astrocytes, and oligodendrocytes. However, current limitations exist regarding the use of these neural stem cells, as continued growth in vitro tends to lead to a reduced neurogenic ability of these cells, and eventually senescence. Thus, a greater focus has been placed on the use of neural stem cells as a vehicle for neuroprotection and/or neuroregeneration of endogenous neurons upon transplantation, rather than as a source of replacement neurons. While such an approach could conceivable be utilized for nearly any neurodegenerative disorder, greater emphasis has been placed on certain disorders affecting specific cell types of the nervous system, including amyotrophic lateral sclerosis (ALS), Parkinson's disease (PD) and age-related macular degeneration (AMD).

3.1 Neural Stem Cells for Motor Neuron Disorders

Amytrophic Lateral Sclerosis (ALS) is characterized by a progressive and irreversible degeneration of motor neurons in the brainstem, cerebral cortex, and spinal cord leading to muscle weakness, dysfunction, and loss of mobility. ALS has a prevalence of approximately 1:50,000 and typically leads to death within 5 years of diagnosis (Cronin et al. 2007). The incidence of ALS is mainly sporadic while 10 % of the disease phenotypes are familial or inherited (FALS). In about 20 % of FALS cases, mutations in superoxide dismutase (SOD1) gene have been established to increased accumulation of SOD1 aggregates causing toxicity to motor neurons via a number of diverse mechanisms (Rosen et al. 1993; Shaw and Valentine 2007). While research in this field has been extremely promising, the SOD1 mutation is not responsible for sporadic ALS, which occurs in a greater percentage of the population (Lindvall et al. 2012). In either case, progression of the disease phenotype is debatable; however, other pathological mechanisms implicated include dysfunction of mitochondria, activation of apoptosis pathways, microglial and astrocyte interactions and defects in calcium homeostasis (Boillee et al. 2006; von Lewinski and Keller 2005).

Spinal muscular atrophy (SMA) is another motor neuron degenerative disease, characterized by the loss of anterior horn cells of the spinal cord, leading to degeneration of alpha motor neurons and resulting in muscular atrophy and paralysis (Ebert and Svendsen 2010; Lefebvre et al. 1995). Unlike late-onset diseases like ALS, SMA is particularly fatal to infants and children. SMA is an autosomal recessive disorder, leading to deletions or mutations in the survival of motor neuron gene (SMN), subsequently affecting the 20 kb SMN1 protein encoded by that gene. In humans, two transcripts of the gene have been found, SMN1 and SMN2, which differ by a single nucleotide. Mutations in SMN1 have severe implications as they often affect production of full length protein, while mutations in SMN2 leads to

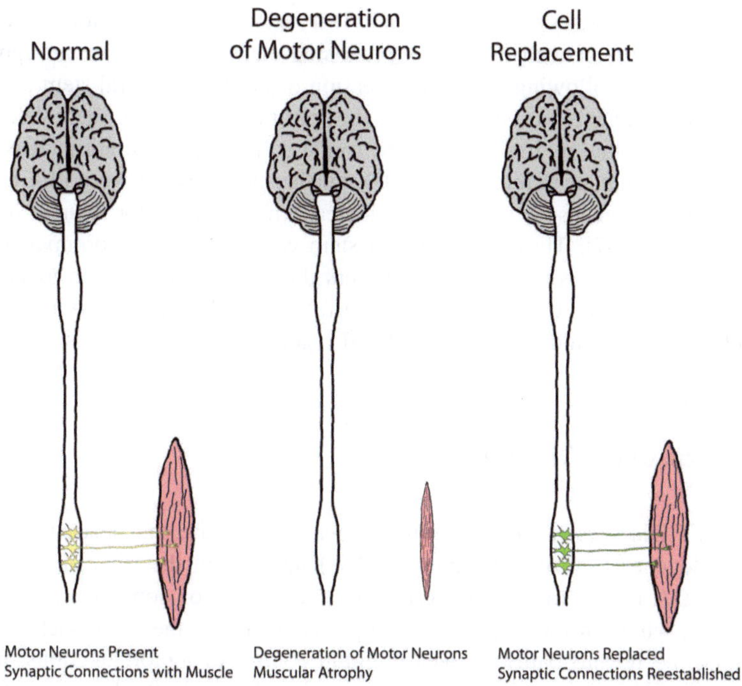

Fig. 1 Neural regeneration strategies for motor neurons. Under normal conditions, spinal motor neurons serve to relay information from the central nervous system to peripheral muscle targets. However, due to degenerative diseases such as ALS or SMA, these motor neurons are lost, and thus connectivity with peripheral muscles is absent. Current strategies for intervention include the prospects of stem cells for replacement of degenerated motor neurons, followed by the re-establishment of functional connectivity with peripheral muscles

production of low levels of the protein, which have been found capable of reducing the severity of the disease phenotype (Lorson et al. 1999).

Both ALS and SMA have been a focus of neural stem cell strategies, as motor neurons may potentially be generated for replacement by neural stem cells (Fig. 1) (Jordan et al. 2009) or at the least, these neural stem cells could be utilized to rescue endogenous motor neurons from disease-related degeneration (Hefferan et al. 2012; Klein et al. 2005; Lepore et al. 2011; Suzuki et al. 2007). Upon transplantation into the spinal cord, neural stem cells typically exhibit a bias toward gliogenic differentiation, and it is hoped that these glial cells would offer some degree of neuroprotection. Indeed, studies have shown that under certain conditions, spinal motor neurons are rescued near the site of neural stem cell transplantation (Hefferan et al. 2012), while other studies have indicated that these neural stem cells may survive and integrate into the host nervous system, but offer little neuroprotection to endogenous motor neurons.

Given the inconclusive results demonstrating neuroprotection provided by neural stem cells, an alternate approach would involve using neural stem cells that are engineered to overexpress secreted survival factors as a vehicle for the delivery of neuroprotective growth factors. A promising candidate for such an approach exists in GDNF. Following the overexpression of GDNF in neural stem cells and subsequent transplantation of these cells into rat models of ALS, these cells were capable of surviving, migrating along the spinal column, and secreting GDNF into the microenvironment (Klein et al. 2005). Furthermore, the delivery of GDNF to the ALS rat model also resulted in an increase in spinal motor neuron survival (Suzuki et al. 2007). Thus, while neural stem cells may not be optimal for cell replacement for neurodegenerative disorders due to a predisposition to generate glial cells, these cells still constitute a powerful approach as a delivery mechanism for factors that may aid in the survival and regeneration of host cells.

3.2 Parkinson's Disease

Parkinson's Diseases (PD) is a neurodegenerative disease that leads to the progressive degeneration of dopaminergic (DA) neurons in the substantia nigra of midbrain. DA neurons are characterized by the release of dopamine as a neurotransmitter, which in turn regulates multiple pathways in the brain such as those governing the reward pathway, pleasure, and control of motor functions. In PD, production of dopamine is reduced, associated with an eventual degeneration of DA neurons (Davie 2008). The symptoms of this disease include rigidity, slow movement, tremors, affected walking and posture, and sometimes dementia. Similar to ALS, the disease occurs in two forms: familial and sporadic form; some forms of familial PD are associated with mutations of α-synuclein protein leading to its accumulation in the form of clusters called Lewy bodies while mutations in Parkin lead to PD in the absence of Lewy bodies (Chung et al. 2001).

As opposed to some other neurodegenerative diseases, PD is confined to a single subtype of neurons, and external supplementation of dopamine has been proven to be effective for a short timeframe. Hence, current therapies for PD include L-3, 4-dihydroxyphenylalanine (L-DOPA) administration to supplement the loss of dopamine through external sources (Purves et al. 2008). However, the inability of L-DOPA to cross the blood–brain barrier coupled with inconsistent levels of dopamine after the drug levels off leads to re-emergence of symptoms. Hence, although drug-based treatments are helpful, they do not provide long-term relief either by neuroprotection of dying DA neurons or by replacement of the damaged DA neurons. Therefore, efforts have focused upon the transplantation of neurons that can integrate into the existing circuitry and deliver dopamine or transplant cells that maintain dopamine concentration (Tsui and Isacson 2011; Wijeyekoon and Barker 2009).

The transplantation of embryonic mesencephalic tissue in PD patients has been shown to reinnervate and restore functionality to the striatum leading to increased dopamine release (Lindvall and Bjorklund 2011). These grafts have been shown to functionally integrate in patient's brain and retain functionality for up to 10 years

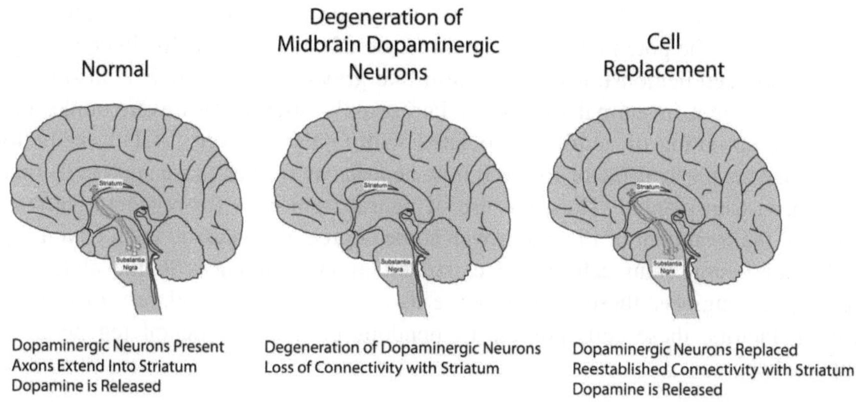

Fig. 2 Neural regeneration strategies for dopaminergic neurons. In the midbrain, dopaminergic neurons are typically found within the substantia nigra and extend axons into the striatum, where dopamine is released. Following the degeneration of these dopaminergic neurons due to diseases such as PD, striatal neurons lose their dopaminergic input. Through the use of stem cells, it is hoped that midbrain dopaminergic neurons can be replaced and eventually restore connectivity with the striatum

after transplantation (Piccini et al. 1999). Unfortunately, not all results of such experiments have been so dramatic. In some clinical trials, patients saw little or no recovery after 2 years (Freed et al. 2001; Olanow et al. 2003). A number of factors such as immune rejection and extreme severity of disease have been shown to affect reproducibility of these experiments. Also, embryonic mesencephalic tissue is limited in number and is not easily accessible, which further complicates treatment options via this tissue source. Nevertheless, these experiments helped demonstrate that cell replacement experiments could potentially succeed in human PD patients and laid the groundwork for stem cell research in this area.

Neural stem cells thus offer a unique opportunity to provide an expandable population of cells that could potentially be utilized for transplantation in animal models of PD (Fig. 2). Experimentally, it has been demonstrated that the transplantation of neural stem cells into the striatum of the 6-Hydroxydopamine (6-OHDA)-lesioned rat model of PD results in long-term survival of these cells, along with modest neurite outgrowth from the transplanted cells (Armstrong et al. 2003; Harrower et al. 2006). However, while the results of these studies were promising, these cells failed to differentiate into DA neurons, which would be necessary to replace those cells that have been lost. Thus, alternate approaches have focused upon the ability of neural stem cells to serve as a vehicle for growth factor delivery to the striatum in an effort to rescue host DA neurons, thereby slowing or preventing the progression of the disease (Akerud et al. 2001; Behrstock et al. 2006; Ebert et al. 2008). The growth factor GDNF has been identified as an optimal candidate for growth factor-based cell rescue, as previous studies have demonstrated that neural stem cells overexpressing GDNF can enhance the survival of host DA neurons (Akerud et al. 2001; Behrstock et al. 2006), and even restore some degree of functionality to the host neurons (Ebert et al. 2008), thereby demonstrating the potential for regenerative use of neural stem cells. Interestingly, it has also been demonstrated that insulin-like growth factor 1 (IGF-1) also has a functional effect in animal models of Parkinson's disease. However, unlike the increased sprouting of neurites associated with GDNF delivery, IGF-1 seems to function as a survival factor for both host neurons as well as those stem cells that have been transplanted. Thus, the use of neural stem cells for regenerative purposes in the treatment of PD has proven some degree of success to date. Given the ability of neural stem cells to be expanded in vitro, along with the ability to genetically engineer these cells to serve as a vehicle for the delivery of critical growth factors, these cells possess tremendous promise for neural regenerative approaches.

4 Pluripotent Stem Cells for Neural Regeneration

As opposed to neural stem cells which are considered to be multipotent due to their ability to form numerous cell types of the nervous system, the advent of pluripotent stem cell research affords new opportunities to derive cell types of interest from a more primitive stem cell population that possesses the ability to differentiate into any cell type of the body. The isolation of embryonic stem cells (ESCs) from the inner cell mass of a mouse blastocyst, which have the ability to maintain their pluripotent state in vitro, marked the first demonstration of pluripotent stem cells (Evans and Kaufman 1981). Nearly 20 years later, similar studies were successful in deriving the first human embryonic stem cells (hESCs) (Thomson et al. 1998).

Based upon the ability to study ESCs of both mouse and human origin, researchers were able to better understand the underlying mechanisms responsible for the maintenance of pluripotency. As a result, it was discovered that it was possible to reprogram somatic cells such as fibroblasts via the introduction of a subset of the transcription factors *Oct4, Sox2, Nanog, Lin28, c-myc*, and/or *Klf4* to generate a new source of pluripotent stem cells known as induced pluripotent stem cells (iPSCs) (Fig. 3). The implications of this discovery were profound, as it eliminated the embryonic origin of stem cells and thus the ethical considerations surrounding ESCs, while also providing the ability to generate pluripotent stem cells from almost any somatic cell of the body, offering prospects for patient- and disease-specific stem cells, as well as personalized medicine.

Pluripotent stem cells, both ESCs and iPSCs, have several features that have led to widespread interest in them as applied to the field of neural regeneration. First, these cells are pluripotent, meaning that they can be cultured and expanded indefinitely, and can also be directed to differentiate into any cell type of the body, including all of the cells of the nervous system. Second, the ability of these cells to recapitulate development has been established and hence, they serve as an excellent in vitro system for developmental studies (Hu et al. 2009; Meyer et al. 2009; Pankratz et al. 2007). Third, with the advent of iPSCs, somatic cells can be

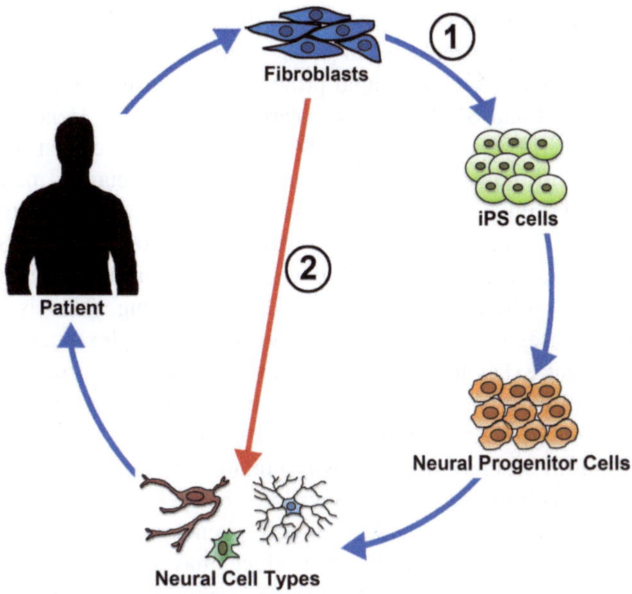

Fig. 3 Paradigm for cellular reprogramming. Through methods of cellular reprogramming, somatic cells such as skin fibroblasts can be isolated from individual patients and used to generate replacement neurons. Initial efforts were focused upon the reprogramming of these fibroblasts into induced pluripotent stem cells (pathway 1), followed by the directed differentiation of these pluripotent stem cells into a neural lineage. More recently, efforts have focused upon the direct reprogramming of somatic cells to a specified neural phenotype (pathway 2), thereby bypassing the pluripotent intermediary

reprogrammed to form stem cells, which could further facilitate cell replacement for neural degeneration from autologous sources, reducing chances of graft rejection. iPSCs have also been demonstrated to retain disease phenotypes in vitro and hence could be used for disease modeling or drug screening (Ebert et al. 2009; Egawa et al. 2012; Meyer et al. 2011). Lastly, a variety of protocols for the differentiation of stem cell-derived neurons in culture have been established (Chambers et al. 2009; Hu et al. 2009; Lee et al. 2007; Li et al. 2005; Perrier et al. 2004; Yan et al. 2005). Due to these unique advantages, significant efforts have been made to use these cells to develop neural regenerative strategies for a variety of neurodegenerative disorders including ALS, SMA, and PD, as well as AMD.

4.1 Motor Neuron Disorders

Disorders affecting motor neurons such as ALS and SMA lead to the degeneration of these cells, and therefore decreased motor function. In the early stages of disease progression, neural regeneration may be accomplished via rescue of existing motor neurons. At later stages of the disease however, motor neurons are already lost and must be replaced in order to reestablish functionality (Fig. 1). Owing to the chronic nature and rapid progression of disease phenotype, current treatments for ALS and SMA have been largely ineffective. Hence, human pluripotent stem cells (hPSCs), including hESCs and hiPSCs have been suggested as a potential source of replacement cells for lost motor neurons, particularly to facilitate cell replacement and support, as well as a novel tool for in vitro studies of neurodegeneration and regeneration for a variety of reasons. First, hPSCs possess the unique ability to recapitulate human development in vitro. Hence, they serve as valuable tools for disease modeling and drug screening. This is particularly important in diseases such as ALS which possess a complex phenotype and no effective treatment exists to date. hiPSCs have been derived from ALS patients that mirror the disease phenotype and efforts are being undertaken to understand the disease pathology (Dimos et al. 2008; Park et al. 2008). Additionally, patient-specific lines of hiPSCs have been shown to retain the disease phenotype in vitro. Utilizing such cells, drug screening assays have been performed to identify novel compounds that could rescue the ALS disease phenotype, leading to the identification of anacardic acid as a compound which may help rescue ALS-affected motor neurons (Egawa et al. 2012). Such experiments further reinforce the ability of hiPSCs to help decipher disease pathogenesis and screen for drugs for use in vivo to potentially reverse the disease phenotype.

Similar studies have also been performed using other motor neuron disorders such as SMA. Utilizing hiPSCs derived from an SMA patient, it was demonstrated that these cells exhibited a reduced ability to differentiate into motor neurons, hence recapitulating the disease phenotype (Ebert et al. 2009). hiPSCs derived from SMA patients may also provide the opportunity to study the sequela of the disease and its underlying mechanisms. Studies have demonstrated the role of apoptotic pathways resulting in motor neuron death, suggesting apoptotic inhibitors could reduce the severity of the disease. Patient-derived iPSCs hold promise as excellent surrogates for drug screening, as well as determining necessary drug concentrations to reduce the severity of the phenotype.

Additionally, hPSCs provide a vital source for cell replacement. A number of protocols have been published in the last decade demonstrating the derivation of functional motor neurons from hPSCs (Dimos et al. 2008; Hu et al. 2009; Hu and Zhang 2009; Li et al. 2005). Such protocols have involved the use of regionalizing agents, such as retinoic acid to specify a caudal fate as well as sonic hedgehog (SHH) to specify a ventral fate (Dimos et al. 2008; Hu et al. 2009; Lee et al. 2007; Li et al. 2005; Shin et al. 2005).

4.2 Parkinson's Disease

Similar to the motor neuron diseases mentioned above, PD will also require the replacement of neurons once they have been lost (Fig. 2). In the last decade, many protocols demonstrating derivation of functional DA neurons from hPSCs have been published (Cho et al. 2008; Cooper et al. 2010; Hu and Zhang 2009; Perrier et al. 2004; Soldner et al. 2009; Sonntag et al. 2007; Swistowski et al. 2010). Similar to other differentiation protocols, differentiation was aided by the use of exogenous factors such as FGF8 to provide a midbrain identity, and later supplemented with factors such as SHH. Other protocols have also focused on the inhibition of SMAD signaling coupled with addition of factors vital to the production of DA neurons, such as SHH, WNT inhibitors, BDNF, GDNF, cAMP and TGFβ3 (Kriks et al. 2011). Mature cells were generated after a minimum of 1 month and midbrain DA neurons were identified by the presence of markers such as βIII-tubulin, tyrosine hydroxylase (TH), LMX1A, NURR, and EN-1.

Additionally, other studies have demonstrated that upon transplantation, hPSCs that were differentiated to a DA fate in vitro, could reverse PD symptoms in animal models (Ben-Hur et al. 2004; Hargus et al. 2010; Roy et al. 2006; Yang et al. 2008). A recent breakthrough was the derivation of a high number of functional DA neurons from hESCs using the smad-inhibition protocol (Kriks et al. 2011). Upon transplantation in rat and mice PD models, a high number of DA neurons survived in the host for at least 4–5 months. Furthermore, the ability of dopaminergic neurons to successfully innervate the striatum and extend axons for at least 3 mm, along with the ability to reverse PD symptoms in animal models, was also demonstrated.

4.3 Retinal Degenerative Disorders

As a part of the central nervous system, the retina functions as an intricate network of neurons whose function it is to receive light stimuli and convert it to an electrical signal that is propagated and interpreted by the superior colliculus in the brain (Dowling and Werblin 1971). The retina is a complex yet well-defined structure composed of six different types of cells: ganglion cells, amacrine cells, bipolar cells, horizontal cells, muller glia cell and photoreceptors (Dowling 1970; Kolb et al. 2001). The retinal pigmented epithelium (RPE) helps to support and maintain the integrity of photoreceptors and their functions (Strauss 1995). Damage to RPE or any other cell type of the retina negatively affects the visual pathway, reducing vision and perhaps eventually leading to blindness.

Among retinal degenerative disorders, Age-related Macular Degeneration (AMD) is one of the leading causes of blindness (Ambati et al. 2003; Gehrs et al. 2006). This disease progresses through an initial stage characterized by the formation of ocular deposits adjacent to RPE called 'drusens' (Hageman et al. 2001). These deposits, over time, lead to degradation of the RPE, with subsequent damage to the photoreceptors ultimately leading to blindness (Bressler et al. 1988; Del Priore et al. 2002). The disease proceeds through two known forms: wet AMD, which involves neovascularization involving extensive and abnormal growth of blood vessels, and dry AMD, which does not involve neovascularization (Gehrs et al. 2006).

Successful treatment of AMD requires a reduction of the disease phenotype coupled with rescue or replacement of the damaged photoreceptors, leading to significant improvements in vision (Bharti et al. 2011). Subretinal transplantation of RPE has been considered a logical move as it involves a transplantation of a single monolayer of cells which are not directly involved or interlaced with a neural network (Gouras and Algvere 1996). Also, the proximity of the retina promotes interactions with the RPE via the subretinal space.

Recently, efforts to replace damaged cells in AMD have used stem cells, including hESCs and hiPSCs (Carr et al. 2009; Lamba et al. 2010). The ability of these cells to generate retinal cell types in a stepwise manner that mimics human retinogenesis has been demonstrated by several groups (Hirami et al. 2009; Idelson et al. 2009; Lamba et al. 2006; Mellough et al. 2012; Meyer et al. 2009, 2011; Osakada et al. 2009a; b). These protocols usually first involve the conversion of these cells to an anterior neural intermediate stage, followed by their subsequent differentiation into cells of the neural retina or RPE. Retinal cell types produced in this fashion have also been demonstrated to be functional in vitro, particularly photoreceptors (Meyer et al. 2011) and RPE (Buchholz et al. 2009; Idelson et al. 2009).

The subretinal transplantation of hPSC-derived RPE is widely believed to possess the greatest likelihood for integration and functional improvement in AMD. Transplantation of hESC-derived RPE into Royal College of Surgeons (RCS) rats has been demonstrated to lead to both the rescue of photoreceptors and improvements in visual acuity (Lund et al. 2006). Similar experiments involving transplantation of hiPSC-derived RPE helped to rescue photoreceptors in the RCS rat in vivo (Carr et al. 2009), while maintaining visual function.

In the last few years, the technology to derive functional RPE cells from hPSCs has been perfected to generate large amounts of RPE with morphological and functional semblance to human RPE (Kokkinaki et al. 2011; Rowland et al. 2012a). Based on promising early results from the RCS rat (Carr et al. 2009; Lu et al. 2009; Lund et al. 2006), the field is now progressing toward clinical trials for AMD where hESC-derived RPEs are being utilized to test restoration of vision (Schwartz et al. 2012). This study demonstrates that hESCs can be generated under optimum conditions and safely transplanted into human subjects without detrimental effects such as teratoma formation or graft rejection. However, further studies will be needed to demonstrate that transplanted RPE cells can successfully restore visual function in patients (Du et al. 2011; Osakada et al. 2010; Rowland et al. 2012b).

5 Reprogramming of Somatic Cells to Neural Lineages

When mouse ESCs were first established in 1981, followed by iPSCs in 2006, they represented novel and promising therapeutic cellular agents for disorders of neural degeneration. The hallmark features of self-renewal and pluripotency make these cells ideal candidates for the replacement, as well as the study, of the myriad of specialized cell types that can be deleteriously affected or lost in disorders of the nervous system. The demonstration of this cellular plasticity, particularly in somatic cell-derived iPSCs, opened the door for other promising cellular replacement strategies. With the lentiviral delivery of key pluripotency-regulating transcription factors, the reprogramming of a terminally differentiated, mammalian somatic cell to a pluripotent stem cell had been demonstrated, characterized, and reproduced many times in the derivation of various iPSC lines. This was considered by many to stand in the face of a major tenet in biology. The next question for the field became: Could this pluripotent intermediary be bypassed, i.e., rather than utilizing an indirect reprogramming pathway via iPSCs, could terminally differentiated, mammalian cells be *directly reprogrammed* to another cell type, in particular, a neural fate?

5.1 A Brief History of Direct Reprogramming

Among the first demonstrations of cellular reprogramming, somatic cell nuclear transfer (SCNT) has been used as a means to reprogram postnatal and cultured cells to a totipotent state in many systems (Campbell et al. 1996; Gurdon et al. 1958; Gurdon and Uehlinger 1966; McGrath and Solter 1983; Prather and First 1990). This technique, first used in *Rana pipiens* (Briggs and King 1952), usually involves the removal of the nucleus from an unfertilized oocyte, which is then replaced by the transplantation of the diploid nucleus of a somatic cell. The egg cell then 'reprograms' the somatic nucleus, cellular division commences, and a new organism results. More recently, human oocytes were successfully used to reprogram skin-derived somatic cells to the blastocyst stage, from which they were able to derive stable, triploid human cell lines that resemble ESCs. This end was finally accomplished by *not* removing the oocyte genome as in previous protocols, and merely adding the diploid nucleus from the somatic cell to the egg (Noggle et al. 2011). While many pathways—including chromatin remodeling—are implicated to be involved, the precise molecular mechanisms regulating this process have yet to be delineated. In between the leap in cloning from frog (Gurdon et al. 1958) to sheep (Campbell et al. 1996), it was discovered that the delivery of a single transcription factor, *MyoD,* was sufficient to convert fibroblasts into muscle cells (Davis et al. 1987). Another milestone in the field came as the first transdifferentiation of cells with therapeutic applicability, when insulin-producing cells in the pancreas were obtained via in vivo transdifferentiation (Zhou et al. 2008). This study demonstrated that these cells alleviated hyperglycemia in a mouse model of Type I diabetes. Thus, when iPSCs were successfully reprogrammed from mouse fibroblasts, a growing body of evidence was suggesting that mammalian somatic cells could potentially be similarly *directly* reprogrammed for regenerative purposes (Fig. 3).

5.2 A Paradigm for Direct Reprogramming

Often, preliminary reports regarding direct reprogramming have utilized the mouse system. However, considering the size of the mouse genome ($\sim$25,000 genes), the odds of choosing and testing 'the right' gene, or worse, right set of genes for reprogramming to a particular cell fate are poor at best, though a quickly growing number of groups have been successful to this end for a variety of cell types (Huang et al. 2011; Ieda et al. 2010; Sekiya and Suzuki 2011; Son et al. 2011; Szabo et al. 2010; Vierbuchen et al. 2010). One way to reconcile the stunning efficiency of this process with its inherent rarity is to view the gene regulatory network as an integrated, dynamic system that has gene expression profile patterns that act as stable 'attractor states' (Huang 2009; Zhou and Huang 2011). A model such as this would also explain the innate stability of iPSCs in their pluripotent state, as well as that of terminally differentiated cell types. This model further predicts that the reprogramming process does not have to be externally regulated throughout the entire process. Rather, the cell only needs to be pushed far enough out of its current 'attractor state' gene expression profile that it is closer to the gene expression profile of the desired cell type, as well as farther away from any other attractor state. And finally, this model correctly predicts that the closer the gene expression profiles of the cell of origin and the final cell type, the easier this transition will be.

5.3 Methods for Reprogramming to a Neural Fate

Skin-derived fibroblasts may share an ectodermal lineage with neural cells, and are readily accessible in large numbers, making them ideal candidates for scientists and/or clinicians interested in studying or treating neurodegenerative disorders. Reprogramming across major lineage boundaries has also been demonstrated, using hepatocytes (Huang et al. 2011; Marro et al. 2011; Sekiya and Suzuki 2011) or germ cells (Tursun et al. 2011) to derive functional induced neuronal cells (iN cells), though one comparison of the transcriptional networks during fibroblast- and hepatocyte-derived reprogramming did reveal timing differences between the two processes (Marro et al. 2011). Furthermore, fibroblast cultures are most likely heterogeneous in their gene expression and harbor multipotent, neural crest-derived stem cells, lending themselves nicely to applications of reprogramming to the neural lineage (Bayreuther et al. 1988). While efficiency is an important variable, so is the ease with which the starting tissue can be obtained. Blood draws and skin-punch biopsies can provide large numbers of mononuclear blood cells or skin fibroblast cells respectively, that are both obtained and stored with relative ease, soon affording clinicians, basic researchers, and patients alike opportunities that were never previously available, or even though possible.

Since the landmark publications demonstrating indirect and direct cellular reprogramming, many groups have been working to apply these principles for regenerative purposes in various ways. One such example is the many alternative induction methods that have since been published to overcome the risk of mutagenesis introduced by retro- or lentiviruses. These include synthetic messenger ribonucleic acid (mRNA) (Warren et al. 2010) and microribonucleic acid (miRNA) systems (Anokye-Danso et al. 2011; Yoo et al. 2011), as well as using small molecules, or a combination of these methods (Ambasudhan et al. 2011; Pang et al. 2011; Zhu et al. 2010) to induce neural conversion. Another very promising method comes from the combination of a tetracycline-induced system with the Piggyback (PB) transposon system. After somatic cells are reprogrammed, integrated DNA is cut out by the transposase enzyme (Kaji et al. 2009; Woltjen et al. 2009). Similar to the increase in studies in gene delivery or induction methods, these discoveries have also lead to an explosion in the field of direct reprogramming to specific neural and neuronal cell types, with hopes ultimately aimed to model and remedy various aspects of neurodegeneration.

5.4 Direct Reprogramming to Neural Cell Types

The elegant, initial demonstration of direct cellular reprogramming to a neural cell type utilized mouse embryonic fibroblasts (MEFs) and postnatal fibroblasts to derive neurons (Veirbuchen et al. 2010), with a conversion efficiency near 20 % at 2 weeks post-infection, indicating a markedly more efficient and rapid conversion than that seen in mouse iPSC establishment. MEF cells were initially used due to their previously demonstrated efficiency in iPSC reprogramming (Takahashi and Yamanaka 2006). Also similar to iPSC generation, a lentiviral system was used in which pools of candidate transcription factors were delivered to cells that were then screened for neuronal conversion. Through these experiments, a combination of three proneural transcription factors—Brn2, Ascl1, and Mytl1 (BAM)—were identified that were sufficient to convert mouse fibroblasts into functional (excitatory glutamatergic and inhibitory GABAergic) iN cells in vitro. This group, along with the Abeliovich group, went on to test if the same three factors could also convert human fetal fibroblasts to iN cells. However, this BAM combination of factors proved insufficient for human cells, as pronounced cell death and immature cell phenotypes were reported (Pang et al. 2011; Qiang et al. 2011). However, with the addition of the Basic helix loop helix (bHLH) transcription factor NeuroD1 to the BAM cocktail, they were able to measure action potentials, detect neuronal markers, and report spontaneous and evoked postsynaptic currents after 5–6 weeks when human fetal fibroblast-derived iN cells were co-cultured with mouse primary cortical neurons (Pang et al. 2011). This timing is similar to human ESC-derived neurons, suggesting that a longer period of maturation is an inherent property of the human system when compared to mouse. Skin-derived fibroblasts from patients with Familial Alzheimer's Disease (FAD) were used to derive iN cells exhibiting disease-specific attributes, providing significant proof-of-principle that iN cells can be used to model human neurodegenerative diseases (Qiang et al. 2011). In this study, they used a five-factor system to generate iN cells, providing yet more evidence for an attractor-state model of cellular differentiation, as the number of pathways to a neuronal identity continues to increase.

5.5 Direct Reprogramming to Specific Neuronal Subtypes

The ability to specify distinct neuronal subtypes in vitro has obvious developmental, clinical, and experimental applications for neural de- and regeneration. DA neurons were among the first neuronal subtype to be directly induced from both mouse and human fibroblasts in 2011 (Caiazzo et al. 2011; Pfisterer et al. 2011). In both studies, actively spiking, TH-positive, induced dopaminergic neuronal cells (iDA neuronal cells) were characterized, though pools of different factors were used to achieve these results, and only the cells induced by the expression of Ascl1, Nurr1, and Lmx1a were demonstrated to be dopamine releasing (Caiazzo et al. 2011). The delivery of this pool of factors resulted in iDA neuronal cells derived from both healthy patients and PD patients. Such an approach has profound implications for Parkinson's-related degeneration of DA neurons, as these reprogrammed cells could conceivably be used to replace those that have been lost.

Alternate efforts have sought to recapitulate the specific midbrain identity of DA neurons lost in PD, with findings that yet another pool of factors was capable of inducing dopaminergic neuronal cells. However, a panel of DA and pan-neuronal genes showed expression levels in the iDA cells fell short of those found in embryonic or adult midbrain DA neurons, though these iDA cells were found capable of partially restoring dopamine function upon transplantation in a mouse model of PD (Kim et al. 2011b).

Beyond the ability to derive DA neuron cells through reprogramming strategies, with implications for PD, efforts have also focused on disorders affecting spinal motor neurons such as ALS (Lou Gehrig's disease) or SMA. Recent studies have indicated that motor neurons can be directly induced from fibroblasts [Induced motor neuron (iMN) cells], using forced expression of a pool of transcription factors that included the aforementioned BAM cocktail of transcription factors along with motor neuron-specific factors (Son et al. 2011). When these iMN cells were co-cultured with myotubes, they sent projections along the length of myotubes that induced contractions. Similar to embryonic and ESC-derived motor neurons, these iMN cells were selectively sensitive to toxicity caused by mutant glial cells from a mouse model of ALS in co-culture, demonstrating their utility both as a phenocopy of motor neurons in disease-states such as SMA or ALS, as well as for studying cell-autonomous contributions to degeneration (Di Giorgio et al. 2007; Nagai et al. 2007). Furthermore, when transplanted into the chick spinal cord, most iMN cells engrafted in the ventral horns with axons extending to the ventral root, demonstrating their ability to survive and integrate in vivo. These landmark proof-of-principle studies will no doubt embolden researchers and clinicians as they seek to develop neural regenerative strategies for a broad spectrum of degenerative disorders.

5.6 Future Implications and Questions for Reprogramming as Applied to Regeneration

The many studies discussed in this chapter made significant discoveries that markedly changed, and will continue to shape, our current understanding of regeneration. As the field of reprogramming has exploded in the last decades, an increasingly apparent need for standardization exists. Currently variations in induction methods, nomenclature, efficiency calculations, quantification methods for the extent of reprogramming, and the methods used to demonstrate functionality of derived cells currently make direct comparisons difficult. Marius Wernig's group, who was first to develop direct reprogramming for iN cells in 2010, recently proposed a panel of criteria to be used to define iN cells with various degrees of reprogramming (Yang et al. 2011). The increasingly stringent criteria are also roughly the order of appearance of aforementioned neuronal properties in both reprogrammed iN cells and neurons during development. Broadly, they include the stepwise appearance of neuronal traits from (1) common morphological features, to (2) unique membrane characteristics, and finally (3) output function. For instance, a characteristic neuronal morphology is the first measureable changed observed, using the specific criteria of complex dendritic arborization, while synaptic plasticity, as demonstrated by short-term facilitation/depression, is the final property to appear. Between these two endpoints exist iN cells with varying degrees of reprogramming, and the authors have offered quantifiable criteria to define the extent of this reprogramming. They also point out the subtle conceptual difference between "partially reprogrammed iN cells" and "immature iN cells," though until mechanisms underlying this process are further elucidated, distinguishing between the two is difficult.

Furthermore, issues of safety and efficiency continue to be a concern. As discussed previously, many recent studies have developed tools aimed at avoiding/removing genome integration events caused by the retro- or lentiviral delivery of genes. This field is advancing rapidly and the affordability of these technologies will continue to increase, as will their efficiencies. The current four-factor lentiviral induction method used for the human system only had an overall efficiency of 2–4 % (Pang et al. 2011), 10-fold lower than the three-factor system used for mouse (Vierbuchen et al. 2010), though direct somatic cell reprogramming methods have generally so far seen higher conversion efficiencies than those observed in iPSC line establishment. As methods are optimized and new genes and compounds are tested, these numbers are expected to only further increase.

As methods for both indirect and direct reprogramming continue to improve, and the mechanisms for each are further delineated, many differences between the two may shrink in significance. However, one major difference between indirect and direct reprogramming is the lack of required cell proliferation in direct reprogramming. This absence could prove to be a detriment for regenerative applications of direct reprogramming, as large numbers of cells are required for transplantation in cellular replacement strategies, and the ability of reprogrammed cells to proliferate in vitro may prove beneficial, or even necessary. As direct reprogramming is currently a much less arduous process than that used to first establish iPSCs and then redifferentiate them, efforts have also focused on directly reprogramming somatic cells into lineage-specific stem cells, such as neural stem cells, in one step. It was indeed demonstrated that functional, bipotent, induced neural progenitor cells (iNPCs) could be derived from mouse fibroblasts by first inducing the overexpression of a set of the pluripotency factors used to establish iPSCs (Oct4, Kl4, Sox2, and c-myc) for 3–6 days, and then allowing the cells to expand in neural reprogramming media supplemented with FGF2, EGF, and FGF4 to support NPCs (Kim et al. 2011a). These iNPCs spontaneously differentiated into multiple neuronal cell types, as well as astrocytes, demonstrating at least a bipotent progenitor. Although these cells did not expand well in culture, they represent another possible unique application of reprogramming for regenerative purposes. As the field of reprogramming is a relatively young one, the current explosion in publications on this topic will continue to deepen our understanding of this dynamic and responsive process. The applicability of those discoveries, to neural regeneration and other biological processes, seems only limited by our imaginations.

References

Akerud P, Canals JM, Snyder EY, Arenas E (2001) Neuroprotection through delivery of glial cell line-derived neurotrophic factor by neural stem cells in a mouse model of Parkinson's disease. J Neurosci 21:8108–8118

Ambasudhan R, Talantova M, Coleman R, Yuan X, Zhu S, Lipton SA, Ding S (2011) Direct reprogramming of adult human fibroblasts to functional neurons under defined conditions. Cell Stem Cell 9:113–118

Ambati J, Ambati BK, Yoo SH, Ianchulev S, Adamis AP (2003) Age-related macular degeneration: etiology, pathogenesis, and therapeutic strategies. Surv Ophthalmol 48:257–293

Anokye-Danso F, Trivedi CM, Juhr D, Gupta M, Cui Z, Tian Y, Zhang Y, Yang W, Gruber PJ, Epstein JA et al (2011) Highly efficient miRNA-mediated reprogramming of mouse and human somatic cells to pluripotency. Cell Stem Cell 8:376–388

Armstrong RJ, Tyers P, Jain M, Richards A, Dunnett SB, Rosser AE, Barker RA (2003) Transplantation of expanded neural precursor cells from the developing pig ventral mesencephalon in a rat model of Parkinson's disease. Exp Brain Res 151:204–217

Bahr M, Bonhoeffer F (1994) Perspectives on axonal regeneration in the mammalian CNS. Trends Neurosci 17:473–479

Bailey SB, Eichler ME, Villadiego A, Rich KM (1993) The influence of fibronectin and laminin during Schwann cell migration and peripheral nerve regeneration through silicon chambers. J Neurocytol 22:176–184

Bayreuther K, Rodemann HP, Hommel R, Dittmann K, Albiez M, Francz PI (1988) Human skin fibroblasts in vitro differentiate along a terminal cell lineage. Proc Natl Acad Sci U S A 85:5112–5116

Behrstock S, Ebert A, McHugh J, Vosberg S, Moore J, Schneider B, Capowski E, Hei D, Kordower J, Aebischer P et al (2006) Human neural progenitors deliver glial cell line-derived neurotrophic factor to parkinsonian rodents and aged primates. Gene Ther 13:379–388

Ben-Hur T, Idelson M, Khaner H, Pera M, Reinhartz E, Itzik A, Reubinoff BE (2004) Transplantation of human embryonic stem cell-derived neural progenitors improves behavioral deficit in Parkinsonian rats. Stem Cells 22:1246–1255

Bharti K, Miller SS, Arnheiter H (2011) The new paradigm: retinal pigment epithelium cells generated from embryonic or induced pluripotent stem cells. Pigment Cell Melanoma Res 24:21–34

Boillee S, Vande Velde C, Cleveland DW (2006) ALS: a disease of motor neurons and their nonneuronal neighbors. Neuron 52:39–59

Bressler NM, Bressler SB, Fine SL (1988) Age-related macular degeneration. Surv Ophthalmol 32:375–413

Briggs R, King TJ (1952) Transplantation of living nuclei from blastula cells into enucleated Frogs' eggs. Proc Natl Acad Sci U S A 38:455–463

Buchholz DE, Hikita ST, Rowland TJ, Friedrich AM, Hinman CR, Johnson LV, Clegg DO (2009) Derivation of functional retinal pigmented epithelium from induced pluripotent stem cells. Stem Cells 27:2427–2434

Caiazzo M, Dell'Anno MT, Dvoretskova E, Lazarevic D, Taverna S, Leo D, Sotnikova TD, Menegon A, Roncaglia P, Colciago G et al (2011) Direct generation of functional dopaminergic neurons from mouse and human fibroblasts. Nature 476:224–227

Campbell KH, McWhir J, Ritchie WA, Wilmut I (1996) Sheep cloned by nuclear transfer from a cultured cell line. Nature 380:64–66

Carr AJ, Vugler AA, Hikita ST, Lawrence JM, Gias C, Chen LL, Buchholz DE, Ahmado A, Semo M, Smart MJ et al (2009) Protective effects of human iPS-derived retinal pigment epithelium cell transplantation in the retinal dystrophic rat. PLoS ONE 4:e8152

Chambers SM, Fasano CA, Papapetrou EP, Tomishima M, Sadelain M, Studer L (2009) Highly efficient neural conversion of human ES and iPS cells by dual inhibition of SMAD signaling. Nat Biotechnol 27:275–280

Cho MS, Lee YE, Kim JY, Chung S, Cho YH, Kim DS, Kang SM, Lee H, Kim MH, Kim JH et al (2008) Highly efficient and large-scale generation of functional dopamine neurons from human embryonic stem cells. Proc Natl Acad Sci U S A 105:3392–3397

Chung KK, Zhang Y, Lim KL, Tanaka Y, Huang H, Gao J, Ross CA, Dawson VL, Dawson TM (2001) Parkin ubiquitinates the alpha-synuclein-interacting protein, synphilin-1: implications for Lewy-body formation in Parkinson disease. Nat Med 7:1144–1150

Cooper O, Hargus G, Deleidi M, Blak A, Osborn T, Marlow E, Lee K, Levy A, Perez-Torres E, Yow A et al (2010) Differentiation of human ES and Parkinson's disease iPS cells into ventral midbrain dopaminergic neurons requires a high activity form of SHH, FGF8a and specific regionalization by retinoic acid. Mol Cell Neurosci 45:258–266

Craveiro LM, Hakkoum D, Weinmann O, Montani L, Stoppini L, Schwab ME (2008) Neutralization of the membrane protein Nogo-A enhances growth and reactive sprouting in established organotypic hippocampal slice cultures. Eur J Neurosci 28:1808–1824

Cronin S, Hardiman O, Traynor BJ (2007) Ethnic variation in the incidence of ALS: a systematic review. Neurology 68:1002–1007

Davie CA (2008) A review of Parkinson's disease. Br Med Bull 86:109–127

Davies SJ, Fitch MT, Memberg SP, Hall AK, Raisman G, Silver J (1997) Regeneration of adult axons in white matter tracts of the central nervous system. Nature 390:680–683

Davies SJ, Goucher DR, Doller C, Silver J (1999) Robust regeneration of adult sensory axons in degenerating white matter of the adult rat spinal cord. J Neurosci 19:5810–5822

Davis RL, Weintraub H, Lassar AB (1987) Expression of a single transfected cDNA converts fibroblasts to myoblasts. Cell 51:987–1000

Del Priore LV, Kuo YH, Tezel TH (2002) Age-related changes in human RPE cell density and apoptosis proportion in situ. Invest Ophthalmol Vis Sci 43:3312–3318

Di Giorgio FP, Carrasco MA, Siao MC, Maniatis T, Eggan K (2007) Non-cell autonomous effect of glia on motor neurons in an embryonic stem cell-based ALS model. Nat Neurosci 10:608–614

Dimos JT, Rodolfa KT, Niakan KK, Weisenthal LM, Mitsumoto H, Chung W, Croft GF, Saphier G, Leibel R, Goland R et al (2008) Induced pluripotent stem cells generated from patients with ALS can be differentiated into motor neurons. Science 321:1218–1221

Dowling JE (1970) Organization of vertebrate retinas. Invest Ophthalmol 9:655–680

Dowling JE, Werblin FS (1971) Synaptic organization of the vertebrate retina. Vis Res Suppl 3:1–15

Du H, Lim SL, Grob S, Zhang K (2011) Induced pluripotent stem cell therapies for geographic atrophy of age-related macular degeneration. Semin Ophthalmol 26:216–224

Ebert AD, Svendsen CN (2010) Stem cell model of spinal muscular atrophy. Arch Neurol 67:665–669

Ebert AD, Beres AJ, Barber AE, Svendsen CN (2008) Human neural progenitor cells over-expressing IGF-1 protect dopamine neurons and restore function in a rat model of Parkinson's disease. Exp Neurol 209:213–223

Ebert AD, Yu J, Rose FF Jr, Mattis VB, Lorson CL, Thomson JA, Svendsen CN (2009) Induced pluripotent stem cells from a spinal muscular atrophy patient. Nature 457:277–280

Egawa N, Kitaoka S, Tsukita K, Naitoh M, Takahashi K, Yamamoto T, Adachi F, Kondo T, Okita K, Asaka et al (2012) Drug screening for ALS using patient-specific induced pluripotent stem cells. Sci Transl Med 4 145ra104

Evans MJ, Kaufman MH (1981) Establishment in culture of pluripotential cells from mouse embryos. Nature 292:154–156

Fawcett JW (2006) Overcoming inhibition in the damaged spinal cord. J Neurotrauma 23:371–383

Fawcett JW, Asher RA (1999) The glial scar and central nervous system repair. Brain Res Bull 49:377–391

Freed CR, Greene PE, Breeze RE, Tsai WY, DuMouchel W, Kao R, Dillon S, Winfield H, Culver S, Trojanowski JQ et al (2001) Transplantation of embryonic dopamine neurons for severe Parkinson's disease. N Engl J Med 344:710–719

Frostick SP, Yin Q, Kemp GJ (1998) Schwann cells, neurotrophic factors, and peripheral nerve regeneration. Microsurgery 18:397–405

Gehrs KM, Anderson DH, Johnson LV, Hageman GS (2006) Age-related macular degeneration–emerging pathogenetic and therapeutic concepts. Ann Med 38:450–471

Gouras P, Algvere P (1996) Retinal cell transplantation in the macula: new techniques. Vis Res 36:4121–4125

Grandpre T, Strittmatter SM (2001) Nogo: a molecular determinant of axonal growth and regeneration. Neuroscientist 7:377–386

Gurdon JB, Uehlinger V (1966) "Fertile" intestine nuclei. Nature 210:1240–1241

Gurdon JB, Elsdale TR, Fischberg M (1958) Sexually mature individuals of *Xenopus laevis* from the transplantation of single somatic nuclei. Nature 182:64–65

Hageman GS, Luthert PJ, Victor Chong NH, Johnson LV, Anderson DH, Mullins RF (2001) An integrated hypothesis that considers drusen as biomarkers of immune-mediated processes at the RPE-Bruch's membrane interface in aging and age-related macular degeneration. Prog Retin Eye Res 20:705–732

Hargus G, Cooper O, Deleidi M, Levy A, Lee K, Marlow E, Yow A, Soldner F, Hockemeyer D, Hallett PJ et al (2010) Differentiated Parkinson patient-derived induced pluripotent stem cells grow in the adult rodent brain and reduce motor asymmetry in Parkinsonian rats. Proc Natl Acad Sci U S A 107:15921–15926

Harrower TP, Tyers P, Hooks Y, Barker RA (2006) Long-term survival and integration of porcine expanded neural precursor cell grafts in a rat model of Parkinson's disease. Exp Neurol 197:56–69

Hawryluk GW, Rowland J, Kwon BK, Fehlings MG (2008) Protection and repair of the injured spinal cord: a review of completed, ongoing, and planned clinical trials for acute spinal cord injury. Neurosurg Focus 25:E14

Hefferan MP, Galik J, Kakinohana O, Sekerkova G, Santucci C, Marsala S, Navarro R, Hruska-Plochan M, Johe K, Feldman E et al (2012) Human neural stem cell replacement therapy for amyotrophic lateral sclerosis by spinal transplantation. PLoS ONE 7:e42614

Hirami Y, Osakada F, Takahashi K, Okita K, Yamanaka S, Ikeda H, Yoshimura N, Takahashi M (2009) Generation of retinal cells from mouse and human induced pluripotent stem cells. Neurosci Lett 458:126–131

Horner PJ, Gage FH (2000) Regenerating the damaged central nervous system. Nature 407:963–970

Horner PJ, Gage FH (2002) Regeneration in the adult and aging brain. Arch Neurol 59:1717–1720

Hu BY, Zhang SC (2009) Differentiation of spinal motor neurons from pluripotent human stem cells. Nat Protoc 4:1295–1304

Hu BY, Du ZW, Zhang SC (2009) Differentiation of human oligodendrocytes from pluripotent stem cells. Nat Protoc 4:1614–1622

Huang S (2009) Reprogramming cell fates: reconciling rarity with robustness. BioEssays 31:546–560

Huang P, He Z, Ji S, Sun H, Xiang D, Liu C, Hu Y, Wang X, Hui L (2011) Induction of functional hepatocyte-like cells from mouse fibroblasts by defined factors. Nature 475:386–389

Huebner EA, Strittmatter SM (2009) Axon regeneration in the peripheral and central nervous systems. Results Probl Cell Differ 48:339–351

Idelson M, Alper R, Obolensky A, Ben-Shushan E, Hemo I, Yachimovich-Cohen N, Khaner H, Smith Y, Wiser O, Gropp M et al (2009) Directed differentiation of human embryonic stem cells into functional retinal pigment epithelium cells. Cell Stem Cell 5:396–408

Ieda M, Fu JD, Delgado-Olguin P, Vedantham V, Hayashi Y, Bruneau BG, Srivastava D (2010) Direct reprogramming of fibroblasts into functional cardiomyocytes by defined factors. Cell 142:375–386

Jordan PM, Ojeda LD, Thonhoff JR, Gao J, Boehning D, Yu Y, Wu P (2009) Generation of spinal motor neurons from human fetal brain-derived neural stem cells: role of basic fibroblast growth factor. J Neurosci Res 87:318–332

Kaji K, Norrby K, Paca A, Mileikovsky M, Mohseni P, Woltjen K (2009) Virus-free induction of pluripotency and subsequent excision of reprogramming factors. Nature 458:771–775

Kim J, Efe JA, Zhu S, Talantova M, Yuan X, Wang S, Lipton SA, Zhang K, Ding S (2011a) Direct reprogramming of mouse fibroblasts to neural progenitors. Proc Natl Acad Sci U S A 108:7838–7843

Kim J, Su SC, Wang H, Cheng AW, Cassady JP, Lodato MA, Lengner CJ, Chung CY, Dawlaty MM, Tsai LH et al (2011b) Functional integration of dopaminergic neurons directly converted from mouse fibroblasts. Cell Stem Cell 9:413–419

Klein SM, Behrstock S, McHugh J, Hoffmann K, Wallace K, Suzuki M, Aebischer P, Svendsen CN (2005) GDNF delivery using human neural progenitor cells in a rat model of ALS. Hum Gene Ther 16:509–521

Kokkinaki M, Sahibzada N, Golestaneh N (2011) Human induced pluripotent stem-derived retinal pigment epithelium (RPE) cells exhibit ion transport, membrane potential, polarized vascular endothelial growth factor secretion, and gene expression pattern similar to native RPE. Stem Cells 29:825–835

Kolb H, Nelson R, Ahnelt P, Cuenca N (2001) Cellular organization of the vertebrate retina. Prog Brain Res 131:3–26

Kriks S, Shim JW, Piao J, Ganat YM, Wakeman DR, Xie Z, Carrillo-Reid L, Auyeung G, Antonacci C, Buch A et al (2011) Dopamine neurons derived from human ES cells efficiently engraft in animal models of Parkinson's disease. Nature 480:547–551

Laabs T, Carulli D, Geller HM, Fawcett JW (2005) Chondroitin sulfate proteoglycans in neural development and regeneration. Curr Opin Neurobiol 15:116–120

Lamba DA, Karl MO, Ware CB, Reh TA (2006) Efficient generation of retinal progenitor cells from human embryonic stem cells. Proc Natl Acad Sci U S A 103:12769–12774
Lamba DA, McUsic A, Hirata RK, Wang PR, Russell D, Reh TA (2010) Generation, purification and transplantation of photoreceptors derived from human induced pluripotent stem cells. PLoS ONE 5:e8763
Lee H, Shamy GA, Elkabetz Y, Schofield CM, Harrsion NL, Panagiotakos G, Socci ND, Tabar V, Studer L (2007) Directed differentiation and transplantation of human embryonic stem cell-derived motoneurons. Stem Cells 25:1931–1939
Lefebvre S, Burglen L, Reboullet S, Clermont O, Burlet P, Viollet L, Benichou B, Cruaud C, Millasseau P, Zeviani M et al (1995) Identification and characterization of a spinal muscular atrophy-determining gene. Cell 80:155–165
Lepore AC, O'Donnell J, Kim AS, Williams T, Tuteja A, Rao MS, Kelley LL, Campanelli JT, Maragakis NJ (2011) Human glial-restricted progenitor transplantation into cervical spinal cord of the SOD1 mouse model of ALS. PLoS ONE 6:e25968
Li XJ, Du ZW, Zarnowska ED, Pankratz M, Hansen LO, Pearce RA, Zhang SC (2005) Specification of motoneurons from human embryonic stem cells. Nat Biotechnol 23:215–221
Lindvall O, Bjorklund A (2011) Cell therapeutics in Parkinson's disease. Neurotherapeutics 8:539–548
Lindvall O, Barker RA, Brustle O, Isacson O, Svendsen CN (2012) Clinical translation of stem cells in neurodegenerative disorders. Cell Stem Cell 10:151–155
Lorson CL, Hahnen E, Androphy EJ, Wirth B (1999) A single nucleotide in the SMN gene regulates splicing and is responsible for spinal muscular atrophy. Proc Natl Acad Sci U S A 96:6307–6311
Lu B, Malcuit C, Wang S, Girman S, Francis P, Lemieux L, Lanza R, Lund R (2009) Long-term safety and function of RPE from human embryonic stem cells in preclinical models of macular degeneration. Stem Cells 27:2126–2135
Lund RD, Wang S, Klimanskaya I, Holmes T, Ramos-Kelsey R, Lu B, Girman S, Bischoff N, Sauve Y, Lanza R (2006) Human embryonic stem cell-derived cells rescue visual function in dystrophic RCS rats. Cloning Stem Cells 8:189–199
Marro S, Pang ZP, Yang N, Tsai MC, Qu K, Chang HY, Sudhof TC, Wernig M (2011) Direct lineage conversion of terminally differentiated hepatocytes to functional neurons. Cell Stem Cell 9:374–382
McGrath J, Solter D (1983) Nuclear transplantation in the mouse embryo by microsurgery and cell fusion. Science 220:1300–1302
Mellough CB, Sernagor E, Moreno-Gimeno I, Steel DH, Lako M (2012) Efficient stage specific differentiation of human pluripotent stem cells towards retinal photoreceptor cells. Stem Cells 30:673–686.
Meyer JS, Shearer RL, Capowski EE, Wright LS, Wallace KA, McMillan EL, Zhang SC, Gamm DM (2009) Modeling early retinal development with human embryonic and induced pluripotent stem cells. Proc Natl Acad Sci U S A 106:16698–16703
Meyer JS, Howden SE, Wallace KA, Verhoeven AD, Wright LS, Capowski EE, Pinilla I, Martin JM, Tian S, Stewart R et al (2011) Optic vesicle-like structures derived from human pluripotent stem cells facilitate a customized approach to retinal disease treatment. Stem Cells 29:1206–1218
Montani L, Gerrits B, Gehrig P, Kempf A, Dimou L, Wollscheid B, Schwab ME (2009) Neuronal Nogo-A modulates growth cone motility via Rho-GTP/LIMK1/cofilin in the unlesioned adult nervous system. J Biol Chem 284:10793–10807
Nagai M, Re DB, Nagata T, Chalazonitis A, Jessell TM, Wichterle H, Przedborski S (2007) Astrocytes expressing ALS-linked mutated SOD1 release factors selectively toxic to motor neurons. Nat Neurosci 10:615–622
Noggle S, Fung HL, Gore A, Martinez H, Satriani KC, Prosser R, Oum K, Paull D, Druckenmiller S, Freeby M et al (2011) Human oocytes reprogram somatic cells to a pluripotent state. Nature 478:70–75

Olanow CW, Goetz CG, Kordower JH, Stoessl AJ, Sossi V, Brin MF, Shannon KM, Nauert GM, Perl DP, Godbold J et al (2003) A double-blind controlled trial of bilateral fetal nigral transplantation in Parkinson's disease. Ann Neurol 54:403–414

Osakada F, Ikeda H, Sasai Y, Takahashi M (2009a) Stepwise differentiation of pluripotent stem cells into retinal cells. Nat Protoc 4:811–824

Osakada F, Jin ZB, Hirami Y, Ikeda H, Danjyo T, Watanabe K, Sasai Y, Takahashi M (2009b) In vitro differentiation of retinal cells from human pluripotent stem cells by small-molecule induction. J Cell Sci 122:3169–3179

Osakada F, Hirami Y, Takahashi M (2010) Stem cell biology and cell transplantation therapy in the retina. Biotechnol Genet Eng Rev 26:297–334

Pang ZP, Yang N, Vierbuchen T, Ostermeier A, Fuentes DR, Yang TQ, Citri A, Sebastiano V, Marro S, Sudhof TC et al (2011) Induction of human neuronal cells by defined transcription factors. Nature 476:220–223

Pankratz MT, Li XJ, Lavaute TM, Lyons EA, Chen X, Zhang SC (2007) Directed neural differentiation of human embryonic stem cells via an obligated primitive anterior stage. Stem Cells 25:1511–1520

Park IH, Arora N, Huo H, Maherali N, Ahfeldt T, Shimamura A, Lensch MW, Cowan C, Hochedlinger K, Daley GQ (2008) Disease-specific induced pluripotent stem cells. Cell 134:877–886

Perrier AL, Tabar V, Barberi T, Rubio ME, Bruses J, Topf N, Harrison NL, Studer L (2004) Derivation of midbrain dopamine neurons from human embryonic stem cells. Proc Natl Acad Sci U S A 101:12543–12548

Pfisterer U, Kirkeby A, Torper O, Wood J, Nelander J, Dufour A, Bjorklund A, Lindvall O, Jakobsson J, Parmar M (2011) Direct conversion of human fibroblasts to dopaminergic neurons. Proc Natl Acad Sci U S A 108:10343–10348

Piccini P, Brooks DJ, Bjorklund A, Gunn RN, Grasby PM, Rimoldi O, Brundin P, Hagell P, Rehncrona S, Widner H et al (1999) Dopamine release from nigral transplants visualized in vivo in a Parkinson's patient. Nat Neurosci 2:1137–1140

Prather RS, First NL (1990) Nuclear transfer in mammalian embryos. Int Rev Cytol 120:169–190

Purves D, Augustine GJ, Fitzpatrick D, Hall WC, LaMantia AS, McNamara JO, White LE (2008) Neuroscience, 4th edn. Sunderland, Sinauer

Qiang L, Fujita R, Yamashita T, Angulo S, Rhinn H, Rhee D, Doege C, Chau L, Aubry L, Vanti WB et al (2011) Directed conversion of Alzheimer's disease patient skin fibroblasts into functional neurons. Cell 146:359–371

Rosen DR, Siddique T, Patterson D, Figlewicz DA, Sapp P, Hentati A, Donaldson D, Goto J, O'Regan JP, Deng HX et al (1993) Mutations in Cu/Zn superoxide dismutase gene are associated with familial amyotrophic lateral sclerosis. Nature 362:59–62

Rowland TJ, Blaschke AJ, Buchholz DE, Hikita ST, Johnson LV, Clegg DO (2012a) Differentiation of human pluripotent stem cells to retinal pigmented epithelium in defined conditions using purified extracellular matrix proteins. J Tissue Eng Regen Med. doi:10.1002/term.1458

Rowland TJ, Buchholz DE, Clegg DO (2012b) Pluripotent human stem cells for the treatment of retinal disease. J Cell Physiol 227:457–466

Roy NS, Cleren C, Singh SK, Yang L, Beal MF, Goldman SA (2006) Functional engraftment of human ES cell-derived dopaminergic neurons enriched by coculture with telomerase-immortalized midbrain astrocytes. Nat Med 12:1259–1268

Schwartz SD, Hubschman JP, Heilwell G, Franco-Cardenas V, Pan CK, Ostrick RM, Mickunas E, Gay R, Klimanskaya I, Lanza R (2012) Embryonic stem cell trials for macular degeneration: a preliminary report. Lancet 379:713–720

Sekiya S, Suzuki A (2011) Direct conversion of mouse fibroblasts to hepatocyte-like cells by defined factors. Nature 475:390–393

Shaw BF, Valentine JS (2007) How do ALS-associated mutations in superoxide dismutase 1 promote aggregation of the protein? Trends Biochem Sci 32:78–85

Shin S, Dalton S, Stice SL (2005) Human motor neuron differentiation from human embryonic stem cells. Stem Cells Dev 14:266–269
Silver J, Miller JH (2004) Regeneration beyond the glial scar. Nat Rev Neurosci 5:146–156
Smith-Thomas LC, Stevens J, Fok-Seang J, Faissner A, Rogers JH, Fawcett JW (1995) Increased axon regeneration in astrocytes grown in the presence of proteoglycan synthesis inhibitors. J Cell Sci 108(Pt 3):1307–1315
So KF, Aguayo AJ (1985) Lengthy regrowth of cut axons from ganglion cells after peripheral nerve transplantation into the retina of adult rats. Brain Res 328:349–354
Soldner F, Hockemeyer D, Beard C, Gao Q, Bell GW, Cook EG, Hargus G, Blak A, Cooper O, Mitalipova M et al (2009) Parkinson's disease patient-derived induced pluripotent stem cells free of viral reprogramming factors. Cell 136:964–977
Son EY, Ichida JK, Wainger BJ, Toma JS, Rafuse VF, Woolf CJ, Eggan K (2011) Conversion of mouse and human fibroblasts into functional spinal motor neurons. Cell Stem Cell 9:205–218
Sonntag KC, Pruszak J, Yoshizaki T, van Arensbergen J, Sanchez-Pernaute R, Isacson O (2007) Enhanced yield of neuroepithelial precursors and midbrain-like dopaminergic neurons from human embryonic stem cells using the bone morphogenic protein antagonist noggin. Stem Cells 25:411–418
Springer JE (2002) Apoptotic cell death following traumatic injury to the central nervous system. J Biochem Mol Biol 35:94–105
Strauss O (1995) The retinal pigment epithelium. Invest Ophthalmol Vis Sci 36:2327–2331
Suzuki M, McHugh J, Tork C, Shelley B, Klein SM, Aebischer P, Svendsen CN (2007) GDNF secreting human neural progenitor cells protect dying motor neurons, but not their projection to muscle, in a rat model of familial ALS. PLoS ONE 2:e689
Swistowski A, Peng J, Liu Q, Mali P, Rao MS, Cheng L, Zeng X (2010) Efficient generation of functional dopaminergic neurons from human induced pluripotent stem cells under defined conditions. Stem Cells 28:1893–1904
Szabo E, Rampalli S, Risueno RM, Schnerch A, Mitchell R, Fiebig-Comyn A, Levadoux-Martin M, Bhatia M (2010) Direct conversion of human fibroblasts to multilineage blood progenitors. Nature 468:521–526
Takahashi K, Yamanaka S (2006) Induction of pluripotent stem cells from mouse embryonic and adult fibroblast cultures by defined factors. Cell 126:663–676
Terenghi G (1999) Peripheral nerve regeneration and neurotrophic factors. J Anat 194(Pt 1):1–14
Thomson JA, Itskovitz-Eldor J, Shapiro SS, Waknitz MA, Swiergiel JJ, Marshall VS, Jones JM (1998) Embryonic stem cell lines derived from human blastocysts. Science 282:1145–1147
Thornton MR, Mantovani C, Birchall MA, Terenghi G (2005) Quantification of N-CAM and N-cadherin expression in axotomized and crushed rat sciatic nerve. J Anat 206:69–78
Tsui A, Isacson O (2011) Functions of the nigrostriatal dopaminergic synapse and the use of neurotransplantation in Parkinson's disease. J Neurol 258:1393–1405
Tursun B, Patel T, Kratsios P, Hobert O (2011) Direct conversion of *C. elegans* germ cells into specific neuron types. Science 331:304–308
Vierbuchen T, Ostermeier A, Pang ZP, Kokubu Y, Sudhof TC, Wernig M (2010) Direct conversion of fibroblasts to functional neurons by defined factors. Nature 463:1035–1041
von Lewinski F, Keller BU (2005) Ca2 + , mitochondria and selective motoneuron vulnerability: implications for ALS. Trends Neurosci 28:494–500
Warren L, Manos PD, Ahfeldt T, Loh YH, Li H, Lau F, Ebina W, Mandal PK, Smith ZD, Meissner A et al (2010) Highly efficient reprogramming to pluripotency and directed differentiation of human cells with synthetic modified mRNA. Cell Stem Cell 7:618–630
Wijeyekoon R, Barker RA (2009) Cell replacement therapy for Parkinson's disease. Biochim Biophys Acta 1792:688–702
Woltjen K, Michael IP, Mohseni P, Desai R, Mileikovsky M, Hamalainen R, Cowling R, Wang W, Liu P, Gertsenstein M et al (2009) Piggybac transposition reprograms fibroblasts to induced pluripotent stem cells. Nature 458:766–770

Yan Y, Yang D, Zarnowska ED, Du Z, Werbel B, Valliere C, Pearce RA, Thomson JA, Zhang SC (2005) Directed differentiation of dopaminergic neuronal subtypes from human embryonic stem cells. Stem Cells 23:781–790

Yang D, Zhang ZJ, Oldenburg M, Ayala M, Zhang SC (2008) Human embryonic stem cell-derived dopaminergic neurons reverse functional deficit in Parkinsonian rats. Stem Cells 26:55–63

Yang N, Ng YH, Pang ZP, Sudhof TC, Wernig M (2011) Induced neuronal cells: how to make and define a neuron. Cell Stem Cell 9:517–525

Yiu G, He Z (2006) Glial inhibition of CNS axon regeneration. Nat Rev Neurosci 7:617–627

Yoo AS, Sun AX, Li L, Shcheglovitov A, Portmann T, Li Y, Lee-Messer C, Dolmetsch RE, Tsien RW, Crabtree GR (2011) MicroRNA-mediated conversion of human fibroblasts to neurons. Nature 476:228–231

Zhou JX, Huang S (2011) Understanding gene circuits at cell-fate branch points for rational cell reprogramming. Trends Genet 27:55–62

Zhou Q, Brown J, Kanarek A, Rajagopal J, Melton DA (2008) In vivo reprogramming of adult pancreatic exocrine cells to beta-cells. Nature 455:627–632

Zhu S, Li W, Zhou H, Wei W, Ambasudhan R, Lin T, Kim J, Zhang K, Ding S (2010) Reprogramming of human primary somatic cells by OCT4 and chemical compounds. Cell Stem Cell 7:651–655

Zorner B, Schwab ME (2010) Anti-Nogo on the go: from animal models to a clinical trial. Ann N Y Acad Sci 1198(Suppl 1):E22–E34

Teleost Fish as a Model System to Study Successful Regeneration of the Central Nervous System

Günther K. H. Zupanc and Ruxandra F. Sîrbulescu

Abstract Traumatic brain injury and spinal cord injury are devastating conditions that may result in death or long-term disability. A promising strategy for the development of effective cell replacement therapies involves the study of regeneration-competent organisms. Among this group, teleost fish are distinguished by their excellent potential to regenerate nervous tissue and to regain function after injury to the central nervous system. In this chapter, we summarize our current understanding of the cellular processes that mediate this regenerative potential, and we show that several of these processes are shared with the normal development of the intact central nervous system; we describe how the spontaneous self-repair of the teleostean central nervous system leads to functional recovery, at physiological and behavioral levels; we discuss the possible function of molecular factors associated with the degenerative and regenerative processes after injury; and, finally, we speculate on evolutionary aspects of adult neurogenesis and neuronal regeneration, and on how a better understanding of these aspects could catalyze the development of therapeutic strategies to overcome the regenerative limits of the mammalian CNS.

Contents

G. K. H. Zupanc (✉) · R. F. Sîrbulescu
Department of Biology, Northeastern University, 134 Mugar Life Science Building, 360 Huntington Avenue, Boston, MA 02115, USA
e-mail: g.zupanc@neu.edu

Current Topics in Microbiology and Immunology (2013) 367: 193–233
DOI: 10.1007/82_2012_297

© Springer-Verlag Berlin Heidelberg 2012
Published Online: 15 December 2012

1 Introduction

Traumatic injuries are among the most devastating conditions affecting the central nervous system (CNS). Annually, at least 1.7 million people sustain a traumatic brain injury in the United States alone; of these individuals, 52,000 die (Faul et al. 2010). Approximately, 270,000 people are currently living with spinal cord injury in the United States, and an estimated 12,000 new cases occur each year (National Spinal Cord Injury Statistical Center 2012).

At the cellular level, one of the characteristic features of CNS trauma is a massive loss of multiple cell types. As the adult mammalian CNS lacks the ability to regenerate, this cell loss results frequently in long-term disability. Currently, no effective therapy is available to overcome these limits of the mammalian CNS.

To replace the cells lost to injury, two therapeutic approaches have been proposed—transplantation of exogenous neural progenitor cells and enhancement of proliferation of endogenous adult stem cells (for review see Richardson et al. 2010). From a clinical perspective, each of these two types of replacement therapy is still in its infancy, and many basic issues remain to be solved, including a better understanding of the self-repair mechanisms of the CNS (Lindvall and Björklund 2004).

One promising approach to learn more about the intrinsic regenerative potential of the CNS involves the study of regeneration-competent organisms (for review see Stocum 2006). Among these organisms, teleost fish are the best studied. Here, we review the current knowledge of the cellular mechanisms that mediate the excellent regenerative potential of teleosts, how this potential relates to their ability to generate new neurons continuously and in large numbers in the intact CNS, and what molecular factors are implicated in the control of proliferation and further development of the adult-born cells in both the intact and the injured CNS.

2 Adult Neurogenesis in the Central Nervous System of Teleost Fish

2.1 Generation of New Cells

The excellent regenerative potential of teleost fish is intimately linked to their ability to generate large numbers of new neurons in numerous regions of the adult CNS (for reviews see Anderson and Waxman 1985; Hitchcock and Raymond 1992; Zupanc 1999a, 2001, 2008a, b, 2009; Otteson and Hitchcock 2003; Zupanc and Zupanc 2006a; Chapouton et al. 2007; Kaslin et al. 2008; Sîrbulescu and Zupanc 2011; Zupanc and Sîrbulescu 2011). In brown ghost knifefish (*Apteronotus leptorhynchus*), 2 h after administration of a single pulse of the thymidine analogue 5-bromo-2'-deoxyuridine (BrdU) 100,000 cells, on average, are labeled in the brain, corresponding to approximately 0.2 % of the total population of brain cells (Zupanc and Horschke 1995). In zebrafish (*Danio rerio*), 30 min after a single BrdU pulse 6,000 labeled cells, on average, are found in the brain, corresponding to roughly 0.06 % of the total number of brain cells (Hinsch and Zupanc 2007). Using the same labeling protocol as in the brain, approximately 10,000 cells/mm^3 have been estimated to enter the S-phase of mitosis within any 2-h period in the spinal cord of brown ghost knifefish (Sîrbulescu et al. 2009). These estimates suggest that the rate of cell proliferation in the adult teleostean CNS is at least one, if not two, orders of magnitude higher than in the mammalian CNS (Lois and Alvarez-Buylla 1994; Williams 2000; Cameron and McKay 2001; Herculano-Houzel and Lent 2005; Becker et al. 2010).

The vast majority of the adult-born cells in the brain originate from specific proliferation zones, which are frequently associated with ventricular regions (Fig. 1). In the adult mammalian brain, only two brain areas give rise to new neurons: the subgranular zone of the dentate gyrus, from where the new cells migrate a short distance into the granule cell layer of the hippocampus (Altman 1969; Luskin 1993; Lois and Alvarez-Buylla 1994; Lois et al. 1996; Pencea et al. 2001; Sanai et al. 2004; Bédard and Parent 2004; Curtis et al. 2007); and the anterior part of the subventricular zone of the lateral ventricle from where the young cells migrate via the rostral migratory stream into the olfactory bulb (Altman and Das 1965; Kaplan and Bell 1984; Eriksson et al. 1998; Gould et al. 1999; Kornack and Rakic 1999; Seri et al. 2001). By contrast, in the adult teleostean brain dozens of proliferation zones have been identified in a variety of different species (Kranz and Richter 1970a, b; Richter and Kranz 1970a, b; Zupanc and Horschke 1995; Zikopoulos et al. 2000; Ekström et al. 2001; Zupanc et al. 2005; Grandel et al. 2006; Fernández et al. 2011; Teles et al. 2012; Tozzini et al. 2012; Maruska et al. 2012).

In the spinal cord of teleost fish, new cells are generated in both the ependymal layer and the parenchyma (Anderson et al. 1983; Anderson and Waxman 1985; Reimer et al. 2008; Takeda et al. 2008; Sîrbulescu et al. 2009). In the teleostean retina, new neurons and photoreceptors originate from two sources—the ciliary

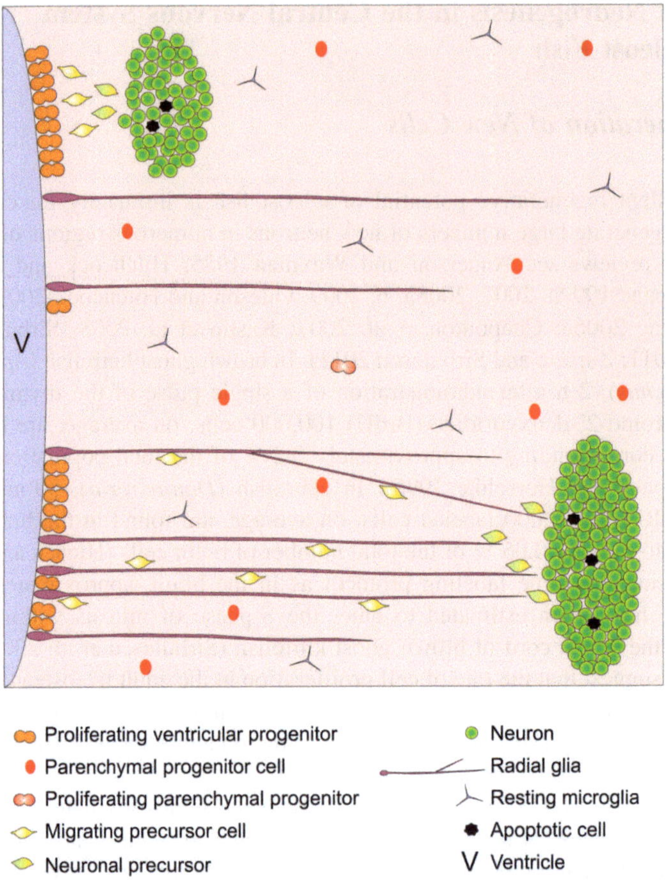

Fig. 1 Generalized schema summarizing the development of adult-born cells in the intact brain. Most of the cells generated in the intact brain are derived from progenitor cells harbored in specific proliferation zones, which are often associated with ventricular areas or derivatives of embryonic ventricles. In some areas of the brain, the progeny of the stem cells/progenitors reside near the proliferation zone where they were born. In other brain areas, the young cells migrate over relatively long distances from the proliferation zone to their target site. Radial glial fibers have been implicated in the guidance of the migrating new cells. Differentiation into neurons or glia starts as early as during the migration of the young cells, and further maturation into specific subtypes of neuronal or glial cells becomes evident after arrival at the target site. At the target site, the number of new cells is regulated through apoptotic cell death. In addition to the mitotically active progenitor cells in the proliferation zones, quiescent progenitor cells exist in the parenchyma. In the intact brain, there is little mitotic activity among this population of progenitor cells. Similarly, most of the microglia remain in a resting state in the absence of injury

marginal zone, which is defined by a small area in the retinal periphery where the retina meets the ciliary epithelium (Johns 1977); and the rod progenitors in the central retina, which are derived from Müller glia through dedifferentiation (Johns 1982; Bernardos et al. 2007).

Examination in vitro of cells isolated from proliferation zones in the dorsal telencephalon and cerebellum has shown that they are capable of self-renewal and generation of different cell types, thus exhibiting genuine characteristics of stem cells (Hinsch and Zupanc 2006). Further examination of such stem cells has revealed a heterogeneity among the different subpopulations in the teleostean brain. One cellular subpopulation displays the characteristics of radial glia (Pellegrini et al. 2007; Ganz et al. 2010; Rothenaigner et al. 2011; Chapouton et al. 2011), whereas a second subpopulation lacks the expression of canonical markers of glial cells (Kaslin et al. 2009; Alunni et al. 2010; Ganz et al. 2010; Rothenaigner et al. 2011). The latter is the case in the zebrafish cerebellum, where progenitors resemble neuroepithelial cells (Kaslin et al. 2009).

2.2 Migration

Based on the distance over which cells migrate after their generation in the adult CNS, two situations can be distinguished (Fig. 1). In certain instances, cells migrate only a short distance to reside near where they were born. In the retina, the new cells that originate from the ciliary marginal zone are continuously added appositionally to the margin of the extant retina (Fig. 2; Johns and Easter 1977; Meyer 1978; Hagedorn and Fernald 1992; Marcus et al. 1999). Similarly, in the optic tectum—the projection target of the retinal ganglion cells in teleosts—the majority of the new cells are generated at the caudal pole, where they remain during their subsequent development (Raymond and Easter 1983; Mansour-Robaey and Pinganaud 1990; Nguyen et al. 1999; Wullimann and Puelles 1999; Ekström et al. 2001; Candal et al. 2005; Zupanc et al. 2005; Grandel et al. 2006). As a result, the optic tectum grows asymmetrically by expanding primarily from its caudal end.

In other instances, cells migrate from the proliferation zones over relatively long distances to specific target areas. In the retina, the rod progenitors arising from proliferating Müller glia in the inner nuclear layer migrate to the outer nuclear layer (Fig. 2; Johns 1982; Julian et al. 1998). In the corpus cerebelli—one subdivision of the teleostean cerebellum—the new cells are generated in specific proliferation zones at and near the midline in the dorsal and ventral molecular layers. Subsequently, the young cells migrate over several hundreds of micrometers along the midline into the granular layer, where they spread out evenly (Zupanc and Horschke 1995; Zupanc et al. 1996, 2005, 2012; Grandel et al. 2006). Thus, in contrast to the optic tectum, the corpus cerebelli grows in a rather symmetric fashion. In the course of their migration, the young cerebellar cells are guided by glial fibrillary acidic protein (GFAP)/vimentin-expressing radial glial fibers, which provide a scaffolding connecting the proliferation zones in the molecular layer and the target area, the granular layer (Zupanc and Clint 2003; Zupanc et al. 2012). There is some indication that a similar mechanism guides

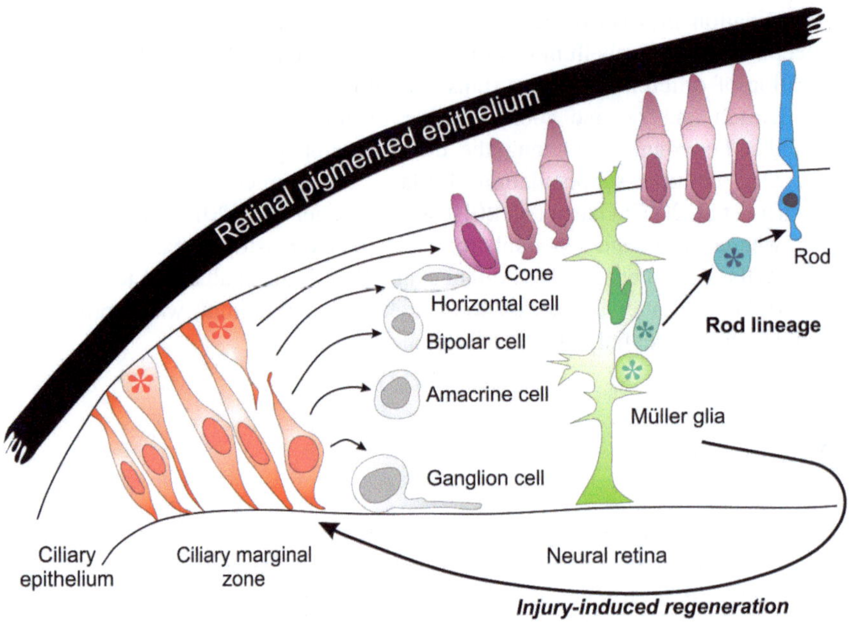

Fig. 2 Development of adult-born cells in the intact and regenerating retina of teleost fish. The ciliary marginal zone (in *red*) between the ciliary epithelium and the neural retina contains multipotent stem cells that span the width of the retinal epithelium. This zone is separated from the retinal pigmented epithelium by a narrow subretinal space. More restricted retinal progenitors (indicated by *asterisk*) within the ciliary marginal zone give rise to retinal ganglion cells, amacrine cells, bipolar cells, horizontal cells, and cone photoreceptors. Müller glia (in *green*) in the differentiated retina create a stem cell niche that supports progenitors which give rise to the rod photoreceptors. Lesions induce the dedifferentiation of Müller glia, thus creating a population of progenitors that form a regenerative niche. The latter resembles the niche defined by the ciliary marginal zone and leads to the generation of all the cell types that have been lost to injury. (Modified after Raymond et al. 2006)

adult-born cells from their proliferation zones to the target areas in the telencephalon (Pellegrini et al. 2007).

2.3 Regulation of Cell Numbers by Apoptotic Cell Death

In the cerebellum, roughly half of the newly generated cells undergo apoptosis a few weeks after they are born (Zupanc et al. 1996; Soutschek and Zupanc 1996; Ott et al. 1997). Most of these cells appear to die shortly after arrival at their target sites. Regulation of the number of newly generated cells through apoptotic cell death is also thought to play a crucial role during development of adult-born cells in other parts of the CNS, including the central posterior/prepacemaker nucleus (Soutschek and Zupanc 1995), the valvula cerbelli pars lateralis (Teles et al. 2012), the retina (Candal et al. 2005; Biehlmaier et al. 2001; Mizuno and Ohtsuka 2008, 2009), and the spinal cord (Sîrbulescu et al. 2009).

2.4 Cellular Differentiation and Integration into Neural Networks

In the cerebellum, the vast majority of the new cells differentiate into granule cell neurons (Zupanc et al. 1996, 2005; Kaslin et al. 2009). In the corpus cerebelli, expression of the neuron-specific marker protein Hu commences when the migrating immature cells reach the granular layer (Zupanc et al. 2012). Retrograde tracing, combined with anti-BrdU immunohistochemistry, has demonstrated that the new granule cells develop proper axonal projections from the granular layer into the associated molecular layer (Zupanc et al. 1996, 2005), thus suggesting integration into the cerebellar neural network.

Quantitative analysis in zebrafish has shown that overall approximately 50 % of all adult-born cells develop into neurons that express the marker protein Hu (Zupanc et al. 2005; Hinsch and Zupanc 2007). Additional molecules co-localizing with BrdU in the zebrafish brain include the neuronal markers acetylated tubulin (Pellegrini et al. 2007), parvalbumin, tyrosine hydroxylase, and serotonin (Grandel et al. 2006), as well as the glial marker S100β (Zupanc et al. 2005; Grandel et al. 2006). Co-localization of any of these markers with BrdU is, however, by far less frequent than the co-localization of BrdU and Hu.

In the spinal cord of adult goldfish, 24 h after a single BrdU pulse, approximately 35 % of the BrdU-positive proliferating cells express the neuronal marker Hu (Takeda et al. 2008). Six weeks after their generation, some of the new neurons produce serotonin, indicating further specialization and possible integration into functional circuits. In the spinal cord of adult zebrafish, olig2-expressing radial glia form a distinct progenitor cell population. These cells appear to give rise only to cells of the oligodendrocyte lineage, but not to neurons or astrocytes (Park et al. 2007).

In the retina, the ciliary marginal zone can generate all types of neurons and glia. The new neurons are incorporated into the retinal circuitry. The rod progenitors, following migration from the inner nuclear layer to the outer nuclear layer, differentiate into mature photoreceptors (Johns 1977, 1982; Raymond et al. 2006).

2.5 Long-Term Persistence

Experiments in which the fate of BrdU-labeled cells was followed over up to 1,010 days have shown that approximately half of the adult-born cells in the brain survive long-term, and probably for the rest of the fish's normal life span (Zupanc et al. 1996, 2005; Ott et al. 1997; Hinsch and Zupanc 2007; Teles et al. 2012). This long-term survival of adult-born cells, together with the continuous generation of new cells, leads to a sustained growth of the brain, as reflected by the increase both in the total number of brain cells and in the size of the brain.

3 Neuronal Regeneration in the Central Nervous System of Teleost Fish

3.1 Lesion Paradigms

To examine nervous tissue regeneration and functional recovery after CNS injuries in teleost fish, a variety of lesion paradigms have been developed (Fig. 3). To the best of our knowledge, the first mentioning of such a paradigm was published in 1922, in form of a brief communication, by Koppányi and Weiss (1922), who transected the spinal cord of European carp (*Carassius vulgaris*), presumably in the cervical or thoracic region. Although the authors did not perform any histological experiments, and thus failed to provide evidence for structural regeneration, they claimed that after 2 months of recovery the swimming movements of the regenerated fish differed "not in the slightest" from those of intact fish. Transection paradigms similar to the one of Koppányi and Weiss have been employed in a number of investigations, particularly to study axonal regeneration and recovery of locomotor function after spinal cord injury (for reviews, see Becker and Becker 2008; Sîrbulescu and Zupanc 2011).

A second lesion paradigm used to study regeneration after spinal cord injury is based on amputation of the tail, including the caudal part of the spinal cord (Waxman and Anderson 1980). This paradigm has been employed in two species of gymnotiform fish, the black ghost knifefish *(Apteronotus albifrons)* and the brown ghost knifefish, primarily to examine neuronal regeneration (for reviews, see Anderson and Waxman 1985; Waxman and Anderson 1986; Sîrbulescu and Zupanc 2011; Zupanc and Sîrbulescu 2011). Since in apteronotids, including these two species, the axonal terminals of spinal motoneurons are modified to form a paired electric organ (de Oliveira-Castro 1955; Bennett 1971; Waxman et al. 1972), amputation of the tail, and thus part of the electric organ, leads to a reduction in the amplitude of the electric organ discharge. This characteristic enables investigators to monitor the partial loss of the electric behavior, and the regain of this behavioral function, which parallels the structural regeneration (Sîrbulescu et al. 2009).

In the brain, the majority of the regeneration studies have focused on the optic tectum, the cerebellum, and the telencephalon. Based on earlier lesion studies that aimed to explore the function of the optic tectum, Walter Kirsche was the first to use this brain system to examine tissue regeneration and functional recovery in teleosts (Kirsche 1960; Kirsche and Kirsche 1961). Lesions are applied by mechanically disrupting varying amounts of tectal tissue.

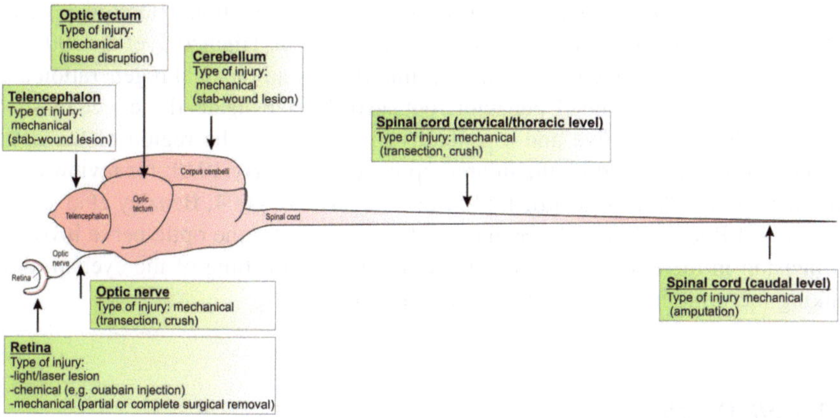

Fig. 3 Lesion paradigms used to study regeneration of the CNS in teleost fish. To indicate the sites of injury, a schematic drawing of the teleostean CNS is shown. Models of traumatic brain injury are mainly based on stab-wound lesions applied to the cerebellum and telencephalon, as well as more extensive tissue lesions in the optic tectum. Models of spinal cord injury involve transection or crush of the cord at cervical or thoracic levels, or amputation of the caudal part of the spinal cord. A variety of lesion paradigms have been applied to the retina, including light/laser-induced ablation, chemical lesions, and mechanical removal of patches of the retina. The effect of such lesions is examined in both the retina and the major projection area of retinal ganglion cells, the optic tectum. Injuries to the optic nerve are applied either through transection of the nerve, or by crushing of the nerve fibers

Cerebellar lesions are applied with a scalpel through the skull into one hemisphere of the dorsalmost subdivision of the cerebellum, the corpus cerebelli (Zupanc et al. 1998; for reviews, see Zupanc and Zupanc 2006a; Zupanc 2008a, b, 2009, 2011; Zupanc and Sîrbulescu 2011). The resulting stab-wound lesion runs from the pial surface through the dorsal molecular layer roughly halfway into the granular layer of the corpus.

In the telencephalon, lesions are generated by inserting a cannula (approximately 26–30 gage) through one nostril and the olfactory bulb into the dorsal telencephalon of one brain hemisphere (Ayari et al. 2010). Alternatively, lesions are applied laterally to the dorsolateral part of one hemisphere of the telencephalon by stabbing with a cannula through the skull (Ayari et al. 2010; Kishimoto et al. 2012).

In cerebellar, telencephalic and tectal models of traumatic brain injury, control tissue can be obtained either from the region corresponding to the lesion site in intact or sham-operated animals, or from the contralateral hemisphere in lesioned animals.

In the retina, a multitude of approaches have been used to induce injuries, including surgical excision of a patch of retina (Hitchcock et al. 1992); intraocular injection of various toxins, such as ouabain (Maier and Wolburg 1979), kainic acid (Negishi et al. 1988), and tunicamycin (Negishi et al. 1991); thermal laser ablation

(Braisted et al. 1994); and ultra-high-intensity light ablation (Bernardos et al. 2007) (for reviews, see Hitchcock and Raymond 1992; Hitchcock et al. 2004).

Severance of the optic nerve is used primarily to study axonal regeneration and recovery of visually guided behavior, but also to investigate the effect of such injury on the degenerative and regenerative processes in the region of the optic tectum that is innervated by the injured optic nerve (Sperry 1948; for reviews, see Stuermer et al. 1992; Bernhardt 1999; Matsukawa et al. 2004; Beazley et al. 2006; Becker and Becker 2007). A common technique to sever the optic nerve involves an incision in the dorsal conjunctiva and a downward-rolling of the eyeball, thus making the optic nerve accessible to crushing or cutting.

3.2 Cell Death

A characteristic feature of the regenerative response of teleost fish after CNS injury is the rapid onset of cell death (Figs. 4, 5a). Using the terminal deoxynucleotidyl transferase dUTP nick end labeling (TUNEL) assay, the first labeled cells can be observed as early as 5 min after application of a stab-wound lesion to the cerebellum of brown ghost knifefish (Zupanc et al. 1998). Thirty minutes after the lesion, the number of labeled cells reaches maximum levels. At 2 days post lesion, the number of TUNEL-positive cells starts to decline, until background levels are reached approximately 20 days after the injury. Further investigation of the morphological appearance of the TUNEL-positive cells using light and electron microscopy indicates that the vast majority of these cells undergo apoptosis, as opposed to necrosis. A similar transient increase in the number of cells undergoing apoptotic cell death has been observed in the dorsal telencephalon (Kroehne et al. 2011), the retina (Vihtelic and Hyde 2000; Yurco and Cameron 2005; Fimbel et al. 2007; Kassen et al. 2009; Bailey et al. 2010), and the spinal cord (Takeda et al. 2008; Sîrbulescu et al. 2009) of teleost fish.

After amputation of the caudal spinal cord, at the interface between the intact tissue and the regenerating tissue, the number of apoptotic cells remains three to four times higher than baseline, for up to 100 days post injury. The persistence of apoptosis in the area of the original lesion is remarkable, as it indicates that this type of cells death may also play a role in the restructuring of the tissue and in the integration of the new cells at this interface. This hypothesis is supported by the observation that the majority of apoptotic cells found at later stages of regeneration are new cells, and that many of the newly differentiated neurons and glia undergo apoptosis in this region of the regenerating spinal cord as late as 150–200 days post lesion (Sîrbulescu and Zupanc 2009).

The predominance of apoptosis in teleosts contrasts with necrosis as the dominant type of cell death in mammals after CNS injury (for reviews see Beattie et al. 2000; Vajda 2002; Liou et al. 2003). Unlike apoptosis, necrosis usually leads to inflammation at the site of the injury (for review see Kerr et al. 1995). This inflammatory response initiates a cascade of events during a secondary (delayed)

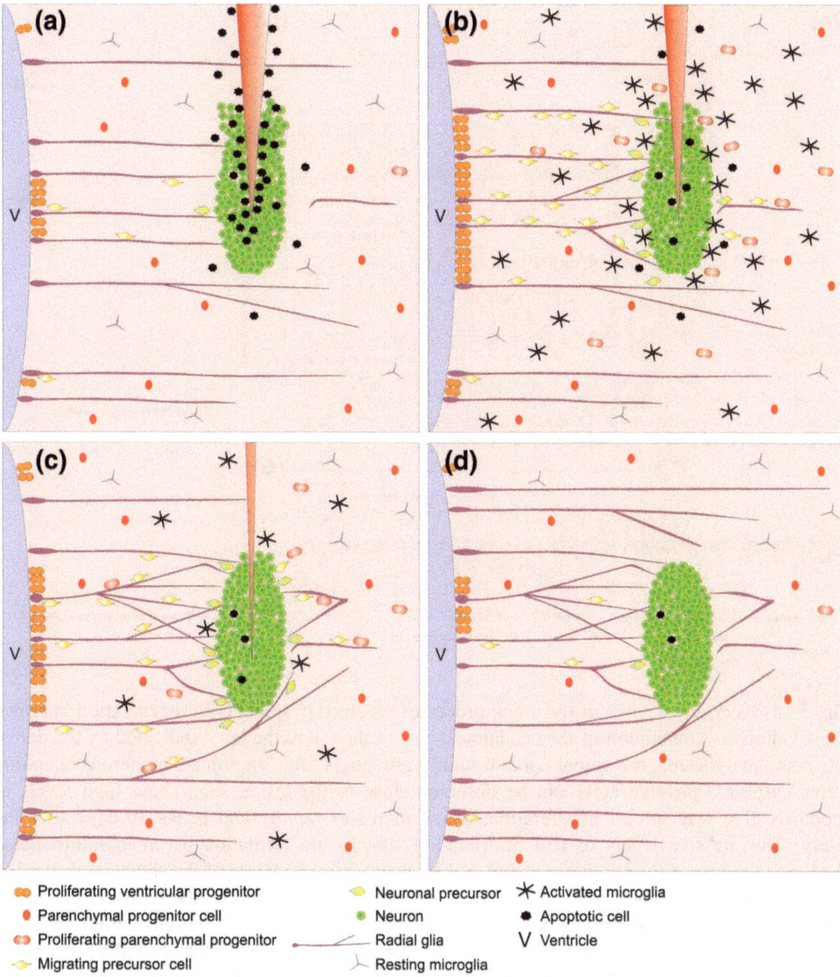

Fig. 4 Sequence of major cellular events mediating brain repair. (**a**) Within a few hours after injury, a large number of cells undergo apoptotic cell death at the lesion site. (**b**) Several days after the lesion, cell proliferation is markedly increased not only among progenitor cells within the ventricular proliferation zones which generate new cells constitutively, but also among the normally quiescent population of progenitor cells in the parenchyma. The new cells derived from these progenitors migrate toward the lesion site where they subsequently differentiate. At the same time, a pronounced increase in the number of activated microglia/macrophages is evident at and near the lesion site. They are thought to remove cellular debris through phagocytic activity. (**c**) Approximately 2 weeks after the lesion, the number of apoptotic cells has almost returned to baseline levels. The numbers of proliferating cells and activated microglia/macrophages have decreased, but are still elevated, as compared to levels in the intact brain. In addition, a meshwork of glial fibers has developed around the injury site. (**d**) Two to three months after the lesion, the cytoarchitecture of the brain area where the lesion occurred is restored. At this time point, only the persistence of the glial meshwork still marks the site of the injury

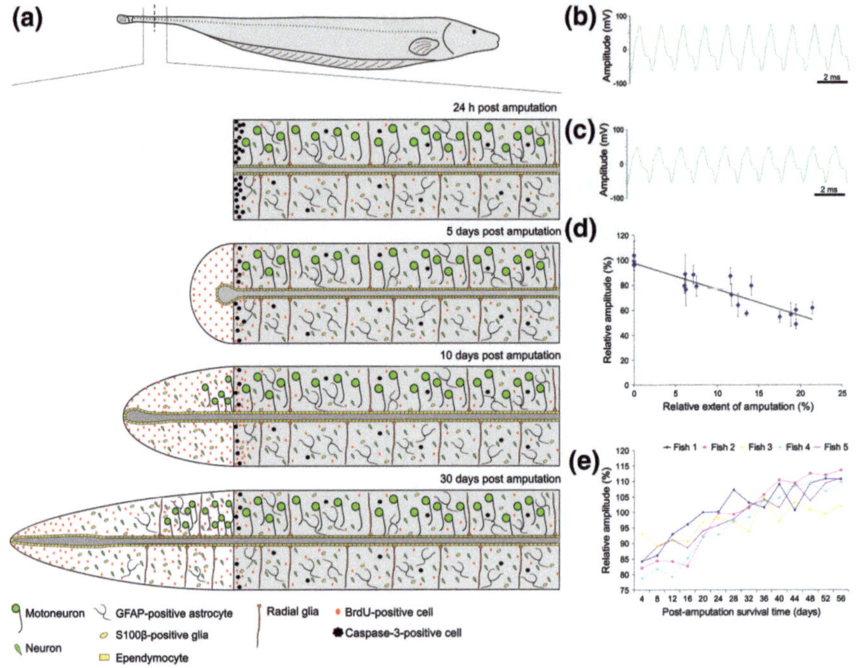

Fig. 5 **a** Overview of some of the major processes involved in spinal cord regeneration in brown ghost knifefish. Amputation of the caudalmost part of the tail at the level indicated by the dotted line completely severs the spinal cord. Within a few hours after the injury, numerous apoptotic active caspase-3-positive cells can be observed close to the lesion site. These high levels of apoptosis give way to cell proliferation, which increases rapidly and peaks 10 days after the injury. The massive levels of cell proliferation lead to the formation of an undifferentiated blastema at the tip of the regenerating tail. Subsequently, the caudal tip of the spinal cord, starting with the ependymal tube, extends into this blastema. The extension of the regenerating spinal cord appears to be supported by cell proliferation both at its caudal end and within the more rostral parenchyma. Differentiation of the newly generated cells into neurons and glia proceeds in a rostro-caudal direction. **b–e** Functional recovery after spinal cord injury. Waveforms of the electric organ discharges before amputation (**b**) and 48 h after amputation (**c**) of 1 cm of tail. The extent to which the amplitude of the discharges drops is directly correlated with the relative extent of tail amputation. Each data point represents one fish (**d**). After amputation, the amplitude of the electric organ discharges gradually recovers, until it reaches baseline levels approximately 30 days later (**e**). (After Sîrbulescu et al. 2009; Sîrbulescu and Zupanc 2011)

phase, which follows the primary (mechanical) phase of the injury. The events that take place during the secondary phase cause progressive cavitation and glial scarring, including upregulation of glial scar-associated molecules that lead to a retraction of axons from the site of the lesion (Balentine 1978; Zhang et al. 1997; Fitch et al. 1999; Horn et al. 2008; for reviews see Reier et al. 1983; Fitch and Silver 2008; Rolls et al. 2009). By contrast, apoptosis is characterized by cell shrinkage, nuclear condensation, and production of membrane-enclosed particles that are digested by other cells. Most significantly, the side-effects that accompany necrosis, such as inflammation of the surrounding tissue, are typically absent in apoptosis (for review see Elmore 2007). Consequently, the elimination of damaged cells through apoptosis, instead of necrosis, is thought to be a key factor contributing to the enormous regenerative capability of the CNS of teleost fish.

3.3 Neuroprotection

In light of the massive wave of cell death triggered by injuries, it is likely that cellular factors exist that protect cells from dying. However, thus far only correlative evidence is available that points to the actual existence of such factors. One such molecular candidate is calbindin-D_{28k}. After application of a stab-wound lesion to the cerebellum, the expression of this vitamin D-dependent calcium-binding protein is transiently increased in granular neurons in the area of the lesion between 16 h and 7 days post injury (Zupanc and Zupanc 2006b). This upregulation of calbindin-D_{28k} might be involved in mitigating the effects of elevated levels of intracellular free Ca^{2+} in the teleostean brain after injury. Such a Ca^{2+} increase has been demonstrated in mammalian models of traumatic brain injury during the secondary phase of tissue damage; it is thought that this increase is a major contributor to the delayed tissue damage and cell death after traumatic brain injury (for review see Weber 2012). Calbindin-D_{28k} has been postulated to exert a neuroprotective function by buffering intracellular free Ca^{2+}. This hypothesis is supported by the findings that calbindin-D_{28k}-expressing neurons exhibit a relative resistance to neurotoxicity induced by glutamate, calcium ionophore, or acidosis (Mattson et al. 1991); and that the rate of survival of neurons can be increased after various types of insults by overexpression of the gene for calbindin-D_{28k} (D'Orlando et al. 2002; Ho et al. 1996; Monje et al. 2001; Phillips et al. 1999).

Another factor potentially involved in the promotion of cellular survival is glutamine synthetase. As revealed by proteome analysis (see Sect. 3.10, Molecular Identification of Regeneration-Associated Factors', below), the abundance of this astrocyte-specific enzyme is increased 3 days after a cerebellar injury (Zupanc et al. 2006). Under normal conditions, glutamine synthetase converts synaptically released glutamate into the non-toxic amino acid glutamine. However, under traumatic conditions, the extracellular level of glutamate is dramatically elevated (Faden et al. 1989; Katayama et al. 1990; Palmer et al. 1994). Then, the amount of glutamine synthetase is insufficient to catalyze the excessive amounts of glutamate released, thus leading to a continuous over-stimulation of glutamatergic synapses. This effect, commonly referred to as excitotoxicity (Olney 1969), is believed to be a major cause of cell death during the secondary phase of tissue damage in the CNS after traumatic injury (Hayes et al. 1992; Young 1992; Weber 2004; Lau and Tymianski 2010). Moreover, the effect of neurodegeneration induced by glutamate excitotoxicity is aggravated by the frequently observed decline in the expression of glutamine synthetase associated with traumatic injury, focal ischemia, and a number of neurological disorders in mammalian systems (Lewis et al. 1989, 1994;

Oliver et al. 1990; Smith et al. 1991; Grosche et al. 1995; Härtig et al. 1995). Thus, the increase in the abundance of glutamine synthetase in the teleostean brain, as opposed to the decrease found in the mammalian brain, after traumatic injury is remarkable, as it may constitute another mechanism that furnishes regeneration-competent organisms with relative protection from cell death.

3.4 Activation of Microglia/Macrophages

Microglia/macrophages have been identified within a few days after lesions in several divisions of the CNS of teleost fish (Fig. 4)—the cerebellum (Zupanc et al. 2003), the dorsal telencephalon (Ayari et al. 2010; Kroehne et al. 2011), and the retina (Craig et al. 2008). In each of these regions, their numbers have been shown to return to control levels by 4 weeks post injury.

The macrophages/microglia are thought to mediate the removal of cellular debris through phagocytotic activity, although direct evidence for such a function in the teleostean CNS is sparse. In the retina, phagocytosis has been shown to play a crucial role during regeneration. Normally, Müller glia engulf the cell bodies of apoptotic photoreceptors. However, if this process is disrupted by inhibiting phagocytosis, both the proliferation of the Müller cells in response to injury and the regeneration of cone photoreceptors are significantly reduced (Bailey et al. 2010).

3.5 Cell Proliferation

The excellent regenerative potential of teleost fish is based not only on the capability to limit the degenerative effects of lesions but also on the ability to generate new cells that replace those lost to injury. In the cerebellum, the dorsal telencephalon, and the spinal cord, the rate of mitosis starts to increase 1 day post lesion and peaks approximately 1 week after the injury, compared to controls (Figs. 4, 5a; Zupanc and Ott 1999; Dervan and Roberts 2003; Takeda et al. 2008; Reimer et al. 2008; Sîrbulescu et al. 2009; Ayari et al. 2010; Kroehne et al. 2011; Kishimoto et al. 2012). In the brain, the number of proliferating cells returns to baseline levels 3–4 weeks post lesion, whereas in the spinal cord, after tail amputation, control levels are still not reached 7 weeks after the injury. This delay is possibly caused by the difference in the extent of the injury—through tail amputation, an entire portion of the spinal cord is removed whereas through stabbing of the respective brain region, a rather limited lesion is created.

Besides the similarity in the time course of the transient upregulation of the proliferative response, a feature shared by different parts of the CNS is the diversity of sources that supply new cells for the repair of the injured tissue. After lesions applied to the cerebellum and the dorsal telencephalon, the cells generated in response to the injury originate from two major sources—the stem cell niches that generate new cells constitutively, and areas in the parenchyma near the injury site, which harbor progenitor cell populations that are quiescent in the intact brain (Zupanc and Ott 1999; Ayari et al. 2010; Kroehne et al. 2011; Kishimoto et al. 2012). Similarly, in the spinal cord cells undergo mitosis after injury throughout the white and gray matters, as well as within the ependymal cell layer surrounding the central canal (Reimer et al. 2008; Takeda et al. 2008; Sîrbulescu et al. 2009). In the retina, mitotic cells appear to be located primarily in the proximity of the injured site (Stenkamp 2007). There is, however, some indication that enhanced proliferative activity can also occur at a distance from the lesion, suggesting the existence of diffusible factors that regulate this response (Yurco and Cameron 2005).

Interestingly, both in the cerebellum and the dorsal telencephalon, it has been shown that a minor population of cells born as early as 2 days *before* the lesion also contribute to the restoration of the damaged tissue (Zupanc and Ott 1999; Kroehne et al. 2011). This observation suggests a direct relationship between the continued cell proliferation in the intact adult brain and the generation of new cells induced by injury. The continuous provision of a pool of undifferentiated cells in the intact brain appears to enable fish to recruit new cells more rapidly and in larger numbers in the event of injury than would be possible by recruiting only those cells that are generated in response to a lesion.

In order to better understand the regenerative potential of teleost fish, an important question concerns the source(s) of the regenerated neurons in the injured CNS. In several regeneration-competent organisms, dedifferentiation of cells in the immediate vicinity of the wound has been shown to represent a major mechanism contributing to the repair of tissues and organs, such as heart (Jopling et al. 2010; Kikuchi et al. 2010), bone (Knopf et al. 2011), and limbs (Kragl et al. 2009). By contrast, the role of dedifferentiation in the process of neuronal regeneration of brain and spinal cord tissue is less clear. Genetic lineage-tracing, combined with lesioning of the dorsal telencephalon, has suggested that most of the regenerating neurons are derived from radial glia-type progenitors, and that dedifferentiation of otherwise non-neurogenic cells plays only a minor role, if any (Kroehne et al. 2011). On the other hand, studies on the retina have shown that, in response to a lesion, the Müller glia can undergo dedifferentiation and give rise to a population of progenitors that serve as the basis for the regeneration of all cell types lost to injury (Yurco and Cameron 2005; Bernardos et al. 2007; Fimbel et al. 2007; Thummel et al. 2008).

3.6 Migration to the Injury Site

Many of the new cells generated distal to the lesion site migrate, within the first few weeks after the injury, from the area where they were born to the area of the wound (Fig. 4). In the corpus cerebelli of brown ghost knifefish, a major source of such cells is the proliferation zone around the midline (Zupanc and Horschke 1995; Zupanc et al. 1996; see Sect. 2.2, 'Migration', above). After application of a stab wound to the dorsal part of one hemisphere of the corpus cerebelli, mitotic activity in this proliferation zone is markedly upregulated. As demonstrated by BrdU pulse-chase experiments, over the following 2 weeks, new cells migrate from this proliferation zone laterally within the dorsal molecular layer to the lesion site (Zupanc and Ott 1999).

The migration of the young cells is paralleled by the appearance of GFAP- and vimentin-expressing radial glial fibers in the dorsal molecular layer, mainly between the injury site and the midline (Clint and Zupanc 2001, 2002). Among other possible functions, these radial glial fibers appear to provide a scaffolding for the migrating young cells, as inferred from the close apposition of BrdU-labeled cells to such fibers (Clint and Zupanc 2001).

A similar directed migration of new cells derived from adult stem cells residing in proliferation zones distal to the injury site has been found in the dorsal telencephalon of zebrafish. After application of a lesion to the dorsolateral portion of one hemisphere, migrating new cells could be traced from the ventricular zone in the ventral telencephalon along a pathway in the subpallial and pallial regions to the injury site (Kishimoto et al. 2012). This destination site is reached within approximately 1 week. Consistent with the notion that the newly generated cells migrate to the injury site, it has been found that the BrdU-labeled cells express polysialylated-neural cell adhesion molecule (PSA-NCAM), a marker for migrating neurons, in the ventricular zone and in the vicinity of the ventricular zone, but not in the area adjacent to the lesion site.

3.7 Neuronal Differentiation

A common feature of any regeneration-competent CNS system examined thus far is that cells lost to injury are replaced by new cells that differentiate into various cell types, including neurons. In the dorsal telencephalon, the young cells start to express the neuronal marker protein Hu as early as 3–4 days after the injury (Ayari et al. 2010; Kroehne et al. 2011; Kishimoto et al. 2012). These new neurons emerge both in the vicinity of the area of the injury and in regions more distal to the injury site (Fig. 4). Over the following few days, the number of Hu-expressing new cells gradually increases, particularly at the injury site. This spatio-temporal pattern suggests that at least some of the young cells acquire immunological properties characteristic of neurons as early as during their migration toward the injury site.

Further examination of these cells at later stages of development has shown that they may differentiate into specific types of neurons, expressing marker proteins characteristic of mature neurons, such as parvalbumin or microtubule-associated protein 2a + 2b (MAP2a + 2b) (Kroehne et al. 2011); or Tbr1, a T-domain transcription factor characteristic of postmitotic glutamatergic neurons found in cortical areas of mammals (Englund et al. 2005) and in the pallium and medial-dorsal-lateral pallium of zebrafish (Kishimoto et al. 2012).

A similar differentiation of new cells into a specific subset of mature neurons has been demonstrated in the cerebellum after application of stab-wound lesions (Zupanc et al. 1996). Since the specific neuronal cell type—granular neurons—was identified by retrograde tracing from the molecular layer, this finding also suggests that the new granule cells have developed proper projections from the granular layer to the associated molecular layer.

In the retina, the progeny of dedifferentiated Müller glia redifferentiate into various cell types, including neurons, bipolar and amacrine cells, as well as Müller glia and cone photoreceptors (Wu et al. 2001; Yurco and Cameron 2005; Fausett and Goldman 2006; Raymond et al. 2006; Bernardos et al. 2007; Fimbel et al. 2007).

After spinal cord injury, the new cells develop into Hu C/D- and serotonin-expressing neurons, and S100β and GFAP-expressing ependymocytes and glial cells (Takeda et al. 2008; Sîrbulescu et al. 2009). After cervical spinal cord transection in zebrafish, new motor neurons are found in the proximity of the lesion site, with approximately 8 % of the mitotic cells differentiating into HB9-positive or islet-1-positive neurons by 2 weeks after injury. By 6–8 weeks after the lesion, the newly formed cells express markers of mature neurons and are decorated with synaptic terminals, indicating successful integration into the spinal cord circuitry (Reimer et al. 2008).

3.8 *Regeneration Defects and Improvement of the Quality of Repair*

Although apparent errors during regeneration in regeneration-competent organisms have rarely been examined systematically, review of the existing literature indicates that such deficiencies might occur more often than commonly assumed. After optic nerve transection in goldfish, new axonal sprouts emerge from the cut nerve stumps (Lanners and Grafstein 1980). Although the gross order of regenerating axons traveling to their synaptic target site, the optic tectum, is reminiscent of the order found in uninjured fish, the fascicle arrangement in the tectal fiber layer is quite erratic (Stuermer and Easter 1984). After destruction of the retina by ouabain injection, even more severe pathway aberrations occur in the course of the regeneration of the retina and of the de novo formation of retinotectal axons (Stuermer et al. 1985). In the retina, these axons exhibit highly abnormal courses, including extensive fascicle crossings, hairpin loops, and circular routes. In the optic tectum, short and long fascicles are not neatly aligned, as observed during normal development, but intermingle and cross each other extensively.

Like the aberrations in the retinotectal projections, errors have also been observed in the regenerated retina itself. After neurotoxic lesions, or surgical excision of small patches of retina, laminar fusions that traverse the inner nuclear layer and the ganglion cell layer are the most noticeable type of error (Raymond et al. 1988; Hitchcock et al. 1992; Mensinger and Powers 2007). These fusions occur frequently at the interface between the regenerated and intact portions of the retina.

After amputation of the caudal part of the tail of brown ghost knifefish, in approximately 5–10 % of the individuals the entire regenerate, including the spinal cord, is torsioned to various degrees (Sîrbulescu and Zupanc 2011). Perhaps related to this observation is the finding that apoptotic cells, identified by caspase-3 immunolabeling, persist at the transition zone between the rostrally located intact spinal cord and the more caudally situated regenerated tissue, at levels up to ten times higher than baseline, for at least 110 days after spinal cord amputation (Sîrbulescu and Zupanc 2009). This phenomenon has been interpreted as an indication of a defective, or incomplete, organization of the transition zone between the intact tissue and the newly generated one. Similarly, in eels it has been reported that the regenerated nervous tissue bridge that spans the gap between the caudal and rostral stumps of the transected spinal cord appears narrower and 'more irregular in shape' (Doyle et al. 2001).

Such observations indicate that, despite the impressive regenerative potential of teleost fish, their ability to repair CNS tissue after injury has certain limits. This notion prompts the question whether regeneration could be improved through experimental manipulation. In brown ghost knifefish, suppression of caspase-3 activation by administration of the apoptosis inhibitor 2,2'-methylenebis(1,3-cyclohexanedione) immediately after amputation of the caudal spinal cord results in a significant increase in the relative number of new cells that differentiate into neurons, in improved survival of the new neurons, and in significantly accelerated functional recovery, as assayed by monitoring the amplitude of the electric organ discharge (Sîrbulescu and Zupanc 2010a).

Improvement of regeneration has also been shown through a different type of experimental manipulation. Raising the ambient water temperature from 22°C to 30°C leads to a decrease in the number of apoptotic cells, an increase in the number of proliferating cells, and improved functional recovery (Sîrbulescu and Zupanc 2010b).

The deficits in regeneration, and the possibility to overcome them by experimental manipulation, enable investigators not only to better understand the mechanisms that mediate the regenerative potential of regeneration-competent organisms, but also to design therapeutic approaches that can be applied to regeneration-incompetent organisms, including humans. For example, the observation that administration of an apoptosis inhibitor immediately after the injury may have a long-term beneficial effect on the 'quality' of the repair of neuronal tissue and on the regain of function prompts testing this effect in a mammalian model system. We propose such a comparative approach as a promising alternative to the traditional paradigms that focus exclusively on the study of regeneration-incompetent organisms to identify potential therapeutic targets.

3.9 Functional Recovery

Lesions of the corpus cerebelli in brown ghost knifefish do not cause any obvious behavioral defects. In the dorsal telencephalon, functional recovery after lesions has not been studied thus far.

In the retina, functional recovery as part of the regenerative process has been examined at both the physiological and behavioral levels, using mainly goldfish as a model system (for reviews see Stenkamp 2007; Fleisch et al. 2011). Removal of parts of the retina by surgical and/or cytotoxic lesioning causes a reduction in the amplitude of different components of the electroretinogram, and a deviation from normal visually mediated reflexive behaviors as revealed through monitoring of the dorsal light reflex and the optokinetic nystagmus. The impairment of the dorsal light reflex is reflected by a tilting of the fish's vertical axis such that the intact eye is positioned away from downwelling light. The partial abolishment of the optokinetic nystagmus becomes evident through a marked reduction in the number of reset eye movements, compared to the untreated control eye after stimulation with high-contrast black-and-white square-wave gratings moved at a certain speed. As the retina regenerates, the components of the electroretinogram gradually reappear, and the two reflexive behaviors are progressively restored (Kästner and Wolburg 1982; Mensinger and Powers 1999, 2007; Lindsey and Powers 2007).

After transection of both optic nerves, fish fail to perform visually mediated behaviors, such as optokinetic reactions and startle responses, thus indicating total blindness. Functional vision is restored within a few weeks, as shown by the reappearance of these behaviors (Sperry 1948). Similarly, color discrimination, as demonstrated by use of a conditioning paradigm, is fully reinstated within a few weeks after transection of the optic nerves, independent of whether the fish learned to discriminate certain colors pre- or postoperatively (Arora and Sperry 1963).

Experiments examining functional recovery after tectal lesions have been initiated by Walter Kirsche in the early 1960s (Kirsche 1960; Kirsche and Kirsche 1961). Unilateral lesions of the optic tectum in Crucian carps (*Carassius carassius*) cause a number of motor aberrations, including sustained lateral body flexure and/or circling movements. The severity of these aberrations correlates positively with the extent of the lesion. In case of complete unilateral lesions of the tectum, tissue regeneration is absent, and functional recovery has not been observed. Partial lesions in young (3–6 cm) fish, leaving intact the proliferation zone located at the caudal-dorsal-medial borders of the tectum, result in a progressive recovery of the behavioral functions. The recovery becomes evident between 30 and 60 days post lesion, and the behavior typically returns to normal by 100 days after the lesion. The functional recovery parallels the reconstitution of the tectal cytoarchitecture, which is preceded by an increase in mitotic activity of the tectal proliferation zone (Richter 1968; Richter and Kranz 1970b, 1977).

Following spinal cord lesions, a large number of investigations have demonstrated an excellent capacity of teleost fish to recover behavioral function. The majority of the early studies have used the return to normal swimming as a measure of spinal cord regeneration (Tuge and Hanzawa 1935, 1937; Bernstein and Bernstein 1967; Bernstein and Gelderd 1970; Coggeshall et al. 1982; Coggeshall and Youngblood 1983). In goldfish, Bernstein (1964) showed functional recovery 60 days after injury by evoking muscle contractions in myotomes located caudally to the lesion site through stimulation of spinal cord regions rostral to the injury. In the same species, the return of the highly stereotypical C-start escape response, as well as the regain of equilibrium and normal feeding behavior, have been used as measures of recovery after a crush lesion of the cervical spinal cord (Zottoli and Freemer 2003). However, the C-start response has not been observed to fully return to preoperative parameters, not even at more than 100 days post lesion. In minnows (*Phoxinus phoxinus*), neural control of the rapid skin color change in response to dark or light backgrounds returns to normal levels approximately 5 months after spinal cord transection (Healey 1962, 1967). In eels, after a similar injury, the tail beat frequency largely returns to baseline levels within 35 days. However, the amplitude of the tail beat is still lower than in intact fish by 45 days post lesion (Doyle et al. 2001). In this species, locomotor recovery can be significantly accelerated if the fish are forced to exercise by being placed in a tank with continuous water current (Doyle and Roberts 2006). After spinal cord transection, lesioned zebrafish can swim only approximately 5 % of the distance covered by controls. However, by 6 weeks post lesion, they swim 57 % of the distance measured in controls within a 5-min interval (Becker et al. 2004). When the zebrafish are challenged with forced swimming against a current, a markedly lower endurance becomes visible in fish that have regenerated, even 10–12 weeks after spinal cord lesions (van Raamsdonk et al. 1998). A possible explanation for this finding is that proximal and distal body regions in lesioned fish do not reach pre-lesion levels of coordination.

In the majority of studies carried out thus far, functional recovery has been assessed after transection lesions which sever axonal tracts at cervical or thoracic levels. Therefore, axonal regeneration, rather than de novo neurogenesis, has been assumed to be the basis of functional recovery. Nevertheless, neuronal regeneration is likely to play an important role in the recovery of functional spinal cord circuits, especially after ablations, as opposed to transection lesions. This has been demonstrated in brown ghost knifefish. Amputation of caudal spinal cord removes a portion of the neurons forming the electric organ, which leads to a proportional reduction in the amplitude of the electric organ discharge (Fig. 5b–d). As neurogenesis proceeds in the regenerating spinal cord, new motoneurons are generated, forming axons which reconstitute the electric organ. Thus, in this teleost the process of neuronal differentiation after spinal cord lesion is paralleled by quantifiable behavioral recovery (Sîrbulescu et al. 2009; Sîrbulescu and Zupanc 2010a, b). The amplitude of the electric organ discharge recovers gradually after a 1-cm tail amputation, reaching baseline levels approximately 40 days after the lesion (Fig. 5e) (Sîrbulescu et al. 2009). The functional recovery becomes evident approximately 8–12 days after the lesion, which corresponds to the time when cell proliferation peaks and neuronal differentiation starts.

3.10 Molecular Identification of Regeneration-Associated Factors

To better understand the molecular mechanisms underlying tissue regeneration in the adult teleostean CNS, it is imperative to identify regeneration-associated genes and proteins on a large scale and to analyze their functions. Our knowledge concerning the multitude of genes and proteins involved in the regenerative processes has been particularly advanced by two experimental approaches, microarrays and proteomics.

Microarray studies have thus far only been carried out in zebrafish, and they have been instrumental in the investigation of gene expression changes that potentially underlie regeneration after lesions to several regions of the CNS. In the retina, such studies have examined changes that occur at various time points between 0 and 21 days after surgical lesions (Cameron et al. 2005), light injury (Kassen et al. 2007; Craig et al. 2008; Calinescu et al. 2009; Qin et al. 2009), optic nerve transection (Saul et al. 2010), or optic nerve crush (Veldman et al. 2007; McCurley and Callard 2010). After spinal cord transection, gene expression has been investigated in the spinal cord caudal to the lesion (Guo et al. 2011), as well as in the brainstem neurons of the nucleus of the medial longitudinal fascicle (Ma et al. 2012).

While some variation exists between studies due to the range of methods and lesion paradigms employed, most of the altered transcripts can be grouped into several major functional categories that reflect remarkably well the cellular phenomena that describe regeneration. Within the first day after injury, a decrease is observed in factors associated with the normal function of affected cells, such as photoreceptor-specific genes encoding opsins, *phosphodiesterase 6c*, *alpha* and *beta transducin* (Craig et al. 2008), and an increase in genes associated with an immune response and/or apoptosis, such as *tumor necrosis factor receptor 21*, *bcl2-associated X protein* (*bax*), *caspase 8*, *caspase 8/FADD like apoptosis regulator*, and *bcl2-associated death promoter* (*bad*) (Kassen et al. 2007; McCurley and Callard 2010), *complement C7 precursor*, *chemokine C-X-C motif receptor 4b* (*cxcr4b*), *lymphocyte cytosolic plastin 1* (*L-plastin*), *perforin 1 precursor*, and *leukocyte surface antigen CD53* (Cameron et al. 2005).

At 1–2 days after injury, increases have been observed in the expression of genes encoding transcription factors, such as *SRY-box containing gene* (*sox11b*), *fos*, *jun*, and *paired box 6a* (*pax6a*) (Craig et al. 2008), and proteins associated with cell cycle progression, such as *mini-chromosome maintenance 3*, *4*, *5*, and *7*, *cyclins B1*, *D1*, *F*, and *E*, and *proliferating cell nuclear antigen* (*pcna*) (Kassen et al. 2007).

At later stages, changes in transcription regulation reflect the predominance of long-term regenerative processes. Some examples include genes associated with continued cell proliferation—*protein regulator of cytokinesis 1, proliferation associated protein 100 (p100), deoxycytidine kinase (dCK), class I c-tubulin, activating transcription factor 3 (atf3), cyclin B1*, and *tumor suppressor p53-*

binding protein (Cameron et al. 2005; Saul et al. 2010; McCurley and Callard 2010); genes associated with cell growth and differentiation, including *engrailed 2b* (*eng2b*) and *zinc finger protein 2* (*zic2*); and genes involved in axonal genesis, including *growth associated-protein43* (*gap43*), *alpha* and *beta tubulin*, and the intermediate filament *plasticin*, which is expressed by regenerating ganglion cells (Cameron et al. 2005; Kassen et al. 2007). Increased tissue remodeling and cell migration are indicated by differential regulation of genes encoding enzymes involved in extracellular matrix processing, such as *matrix metalloproteinase 2* (*mmp2*), *mmp9*, and *mmp13* (Cameron et al. 2005; Kassen et al. 2007). Interestingly, genes encoding some of the secreted growth factors upregulated during retina regeneration, including *midkine*, *progranulin*, and *galectin*, are also found after heart injuries in zebrafish, indicating common elements between various regenerative processes (Qin et al. 2009).

Differential proteome analysis has been performed in brown ghost knifefish on cerebellar tissue collected at two time points—30 min and 3 days after the lesion—reflecting different stages of the regenerative process (Zupanc et al. 2006; Ilieş et al. 2012). Thirty minutes after the lesion, the list of proteins whose abundance is significantly altered in lesioned tissue, compared to intact tissue, is comprised of two major categories (Ilieş et al. 2012). The first category includes proteins characteristic of degenerative processes, such as regulation of apoptotic cell death, destruction of the cytoskeleton, or remodeling of synaptic connections. Examples of upregulated proteins in this category include MHC class I heavy chain, 26S proteasome non-ATPase regulatory subunit 8, and ubiquitin-specific protease 5, while levels of spectrin alpha 2 are decreased. The second category includes proteins associated with regenerative processes, such as regrowth of neurites and promotion of mitotic activity. Examples are alpha-internexin neuronal intermediate filament protein, erythrocyte membrane protein 4.1N, and tubulin alpha-1C chain. The finding of upregulation in the latter proteins is remarkable because it indicates that the course toward regeneration is set very early—within a few minutes after the injury—in the CNS of a regeneration-competent organism.

Three days after lesioning of the cerebellum, the list of proteins that exhibit significant changes in their abundance is dominated by cytoskeletal proteins (such as beta-actin and beta-tubulin), presumably reflecting the repair of injured axons; and by proteins that mediate the correct assembly of structural proteins (such as chaperonin-containing tailless-complex polypeptide 1, subunit epsilon, tropomodulins-3 and -4, and bullous pemphigoid antigen 1) (Zupanc et al. 2006). Furthermore, several proteins potentially involved in neuroprotection exhibit altered abundances at this time point. Whereas the levels of 78,000-Da glucose-regulated protein and glutamine synthetase are increased, the level of cytosolic aspartate aminotransferase is decreased. Similar to the results of microarray studies, which have shown an upregulation in the expression of genes encoding transcription factors 1–2 days after injury (see above), proteomic analysis has also revealed an increase in the abundance of a potential transcriptional regulator, bone marrow zinc finger 2, 3 days after a cerebellar lesion. Taken together, these observations are congruent with the notion that at this stage of regeneration, cells that have survived the primary phase of tissue damage are protected from death through a number of mechanisms, while at the same time processes are initiated that finally lead to the repair of injured cells and to the formation of new cells.

Large-scale approaches, such as microarray and proteomics studies, provide extremely rich datasets comprising numerous genes and proteins potentially involved in regeneration. In order to verify and further examine the role that these candidates play, subsequent in-depth studies need to focus on the spatio-temporal dynamics of one or a few of such molecules. A summary of the molecular factors that have been the subject of such detailed investigations is given in Table 1. This growing list of factors, albeit far from comprehensive, begins to describe the special properties of a molecular microenvironment that supports, rather than inhibits the complex process of regeneration in the CNS of vertebrates.

Immediately after injury, apoptotic cell death increases dramatically, but subsequent rises in the levels of apoptotic regulators, such as phospho-Akt (protein kinase B) and B cell lymphoma 2 (Bcl-2) (Koriyama et al. 2006) facilitate a rapid decrease in the number of apoptotic cells (Zupanc et al. 1998; Sîrbulescu et al. 2009). With a short lag, as compared to the initial wave of apoptosis, a marked increase in the expression of transcription factors, such as ATF3, Sox11b, Pax6, Nkx6.1, Krueppel-like factor (KLF) 6a, and KLF7a, and in growth-associated proteins, such as tubulin alpha 1A (Tuba1a) and GAP-43, indicates the transition toward regenerative processes, including cell proliferation, neuronal differentiation, and axonal regrowth (Veldman et al. 2007; Reimer et al. 2009; Saul et al. 2010). Elevated levels of growth factors such as fibroblast growth factor 3 (FGF-3) and several receptors of the retinoic acid pathway have been proposed to play a role in axonal regeneration (Reimer et al. 2009). Microtubule-associated protein 1B (MAP1B), a protein that has been associated in mammals with regeneration in the peripheral nervous system (Soares et al. 2002), is expressed in adult rainbow trout at high levels in both spinal and supraspinal neurons that show regenerative capacity, as well as in the glial cells of the CNS (Alfei et al. 2004).

The excellent regeneration of teleost fish after CNS injury is facilitated not only through the upregulation of growth-promoting molecules, but also through the absence or active inhibition of molecules that interfere with axonal regrowth. In zebrafish, Nogo-A, one of the main inhibitory proteins in mammalian myelin (Bradbury and McMahon 2006; Yiu and He 2006), lacks the inhibitory N-terminal region (Diekmann et al. 2005). Immunoreactivity to chondroitin sulfate proteoglycans, major components of the glial scar in mammals (Silver and Miller 2004), is not elevated after CNS injuries in zebrafish (Becker and Becker 2007). A recent study has shown that the elevated expression of a regulatory microRNA, miR-133b, plays an important role in spinal cord regeneration in zebrafish by inhibiting the RhoA GTPase, which promotes growth cone collapse, and therefore, would prevent axonal regrowth (Yu et al. 2011b).

Taken together, the studies summarized here are beginning to sketch the fundamental differences that enable teleost fish to regenerate CNS tissue after injury. Nevertheless, considerable work will be required to advance our understanding of the molecular interactions which orchestrate this highly complex process.

Table 1 Molecular factors involved in CNS regeneration of teleost fish

Class	Molecular factor	Factor regulation		Effect on regeneration	Model system		References
		Modulation	Time point/interval		Species	CNS region	
Apoptotic factors/regulators	p-Akt	Increased expression	3–40 days	+	Goldfish (*Carassius auratus*)	Retina	Koriyama et al. (2006)
	p-Bad	Increased expression	3–40 days	±	Goldfish	Retina	Koriyama et al. (2006)
	Bax	No change	0–30 days	–	Goldfish	Retina	Koriyama et al. (2006)
	Bcl-2	Increased expression	10–20 days	+	Goldfish	Retina	Koriyama et al. (2006)
	Caspase-3	Increased expression	12 h–2 days	±	Brown ghost knifefish (*Apteronotus leptorhynchus*)	Spinal cord	Sîrbulescu et al. (2009), Sîrbulescu and Zupanc (2009, 2010a)
		Decreased activity	10–30 days	±	Goldfish	Retina	Koriyama et al. (2006)
Axonal growth inhibitors	Nogo-A	N/A	N/A	Not inhibitory (lacks N-terminus)	Zebrafish (*Danio rerio*)	Spinal cord	Diekmann et al. (2005)
	Sema-3A	Decreased expression	2–7 days	–	Goldfish	Retina	Rosenzweig et al. (2010)

(continued)

Table 1 (continued)

Class	Molecular factor	Factor regulation		Effect on regeneration	Model system		References
		Modulation	Time point/interval		Species	CNS region	
Cytoskeletal and associated structural proteins	MAP1B	Increased expression	N/A	+	Trout (*Oncorhynchus mykiss*)	Spinal cord	Alfei et al. (2004)
	GFAP	Increased expression	40 days; 4 months	±	Brown ghost knifefish; Black ghost knifefish (*Apteronotus albifrons*)	Spinal cord Brain	Anderson et al. (1984), Sîrbulescu et al. (2009), Clint and Zupanc (2001)
	Vimentin	Increased expression	15–100 days	+	Brown ghost knifefish	Brain	Clint and Zupanc (2002)
ECM components	CSPG	No change	N/A	–	Zebrafish	Spinal cord	Becker and Becker (2007)
	Tenascin C	Increased expression	4 h, 2–11 days	+	Zebrafish	Spinal cord	Yu et al. (2011a)
Growth-associated proteins	GAP-43	Increased expression	7–14 days	+	Zebrafish	Spinal cord Retina	Becker et al. (1998) Kusik et al. (2010)
	Tuba1a	Increased expression	2–14 days	+	Zebrafish	Optic nerve	Goldman and daysing (2000), Senut et al. (2004), Veldman et al. (2010)

(continued)

Table 1 (continued)

Class	Molecular factor	Factor regulation		Effect on regeneration	Model system		References
		Modulation	Time point/interval		Species	CNS region	
Growth factors/receptors	BDNF	No change	1–20 days	+	Eel (*Anguilla anguilla*)	Spinal cord	Dalton et al. (2009)
	FGF3	Increased expression	14 days	+	Zebrafish	Spinal cord	Reimer et al. (2009)
	Retinoic acid receptor	Increased expression	14 days	+	Zebrafish	Spinal cord	Reimer et al. (2009)
	Trk B	No change	1–20 days	+	Eel	Spinal cord	Dalton et al. (2009)
Intra-/inter-cellular messengers	cAMP	N/A	N/A	+	Zebrafish	Spinal cord	Bhatt et al. (2004)
	NO	Increased expression	5–40 days	+	Goldfish	Retina	Koriyama et al. (2009)
Micro RNA	miR-133b	Increased expression	6 h–7 days	+	Zebrafish	Spinal cord	Yu et al. (2011b)
mRNA editing enzymes	Apobec2a/2b	Increased expression	6 h–8 days	+	Zebrafish	Retina	Powell et al. (2012)

(continued)

Table 1 (continued)

Class	Molecular factor	Factor regulation		Effect on regeneration	Model system		References
		Modulation	Time point/interval		Species	CNS region	
Transcription regulators	ATF3	Increased expression	24 h–7 days	+	Zebrafish	Retina	Veldman et al. (2007), Saul et al. (2010)
	KLF6a/KLF7a	Increased expression	1–12 days	+	Zebrafish	Retina	Veldman et al. (2007)
	Nkx6.1	Increased expression	14 days	+	Zebrafish	Spinal cord	Reimer et al. (2009)
	Pax6	Increased expression	14 days	+	Zebrafish	Spinal cord	Reimer et al. (2009)
	Sonic hedgehog a	Increased expression	14–42 days	+	Zebrafish	Spinal cord	Reimer et al. (2009)
	Sox11b	Increased expression	11 days	+	Zebrafish	Spinal cord	Guo et al. (2011)

(continued)

Table 1 (continued)

Class	Molecular factor	Factor regulation		Effect on regeneration	Model system		References
		Modulation	Time point/interval		Species	CNS region	
Transmembrane cell adhesion molecules	Contactin-1a	Increased expression	6–14 days	+	Zebrafish	Spinal cord	Schweitzer et al. (2007)
	L 1.1	Increased expression	3–56 days	+	Zebrafish	Spinal cord	Becker et al. (1998, 2004)
	NCAM	No change	7–14 days	+	Zebrafish	Spinal cord	Becker et al. (1998)
	zfNLRR	Increased expression	24 h–9 days	+	Zebrafish	Spinal cord Retina	Bormann et al. (1999)
	P0	Increased expression	2–180 days	+	Zebrafish	Brain Spinal cord	Schweitzer et al. (2003)
Other	Calbindin-D28 k	Increased expression	16 h–7 days	+	Brown ghost knifefish	Brain	Zupanc and Zupanc (2006b)
	Cysteine/Glycine-rich protein 1a	Increased expression	3–21 days	+	Zebrafish	Spinal cord	Ma et al. (2012)
	Somatostatin	Increased expression	24 h–25 days	+	Brown ghost knifefish	Brain	Zupanc (1999b)
	Transglutaminase (neural)	Increased expression	10–40 days	+	Goldfish	Retina	Sugitani et al. (2006)

4 Conclusions

The study of regeneration-competent organisms provides us with the unique opportunity to gain a broad biological understanding of tissue repair in the adult CNS. As shown in this chapter, neuronal regeneration is intimately linked to adult neurogenesis. Every regeneration-competent organism examined thus far also generates new neurons constitutively, both in large numbers and in many regions of the adult CNS. Comparative analysis has suggested that adult neurogenesis is a primitive vertebrate trait (Zupanc 2006b). It is likely that the availability of all the cellular and molecular regulatory mechanisms necessary for the generation of new neurons in the intact CNS has greatly facilitated the repurposing of the cellular machinery for neuronal regeneration with only slight adaptive changes. We hypothesize that such a regenerative ability was also shared by many, if not all early vertebrates.

One group of organisms that have shown extraordinary adaptation under the selective pressure caused by injury are the Gymnotiformes; they include the black and brown ghost knifefish mentioned earlier in this chapter. Many species of this teleostean order are distinguished by their elongated, compressed caudal part of the body ('tail'). In their natural habitat, gymnotiforms often suffer from damage to, or loss of, parts of the tail because of predatory fish specialized in tail-eating (Mago-Leccia et al. 1985; Lundberg et al. 1996; de Santana and Vari 2010). Probably in response to this selective pressure, gymnotiforms have developed the extraordinary ability to regenerate tails, including large parts of the caudal spinal cord, rapidly and with high efficiency—even after repeated loss of the tail.

A broad understanding of the biology of adult neurogenesis and neuronal regeneration will also facilitate the analysis of the selective pressures that have caused the loss of the regenerative potential during the evolution of mammals. It has been hypothesized that in vertebrates with indeterminate growth, such as teleosts, the primary function of adult neurogenesis is to ensure a constant ratio of the number of central neuronal elements within motor/sensory pathways relative to the number of peripheral elements—muscle fibers and receptor cells—when the number of peripheral elements grows ('numerical matching hypothesis'; Zupanc 1999a, 2001, 2006a, b, 2008a, b, 2009). Such a function is consistent with the existence of an enormous neurogenic potential in the teleostean brain, which correlates with continuous peripheral growth and the constant increase in the number of individual muscle fibers or receptor cells (Johns and Easter 1977; Corwin 1981; Zakon 1984; Weatherley and Gill 1985; Koumans and Akster 1995; Zimmerman and Lowery 1999; Rowlerson and Veggetti 2001). By contrast, growth in mammals is primarily the result of an increase in the size, not in the number, of individual peripheral elements (Rowe and Goldspink 1969). As a corollary, the matching hypothesis predicts that, when the mode of muscle growth is shifted from hyperplasia to hypertrophy in the course of the evolution of mammals, the neurogenic potential of CNS structures forming part of the motor pathway was reduced in parallel. Similarly, the number of neurogenic regions and the rate of neurogenesis decreased in brain areas associated with sensory processing when the continuous formation of new sensory cells in the sensory organs during adulthood was gradually abandoned.

Phylogenetic studies have shown that phenotypes are commonly lost by alteration of regulatory mechanisms without extensive modification of the individual developmental processes (such as cellular proliferation, migration, or differentiation) that had produced the lost form. On the other hand, novel environmental factors can initiate the expression of these ancestral elements in novel combinations through the process of developmental recombination to give rise to novel phenotypic traits, followed by the genetic accommodation of change (West-Eberhard 2003, 2005). If this hypothesis is correct, then the loss of adult neurogenesis and neuronal regeneration as phenotypic traits in many regions of the mammalian brain may have left intact some of the genetic or molecular pathways that induced the original phenotype. The existence of such ancestral elements has important implications from a biomedical point of view because then it should be possible to reactivate these elements under certain environmental conditions by turning off or on switches that control them. Indeed, quiescent stem cells have been discovered in the adult mammalian brain that can be activated by certain signals from the cellular environment, such as epidermal growth factor (Reynolds and Weiss 1992; Weiss et al. 1996; Palmer et al. 1999; Kondo and Raff 2000). Clearly, these adult stem cells could provide the substrate for the development of a cell replacement therapy based on the intrinsic ancestral potential of the adult human brain. The study of regeneration-competent organisms could play an important role in identifying the factors required to reactivate this potential.

References

Alfei L, Soares S, Alunni A, Ravaille-Veron M, Von Boxberg Y, Nothias F (2004) Expression of MAP1B protein and its phosphorylated form MAP1B-P in the CNS of a continuously growing fish, the rainbow trout. Brain Res 1009:54–66

Altman J (1969) Autoradiographic and histological studies of postnatal neurogenesis: IV. Cell proliferation and migration in the anterior forebrain, with special reference to persisting neurogenesis in the olfactory bulb. J Comp Neurol 137:433–458

Altman J, Das GD (1965) Autoradiographic and histological evidence of postnatal hippocampal neurogenesis in rats. J Comp Neurol 124:319–336

Alunni A, Hermel JM, Heuzé A, Bourrat F, Jamen F, Joly JS (2010) Evidence for neural stem cells in the medaka optic tectum proliferation zones. Dev Neurobiol 70:693–713

Anderson MJ, Swanson KA, Waxman SG, Eng LF (1984) Glial fibrillary acidic protein in regenerating teleost spinal cord. J Histochem Cytochem 32:1099–1106

Anderson MJ, Waxman SG (1985) Neurogenesis in adult vertebrate spinal cord in situ and in vitro: a new model system. Ann N Y Acad Sci 457:213–233

Anderson MJ, Waxman SG, Laufer M (1983) Fine structure of regenerated ependyma and spinal cord in *Sternarchus albifrons*. Anat Rec 205:73–83

Arora HL, Sperry RW (1963) Color discrimination after optic nerve regeneration in the fish *Astronotus ocellatus*. Dev Biol 7:234–243

Ayari B, El Hachimi KH, Yanicostas C, Landoulsi A, Soussi-Yanicostas N (2010) Prokineticin 2 expression is associated with neural repair of injured adult zebrafish telencephalon. J Neurotrauma 27:959–972
Bailey TJ, Fossum SL, Fimbel SM, Montgomery JE, Hyde DR (2010) The inhibitor of phagocytosis, O-phospho-L-serine, suppresses Müller glia proliferation and cone cell regeneration in the light-damaged zebrafish retina. Exp Eye Res 91:601–612
Balentine JD (1978) Pathology of experimental spinal cord trauma. I. The necrotic lesion as a function of vascular injury. Lab Invest 39:236–253
Beattie MS, Farooqui AA, Bresnahan JC (2000) Review of current evidence for apoptosis after spinal cord injury. J Neurotrauma 17:915–925
Beazley LD, Rodger J, Bartlett CA, Taylor AL, Dunlop SA (2006) Optic nerve regeneration: molecular pre-requisites and the role of training—restoring vision after optic nerve injury. Adv Exp Med Biol 572:389–395
Becker CG, Becker T (2007a) Growth and pathfinding of regenerating axons in the optic projection of adult fish. J Neurosci Res 85:2793–2799
Becker CG, Becker T (2007b) Zebrafish as a model system for successful spinal cord regeneration. In: Becker CG, Becker T (eds) Model organisms in spinal cord regeneration. Wiley, Weinheim, pp 289–319
Becker CG, Becker T (2008) Adult zebrafish as a model for successful central nervous system regeneration. Restor Neurol Neurosci 26:71–80
Becker CG, Lieberoth BC, Morellini F, Feldner J, Becker T, Schachner M (2004) L1.1 is involved in spinal cord regeneration in adult zebrafish. J Neurosci 24:7837–7842
Becker D, Gary DS, Rosenzweig ES, Grill WM, McDonald JW (2010) Functional electrical stimulation helps replenish progenitor cells in the injured spinal cord of adult rats. Exp Neurol 222:211–218
Becker T, Bernhardt RR, Reinhard E, Wullimann MF, Tongiorgi E, Schachner M (1998) Readiness of zebrafish brain neurons to regenerate a spinal axon correlates with differential expression of specific cell recognition molecules. J Neurosci 18:5789–5803
Bédard A, Parent A (2004) Evidence of newly generated neurons in the human olfactory bulb. Dev Brain Res 151:159–168
Bennett MVL (1971) Electric organs. In: Hoar WS, Randall DJ (eds) Fish physiology. Sensory systems and electric organs, vol 5, Academic Press, New York, pp 347–491
Bernardos RL, Barthel LK, Meyers JR, Raymond PA (2007) Late-stage neuronal progenitors in the retina are radial Müller glia that function as retinal stem cells. J Neurosci 27:7028–7040
Bernhardt RR (1999) Cellular and molecular bases of axonal regeneration in the fish central nervous system. Exp Neurol 157:223–240
Bernstein JJ (1964) Relation of spinal cord regeneration to age in adult goldfish. Exp Neurol 9:161–174
Bernstein JJ, Bernstein ME (1967) Effect of glial-ependymal scar and teflon arrest on the regenerative capacity of goldfish spinal cord. Exp Neurol 19:25–32
Bernstein JJ, Gelderd JB (1970) Regeneration of the long spinal tracts in the goldfish. Brain Res 20:33–38
Bhatt DH, Otto SJ, Depoister B, Fetcho JR (2004) Cyclic AMP-induced repair of zebrafish spinal circuits. Science 305:254–258
Biehlmaier O, Neuhauss SC, Kohler K (2001) Onset and time course of apoptosis in the developing zebrafish retina. Cell Tissue Res 306:199–207
Bormann P, Roth LW, Andel D, Ackermann M, Reinhard E (1999) zfNLRR, a novel leucine-rich repeat protein is preferentially expressed during regeneration in zebrafish. Mol Cell Neurosci 13:167–179
Bradbury EJ, McMahon SB (2006) Spinal cord repair strategies: why do they work? Nat Rev Neurosci 7:644–653
Braisted JE, Essman TF, Raymond PA (1994) Selective regeneration of photoreceptors in goldfish retina. Development 120:2409–2419

Calinescu AA, Vihtelic TS, Hyde DR, Hitchcock PF (2009) Cellular expression of *midkine-a* and *midkine-b* during retinal development and photoreceptor regeneration in zebrafish. J Comp Neurol 514:1–10

Cameron DA, Gentile KL, Middleton FA, Yurco P (2005) Gene expression profiles of intact and regenerating zebrafish retina. Mol Vis 11:775–791

Cameron HA, McKay RD (2001) Adult neurogenesis produces a large pool of new granule cells in the dentate gyrus. J Comp Neurol 435:406–417

Candal E, Anadón R, DeGrip WJ, Rodríguez-Moldes I (2005) Patterns of cell proliferation and cell death in the developing retina and optic tectum of the brown trout. Dev Brain Res 154:101–119

Chapouton P, Jagasia R, Bally-Cuif L (2007) Adult neurogenesis in non-mammalian vertebrates. Bioessays 29:745–757

Chapouton P, Webb KJ, Stigloher C, Alunni A, Adolf B, Hesl B, Topp S, Kremmer E, Bally-Cuif L (2011) Expression of *hairy/enhancer of split* genes in neural progenitors and neurogenesis domains of the adult zebrafish brain. J Comp Neurol 519:1748–1769

Clint SC, Zupanc GKH (2001) Neuronal regeneration in the cerebellum of adult teleost fish, *Apteronotus leptorhynchus*: guidance of migrating young cells by radial glia. Dev Brain Res 130:15–23

Clint SC, Zupanc GKH (2002) Up-regulation of vimentin expression during regeneration in the adult fish brain. Neuroreport 13:317–320

Coggeshall RE, Birse SG, Youngblood CS (1982) Recovery from spinal transection in fish. Neurosci Lett 32:259–264

Coggeshall RE, Youngblood CS (1983) Recovery from spinal transection in fish: regrowth of axons past the transection. Neurosci Lett 38:227–231

Corwin JT (1981) Postembryonic production and aging of inner ear hair cells in sharks. J Comp Neurol 201:541–553

Craig SE, Calinescu AA, Hitchcock PF (2008) Identification of the molecular signatures integral to regenerating photoreceptors in the retina of the zebrafish. J Ocul Biol Dis Infor 1:73–84

Curtis MA, Kam M, Nannmark U, Anderson MF, Axell MZ, Wikkelso C, Holtas S, van Roon-Mom WMC, Björk-Eriksson T, Nordborg C, Frisén J, Dragunow M, Faull RLM, Eriksson PS (2007) Human neuroblasts migrate to the olfactory bulb via a lateral ventricular extension. Science 315:1243–1249

Dalton VS, Roberts BL, Borich SM (2009) Brain derived neurotrophic factor and trk B mRNA expression in the brain of a brain stem-spinal cord regenerating model, the European eel, after spinal cord injury. Neurosci Lett 461:275–279

de Oliveira-Castro G (1955) Differentiated nervous fibers that constitute the electric organ of *Sternarchus albifrons*. Linn. An Acad Bras Cienc 27:557–564

de Santana CD, Vari RP (2010) Electric fishes of the genus *Sternarchorhynchus* (Teleostei, Ostariophysi, Gymnotiformes); phylogenetic and revisionary studies. Zool J Linn Soc-Lond 159:223–371

Dervan AG, Roberts BL (2003) Reaction of spinal cord central canal cells to cord transection and their contribution to cord regeneration. J Comp Neurol 458:293–306

Diekmann H, Klinger M, Oertle T, Heinz D, Pogoda HM, Schwab ME, Stuermer CAO (2005) Analysis of the reticulon gene family demonstrates the absence of the neurite growth inhibitor Nogo-A in fish. Mol Biol Evol 22:1635–1648

D'Orlando C, Celio MR, Schwaller B (2002) Calretinin and calbindin D-28 k, but not parvalbumin protect against glutamate-induced delayed excitotoxicity in transfected N18-RE 105 neuroblastoma-retina hybrid cells. Brain Res 945:181–190

Doyle LM, Roberts BL (2006) Exercise enhances axonal growth and functional recovery in the regenerating spinal cord. Neuroscience 141:321–327

Doyle LM, Stafford PP, Roberts BL (2001) Recovery of locomotion correlated with axonal regeneration after a complete spinal transection in the eel. Neuroscience 107:169–179

Ekström P, Johnsson C-M, Ohlin L-M (2001) Ventricular proliferation zones in the brain of an adult teleost fish and their relation to neuromeres and migration (secondary matrix) zones. J Comp Neurol 436:92–110

Elmore S (2007) Apoptosis: a review of programmed cell death. Toxicol Pathol 35:495–516

Englund C, Fink A, Lau C, Pham D, Daza RA, Bulfone A, Kowalczyk T, Hevner RF (2005) Pax6, Tbr2, and Tbr1 are expressed sequentially by radial glia, intermediate progenitor cells, and postmitotic neurons in developing neocortex. J Neurosci 25:247–251

Eriksson PS, Perfilieva E, Bjork-Eriksson T, Alborn AM, Nordborg C, Peterson DA, Gage FH (1998) Neurogenesis in the adult human hippocampus. Nat Med 4:1313–1317

Faden AI, Demediuk P, Panter SS, Vink R (1989) The role of excitatory amino acids and NMDA receptors in traumatic brain injury. Science 244:798–800

Faul M, Xu L, Wald MM, Coronado VG (2010) Traumatic brain injury in the United States: emergency department visits, hospitalizations and deaths 2002–2006. Center for Disease Control and Prevention, National Center for Injury Prevention and Control, Atlanta

Fausett BV, Goldman D (2006) A role for alpha1 tubulin-expressing Müller glia in regeneration of the injured zebrafish retina. J Neurosci 26:6303–6313

Fernández AS, Rosillo JC, Casanova G, Olivera-Bravo S (2011) Proliferation zones in the brain of adult fish *Austrolebias* (*Cyprinodontiform*[sic]*: Rivulidae*): a comparative study. Neuroscience 189:12–24

Fimbel SM, Montgomery JE, Burket CT, Hyde DR (2007) Regeneration of inner retinal neurons after intravitreal injection of ouabain in zebrafish. J Neurosci 27:1712–1724

Fitch MT, Doller C, Combs CK, Landreth GE, Silver J (1999) Cellular and molecular mechanisms of glial scarring and progressive cavitation: in vivo and in vitro analysis of inflammation-induced secondary injury after CNS trauma. J Neurosci 19:8182–8198

Fitch MT, Silver J (2008) CNS injury, glial scars, and inflammation: inhibitory extracellular matrices and regeneration failure. Exp Neurol 209:294–301

Fleisch VC, Fraser B, Allison WT (2011) Investigating regeneration and functional integration of CNS neurons: lessons from zebrafish genetics and other fish species. Biochim Biophys Acta 1812:364–380

Ganz J, Kaslin J, Hochmann S, Freudenreich D, Brand M (2010) Heterogeneity and Fgf dependence of adult neural progenitors in the zebrafish telencephalon. Glia 58:1345–1363

Goldman D, Ding J (2000) Different regulatory elements are necessary for alpha1 tubulin induction during CNS development and regeneration. Neuroreport 11:3859–3863

Gould E, Reeves AJ, Fallah M, Tanapat P, Gross CG, Fuchs E (1999) Hippocampal neurogenesis in adult old world primates. Proc Natl Acad Sci USA 96:5263–5267

Grandel H, Kaslin J, Ganz J, Wenzel I, Brand M (2006) Neural stem cells and neurogenesis in the adult zebrafish brain: origin, proliferation dynamics, migration and cell fate. Dev Biol 295:263–277

Grosche J, Härtig W, Reichenbach A (1995) Expression of glial fibrillary acidic protein (GFAP), glutamine synthetase (GS), and Bcl-2 protooncogene protein by Müller (glial) cells in retinal light damage of rats. Neurosci Lett 185:119–122

Guo Y, Ma L, Cristofanilli M, Hart RP, Hao A, Schachner M (2011) Transcription factor Sox11b is involved in spinal cord regeneration in adult zebrafish. Neuroscience 172:329–341

Hagedorn M, Fernald RD (1992) Retinal growth and cell addition during embryogenesis in the teleost, *Haplochromis burtoni*. J Comp Neurol 321:193–208

Härtig W, Grosche J, Distler C, Grimm D, El-Hifnawi E, Reichenbach A (1995) Alterations of Müller (glial) cells in dystrophic retinae of RCS rats. J Neurocytol 24:507–517

Hayes RL, Jenkins LW, Lyeth BG (1992) Neurotransmitter-mediated mechanisms of traumatic brain injury: acetylcholine and excitatory amino acids. J Neurotrauma 9:S173–S187

Healey EG (1962) Experimental evidence for regeneration following spinal section in the minnow (*Phoxinus phoxinus* L.). Nature 194:395–396

Healey EG (1967) Experimental evidence for the regeneration of nerve fibres controlling colour changes after anterior spinal section in the minnow (*Phoxinus phoxinus* L.). Proc R Soc Lond B 168:57–81

Herculano-Houzel S, Lent R (2005) Isotropic fractionator: a simple, rapid method for the quantification of total cell and neuron numbers in the brain. J Neurosci 25:2518–2521
Hinsch K, Zupanc GKH (2006) Isolation, cultivation, and differentiation of neural stem cells from adult fish brain. J Neurosci Methods 158:75–88
Hinsch K, Zupanc GKH (2007) Generation and long-term persistence of new neurons in the adult zebrafish brain: a quantitative analysis. Neuroscience 146:679–696
Hitchcock P, Ochocinska M, Sieh A, Otteson D (2004) Persistent and injury-induced neurogenesis in the vertebrate retina. Prog Retin Eye Res 23:183–194
Hitchcock PF, Lindsey Myhr KJ, Easter SS Jr, Mangione-Smith R, Dwyer Jones D (1992) Local regeneration in the retina of the goldfish. J Neurobiol 23:187–203
Hitchcock PF, Raymond PA (1992) Retinal regeneration. Trends Neurosci 15:103–108
Ho B-K, Alexianu ME, Colom LV, Mohamed AH, Serrano F, Appel SH (1996) Expression of calbindin-D_{28K} in motoneuron hybrid cells after retroviral infection with calbindin-D_{28K} cDNA prevents amyotrophic lateral sclerosis IgG-mediated cytotoxicity. Proc Natl Acad Sci USA 93:6796–6801
Horn KP, Busch SA, Hawthorne AL, van Rooijen N, Silver J (2008) Another barrier to regeneration in the CNS: activated macrophages induce extensive retraction of dystrophic axons through direct physical interactions. J Neurosci 28:9330–9341
Ilieş I, Zupanc MM, Zupanc GKH (2012) Proteome analysis reveals protein candidates involved in early stages of brain regeneration of teleost fish. Neuroscience 219:302–313
Johns PR (1977) Growth of the adult goldfish eye. III. Source of the new retinal cells. J Comp Neurol 176:343–357
Johns PR (1982) Formation of photoreceptors in larval and adult goldfish. J Neurosci 2:178–198
Johns PR, Easter SS Jr (1977) Growth of the adult goldfish eye: II. Increase in retinal cell number. J Comp Neurol 176:331–342
Jopling C, Sleep E, Raya M, Martí M, Raya A, Izpisúa Belmonte JC (2010) Zebrafish heart regeneration occurs by cardiomyocyte dedifferentiation and proliferation. Nature 464:606–609
Julian D, Ennis K, Korenbrot JI (1998) Birth and fate of proliferative cells in the inner nuclear layer of the mature fish retina. J Comp Neurol 394:271–282
Kaplan MS, Bell DH (1984) Mitotic neuroblasts in the 9-day-old and 11-month-old rodent hippocampus. J Neurosci 4:1429–1441
Kaslin J, Ganz J, Brand M (2008) Proliferation, neurogenesis and regeneration in the non-mammalian vertebrate brain. Philos Trans R Soc Lond B Biol Sci 363:101–122
Kaslin J, Ganz J, Geffarth M, Grandel H, Hans S, Brand M (2009) Stem cells in the adult zebrafish cerebellum: initiation and maintenance of a novel stem cell niche. J Neurosci 29:6142–6153
Kassen SC, Ramanan V, Montgomery JE, C TB, Liu CG, Vihtelic TS, Hyde DR (2007) Time course analysis of gene expression during light-induced photoreceptor cell death and regeneration in *albino* zebrafish. Dev Neurobiol 67:1009–1031
Kassen SC, Thummel R, Campochiaro LA, Harding MJ, Bennett NA, Hyde DR (2009) CNTF induces photoreceptor neuroprotection and Müller glial cell proliferation through two different signaling pathways in the adult zebrafish retina. Exp Eye Res 88:1051–1064
Kästner R, Wolburg H (1982) Functional regeneration of the visual system in teleosts: comparative investigations after optic nerve crush and damage of the retina. Z Naturforsch 37c:1274–1280
Katayama Y, Becker DP, Tamura T, Hovda DA (1990) Massive increases in extracellular potassium and the indiscriminate release of glutamate following concussive brain injury. J Neurosurg 73:889–900
Kerr JFR, Gobé GC, Winterford CM, Harmon BV (1995) Anatomical methods in cell death. In: Schwartz LM, Osborne BA (eds) Cell death. Academic Press, San Diego, pp 1–27
Kikuchi K, Holdway J, Werdich A, Anderson R, Fang Y, Egnaczyk G, Evans T, Macrae CS, DY, Poss K (2010) Primary contribution to zebrafish heart regeneration by *gata4*$^{+}$ cardiomyocytes. Nature 464:601–605

Kirsche W (1960) Zur Frage der Regeneration des Mittelhirnes der Teleostei. Verh Anat Gesell 56:259–270

Kirsche W, Kirsche K (1961) Experimentelle Untersuchungen zur Frage der Regeneration und Funktion des Tectum opticum von *Carassius carassius* L. Z Mikrosk Anat Forsch 67:140–182

Kishimoto N, Shimizu K, Sawamoto K (2012) Neuronal regeneration in a zebrafish model of adult brain injury. Dis Models Mech 5:200–209

Knopf F, Hammond C, Chekuru A, Kurth T, Hans S, Weber CW, Mahatma G, Fisher S, Brand M, Schulte-Merker S, Weidinger G (2011) Bone regenerates via dedifferentiation of osteoblasts in the zebrafish fin. Dev Cell 20:713–724

Kondo T, Raff M (2000) Oligodendrocyte precursor cells reprogrammed to become multipotential CNS stem cells. Science 289:1754–1757

Koppányi T, Weiss P (1922) Funktionelle Regeneration des Rückenmarkes bei Anamniern. Anz Akad Wiss Wien, Math-Naturw Kl 59:206

Koriyama Y, Homma K, Kato S (2006) Activation of cell survival signals in the goldfish retinal ganglion cells after optic nerve injury. Adv Exp Med Biol 572:333–337

Koriyama Y, Yasuda R, Homma K, Mawatari K, Nagashima M, Sugitani K, Matsukawa T, Kato S (2009) Nitric oxide-cGMP signaling regulates axonal elongation during optic nerve regeneration in the goldfish in vitro and in vivo. J Neurochem 110:890–901

Kornack DR, Rakic P (1999) Continuation of neurogenesis in the hippocampus of the adult macaque monkey. Proc Natl Acad Sci USA 96:5768–5773

Koumans JTM, Akster HA (1995) Myogenic cells in development and growth of fish. Comp Biochem Physiol 110A:3–20

Kragl M, Knapp D, Nacu E, Khattak S, Maden M, Epperlein HH, Tanaka EM (2009) Cells keep a memory of their tissue origin during axolotl limb regeneration. Nature 460:60–65

Kranz D, Richter W (1970a) Autoradiographische Untersuchungen über die Lokalisation der Matrixzonen des Diencephalons von juvenilen und adulten *Lebistes reticulatus* (Teleostei). Z mikrosk-anat Forsch 82:42–66

Kranz D, Richter W (1970b) Autoradiographische Untersuchungen zur DNS-Synthese im Cerebellum und in der Medulla oblongata von Teleostiern verschiedenen Lebensalters. Z mikrosk-anat Forsch 82:264–292

Kroehne V, Freudenreich D, Hans S, Kaslin J, Brand M (2011) Regeneration of the adult zebrafish brain from neurogenic radial glia-type progenitors. Development 138:4831–4841

Kusik BW, Hammond DR, Udvadia AJ (2010) Transcriptional regulatory regions of *gap43* needed in developing and regenerating retinal ganglion cells. Dev Dyn 239:482–495

Lanners HN, Grafstein B (1980) Early stages of axonal regeneration in the goldfish optic tract: an electron microscopic study. J Neurocytol 9:733–751

Lau A, Tymianski M (2010) Glutamate receptors, neurotoxicity and neurodegeneration. Pflügers Arch 460:525–542

Lewis GP, Erickson PA, Guérin CJ, Anderson DH, Fisher SK (1989) Changes in the expression of specific Müller cell proteins during long-term retinal detachment. Exp Eye Res 49:93–111

Lewis GP, Guérin CJ, Anderson DH, Matsumoto B, Fisher SK (1994) Rapid changes in the expression of glial cell proteins caused by experimental retinal detachment. Am J Ophthalmol 118:368–376

Lindsey AE, Powers MK (2007) Visual behavior of adult goldfish with regenerating retina. Vis Neurosci 24:247–255

Lindvall O, Björklund A (2004) Cell replacement therapy: helping the brain to repair itself. NeuroRx 1:379–381

Liou AKF, Clark RS, Henshall DC, Yin X-M, Chen J (2003) To die or not to die for neurons in ischemia, traumatic brain injury and epilepsy: a review on the stress-activated signaling pathways and apoptotic pathways. Prog Neurobiol 69:103–142

Lois C, Alvarez-Buylla A (1994) Long-distance neuronal migration in the adult mammalian brain. Science 264:1145–1148

Lois C, Garcia-Verdugo JM, Alvarez-Buylla A (1996) Chain migration of neuronal precursors. Science 271:978–981

Lundberg JG, Cox-Fernandes C, Albert JS, Garcia M (1996) *Magosternarchus*, a new genus with two new species of electric fishes (Gymnotiformes: Apteronotidae) from the Amazon River Basin, South America. Copeia 1996:657–670

Luskin MB (1993) Restricted proliferation and migration of postnatally generated neurons from the forebrain subventricular zone. Neuron 11:173–189

Ma L, Yu YM, Guo Y, Hart RP, Schachner M (2012) Cysteine- and glycine-rich protein 1a is involved in spinal cord regeneration in adult zebrafish. Eur J Neurosci 35:353–365

Mago-Leccia F, Lundberg JG, Baskin JN (1985) Systematics of the South American freshwater fish genus *Adontosternarchus* (Gymnotiformes, Apteronotidae). Contrib Sci (Los Angel, Calif) 358:1–19

Maier W, Wolburg H (1979) Regeneration of the goldfish retina after exposure to different doses of ouabain. Cell Tissue Res 202:99–118

Mansour-Robaey S, Pinganaud G (1990) Quantitative and morphological study of cell proliferation during morphogenesis in the trout visual system. J Hirnforsch 31:495–504

Marcus RC, Delaney CL, Easter SS (1999) Neurogenesis in the visual system of embryonic and adult zebrafish (*Danio rerio*). Vis Neurosci 16:417–424

Maruska KP, Carpenter RE, Fernald RD (2012) Characterization of cell prolfieration throughout the brain of the African cichlid fish *Astatotilapia burtoni* and its regulation by social status. J Comp Neurol 520:3471–3491

Matsukawa T, Arai K, Koriyama Y, Liu Z, Kato S (2004) Axonal regeneration of fish optic nerve after injury. Biol Pharm Bull 27:445–451

Mattson MP, Rychlik B, Chu C, Christakos S (1991) Evidence for calcium-reducing and excito-protective roles for the calcium-binding protein calbindin-D_{28k} in cultured hippocampal neurons. Neuron 6:41–51

McCurley AT, Callard GV (2010) Time course analysis of gene expression patterns in zebrafish eye during optic nerve regeneration. J Exp Neurosci 2010:17–33

Mensinger AF, Powers MK (1999) Visual function in regenerating teleost retina following cytotoxic lesioning. Vis Neurosci 16:241–251

Mensinger AF, Powers MK (2007) Visual function in regenerating teleost retina following surgical lesioning. Vis Neurosci 24:299–307

Meyer RL (1978) Evidence from thymidine labeling for continuing growth of retina and tectum in juvenile goldfish. Exp Neurol 59:99–111

Mizuno TA, Ohtsuka T (2008) Quantitative analysis of activated caspase-3-positive cells and a circadian cycle of programmed cell death in the adult teleost retina. Neurosci Lett 446:16–19

Mizuno TA, Ohtsuka T (2009) Quantitative study of apoptotic cells in the goldfish retina. Zool Sci 26:157–162

Monje ML, Phillips R, Sapolsky R (2001) Calbindin overexpression buffers hippocampal cultures from the energetic impairments caused by glutamate. Brain Res 911:37–42

National Spinal Cord Injury Statistical Center (NSCISC) (2012) Facts and figures at a glance. Updated 2012. Available at www.uscisc.uab.edu. Accessed on Aug 29, 2012

Negishi K, Sugawara K, Shinagawa S, Teranishi T, Kuo CH, Takasaki Y (1991) Induction of immunoreactive proliferating cell nuclear antigen (PCNA) in goldfish retina following intravitreal injection with tunicamycin. Dev Brain Res 63:71–83

Negishi K, Teranishi T, Kato S, Nakamura Y (1988) Immunohistochemical and autoradiographic studies on retinal regeneration in teleost fish. Neurosci Res Suppl 8:S43–S57

Nguyen V, Deschet K, Henrich T, Godet E, Joly JS, Wittbrodt J, Chourrout D, Bourrat F (1999) Morphogenesis of the optic tectum in the medaka (*Oryzias latipes*): a morphological and molecular study, with special emphasis on cell proliferation. J Comp Neurol 413:385–404

Oliver CN, Starke-Reed PE, Stadtman ER, Liu GJ, Carney JM, Floyd RA (1990) Oxidative damage to brain proteins, loss of glutamine synthetase activity, and production of free radicals during ischemia/reperfusion-induced injury to gerbil brain. Proc Natl Acad Sci USA 87:5144–5147

Olney JW (1969) Brain lesions, obesity, and other disturbances in mice treated with monosodium glutamate. Science 164:719–721

Ott R, Zupanc GKH, Horschke I (1997) Long-term survival of postembryonically born cells in the cerebellum of gymnotiform fish, *Apteronotus leptorhynchus*. Neurosci Lett 221:185–188
Otteson DC, Hitchcock PF (2003) Stem cells in the teleost retina: persistent neurogenesis and injury-induced regeneration. Vision Res 43:927–936
Palmer AM, Marion DW, Botscheller ML, Bowen DM, DeKosky ST (1994) Increased transmitter amino acid concentration in human ventricular CSF after brain trauma. Neuroreport 6:153–156
Palmer TD, Markakis EA, Willhoite AR, Safar F, Gage FH (1999) Fibroblast growth factor-2 activates a latent neurogenic program in neural stem cells from diverse regions of the adult CNS. J Neurosci 19:8487–8497
Park HC, Shin J, Roberts RK, Appel B (2007) An olig2 reporter gene marks oligodendrocyte precursors in the postembryonic spinal cord of zebrafish. Dev Dyn 236:3402–3407
Pellegrini E, Mouriec K, Anglade I, Menuet A, Le Page Y, Gueguen MM, Marmignon MH, Brion F, Pakdel F, Kah O (2007) Identification of aromatase-positive radial glial cells as progenitor cells in the ventricular layer of the forebrain in zebrafish. J Comp Neurol 501:150–167
Pencea V, Bingaman KD, Freedman LJ, Luskin MB (2001) Neurogenesis in the subventricular zone and rostral migratory stream of the neonatal and adult primate forebrain. Exp Neurol 172:1–16
Phillips RG, Meier TJ, Giuli LC, McLaughlin JR, Ho DY, Sapolsky RM (1999) Calbindin D_{28K} gene transfer via herpes simplex virus amplicon vector decreases hippocampal damage in vivo following neurotoxic insults. J Neurochem 73:1200–1205
Powell C, Elsaeidi F, Goldman D (2012) Injury-dependent Müller glia and ganglion cell reprogramming during tissue regeneration requires Apobec2a and Apobec2b. J Neurosci 32:1096–1109
Qin Z, Barthel LK, Raymond PA (2009) Genetic evidence for shared mechanisms of epimorphic regeneration in zebrafish. Proc Natl Acad Sci USA 106:9310–9315
Raymond PA, Barthel LK, Bernardos RL, Perkowski JJ (2006) Molecular characterization of retinal stem cells and their niches in adult zebrafish. BMC Dev Biol 6:36
Raymond PA, Easter SS Jr (1983) Postembryonic growth of the optic tectum in goldfish. I. Location of germinal cells and numbers of neurons produced. J Neurosci 3:1077–1091
Raymond PA, Reifler MJ, Rivlin PK (1988) Regeneration of goldfish retina: rod precursors are a likely source of regenerated cells. J Neurobiol 19:431–463
Reier PJ, Stensaas LJ, Guth L (1983) The astrocytic scar as an impediment to regeneration in the central nervous system. In: Kao CC, Bunge RP, Reier PJ (eds) Spinal cord reconstruction. Raven Press, New York, pp 163–195
Reimer MM, Kuscha V, Wyatt C, Sörensen I, Frank RE, Knüwer M, Becker T, Becker CG (2009) Sonic hedgehog is a polarized signal for motor neuron regeneration in adult zebrafish. J Neurosci 29:15073–15082
Reimer MM, Sörensen I, Kuscha V, Frank RE, Liu C, Becker CG, Becker T (2008) Motor neuron regeneration in adult zebrafish. J Neurosci 28:8510–8516
Reynolds BA, Weiss S (1992) Generation of neurons and astrocytes from isolated cells of the adult mammalian central nervous system. Science 255:1707–1710
Richardson RM, Singh A, Sun D, Fillmore HL, Dietrich DW 3rd, Bullock MR (2010) Stem cell biology in traumatic brain injury: effects of injury and strategies for repair. J Neurosurg 112:1125–1138
Richter W (1968) Regeneration im Tectum opticum bei adulten *Lebistes reticulatus* (Peters 1859) (Poeciliidae, Cyprinodontiformes, Teleostei). J Hirnforsch 10:173–186
Richter W, Kranz D (1970a) Autoradiographische Untersuchungen über die Abhängigkeit des ^{3}H-Thymidin-Index vom Lebensalter in den Matrixzonen des Telencephalons von *Lebistes reticulatus* (Teleostei). Z mikrosk-anat Forsch 81:530–554
Richter W, Kranz D (1970b) Die Abhängigkeit der DNS-Synthese in den Matrixzonen des Mesencephalons vom Lebensalter der Versuchstiere (*Lebistes reticulatus*—Teleostei): Autoradiographische Untersuchungen. Z mikrosk-anat Forsch 82:76–92

Richter W, Kranz D (1977) Über die Bedeutung der Zellproliferation für die Hirnregeneration bei niederen Vertebraten: Autoradiographische Untersuchungen. Verh Anat Gesell 71:439–445

Rolls A, Shechter R, Schwartz M (2009) The bright side of the glial scar in CNS repair. Nat Rev Neurosci 10:235–241

Rosenzweig S, Raz-Prag D, Nitzan A, Galron R, Paz M, Jeserich G, Neufeld G, Barzilai A, Solomon AS (2010) Sema-3A indirectly disrupts the regeneration process of goldfish optic nerve after controlled injury. Graefes Arch Clin Exp Ophthalmol 248:1423–1435

Rothenaigner I, Krecsmarik M, Hayes JA, Bahn B, Lepier A, Fortin G, Gotz M, Jagasia R, Bally-Cuif L (2011) Clonal analysis by distinct viral vectors identifies bona fide neural stem cells in the adult zebrafish telencephalon and characterizes their division properties and fate. Development 138:1459–1469

Rowe RWD, Goldspink G (1969) Muscle fibre growth in five different muscles in both sexes of mice. J Anat 104:519–530

Rowlerson A, Veggetti A (2001) Cellular mechanisms of post-embryonic muscle growth in aquaculture species. In: Johnston IA (ed) Muscle development and growth. Academic Press, San Diego, pp 103–140

Sanai N, Tramontin AD, Quinones-Hinojosa A, Barbaro NM, Gupta N, Kunwar S, Lawton MT, McDermott MW, Parsa AT, García-Verdugo JM, Berger MS, Alvarez-Buylla A (2004) Unique astrocyte ribbon in adult human brain contains neural stem cells but lacks chain migration. Nature 427:740–744

Saul KE, Koke JR, García DM (2010) Activating transcription factor 3 (ATF3) expression in the neural retina and optic nerve of zebrafish during optic nerve regeneration. Comp Biochem Physiol A 155:172–182

Schweitzer J, Becker T, Becker CG, Schachner M (2003) Expression of protein zero is increased in lesioned axon pathways in the central nervous system of adult zebrafish. Glia 41:301–317

Schweitzer J, Gimnopoulos D, Lieberoth BC, Pogoda HM, Feldner J, Ebert A, Schachner M, Becker T, Becker CG (2007) Contactin1a expression is associated with oligodendrocyte differentiation and axonal regeneration in the central nervous system of zebrafish. Mol Cell Neurosci 35:194–207

Senut MC, Gulati-Leekha A, Goldman D (2004) An element in the α1-tubulin promoter is necessary for retinal expression during optic nerve regeneration but not after eye injury in the adult zebrafish. J Neurosci 24:7663–7673

Seri B, García-Verdugo JM, McEwen BS, Alvarez-Buylla A (2001) Astrocytes give rise to new neurons in the adult mammalian hippocampus. J Neurosci 21:7153–7160

Silver J, Miller JH (2004) Regeneration beyond the glial scar. Nat Rev Neurosci 5:146–156

Sîrbulescu RF, Ilieş I, Zupanc GKH (2009) Structural and functional regeneration after spinal cord injury in the weakly electric teleost fish, *Apteronotus leptorhynchus*. J Comp Physiol A 195:699–714

Sîrbulescu RF, Zupanc GKH (2009) Dynamics of caspase-3-mediated apoptosis during spinal cord regeneration in the teleost fish, *Apteronotus leptorhynchus*. Brain Res 1304:14–25

Sîrbulescu RF, Zupanc GKH (2010a) Inhibition of caspase-3-mediated apoptosis improves spinal cord repair in a regeneration-competent vertebrate system. Neuroscience 171:599–612

Sîrbulescu RF, Zupanc GKH (2010b) Effect of temperature on spinal cord regeneration in the weakly electric fish, *Apteronotus leptorhynchus*. J Comp Physiol A 196:359–368

Sîrbulescu RF, Zupanc GKH (2011) Spinal cord repair in regeneration-competent vertebrates: adult teleost fish as a model system. Brain Res Rev 67:73–93

Smith CD, Carney JM, Starke-Reed PE, Oliver CN, Stadtman ER, Floyd RA, Markesbery WR (1991) Excess brain protein oxidation and enzyme dysfunction in normal aging and in Alzheimer disease. Proc Natl Acad Sci USA 88:10540–10543

Soares S, von Boxberg Y, Lombard MC, Ravaille-Veron M, Fischer I, Eyer J, Nothias F (2002) Phosphorylated MAP1B is induced in central sprouting of primary afferents in response to peripheral injury but not in response to rhizotomy. Eur J Neurosci 16:593–606

Soutschek J, Zupanc GKH (1995) Apoptosis as a regulator of cell proliferation in the central posterior/prepacemaker nucleus of adult gymnotiform fish, *Apteronotus leptorhynchus*. Neurosci Lett 202:133–136

Soutschek J, Zupanc GKH (1996) Apoptosis in the cerebellum of adult teleost fish, *Apteronotus leptorhynchus*. Dev Brain Res 97:279–286

Sperry RW (1948) Patterning of central synapses in regeneration of the optic nerve in teleosts. Physiol Zool 21:351–361

Stenkamp DL (2007) Neurogenesis in the fish retina. Int Rev Cytol 259:173–224

Stocum DL (2006) Regenerative biology and medicine. Academic Press, Burlington

Stuermer CAO, Bastmeyer M, Bahr M, Strobel G, Paschke K (1992) Trying to understand axonal regeneration in the CNS of fish. J Neurobiol 23:537–550

Stuermer CAO, Easter SS Jr (1984) A comparison of the normal and regenerated retinotectal pathways of goldfish. J Comp Neurol 223:57–76

Stuermer CAO, Niepenberg A, Wolburg H (1985) Aberrant axonal paths in regenerated goldfish retina and tectum opticum following intraocular injection of ouabain. Neurosci Lett 58: 333–338

Sugitani K, Matsukawa T, Koriyama Y, Shintani T, Nakamura T, Noda M, Kato S (2006) Upregulation of retinal transglutaminase during the axonal elongation stage of goldfish optic nerve regeneration. Neuroscience 142:1081–1092

Takeda A, Nakano M, Goris RC, Funakoshi K (2008) Adult neurogenesis with 5-HT expression in lesioned goldfish spinal cord. Neuroscience 151:1132–1141

Teles MC, Sîrbulescu RF, Wellbrock UM, Oliveira RF, Zupanc GKH (2012) Adult neurogenesis in the brain of the Mozambique tilapia, *Oreochromis mosambicus*. J Comp Physiol A 198:427–449

Thummel R, Kassen SC, Enright JM, Nelson CM, Montgomery JE, Hyde DR (2008) Characterization of Müller glia and neuronal progenitors during adult zebrafish retinal regeneration. Exp Eye Res 87:433–444

Tozzini ET, Baumgart M, Battistoni G, Cellerino A (2012) Adult neurogenesis in the short-lived teleost *Nothobranchius furzeri*: localization of neurogenic niches, molecular characterization and effects of aging. Aging Cell 11:241–251

Tuge H, Hanzawa S (1935) Physiology of the spinal fish, with special reference to the postural mechanism. Sci Rep Tohoku Imp Univ, Biol 10:589–606

Tuge H, Hanzawa S (1937) Physiological and morphological regeneration of the sectioned spinal cord in adult teleosts. J Comp Neurol 67:343–365

Vajda FJ (2002) Neuroprotection and neurodegenerative disease. J Clin Neurosci 9:4–8

van Raamsdonk W, Maslam S, de Jong DH, Smit-Onel MJ, Velzing E (1998) Long term effects of spinal cord transection in zebrafish: swimming performances, and metabolic properties of the neuromuscular system. Acta Histochem 100:117–131

Veldman MB, Bemben MA, Goldman D (2010) *Tuba1a* gene expression is regulated by KLF6/7 and is necessary for CNS development and regeneration in zebrafish. Mol Cell Neurosci 43:370–383

Veldman MB, Bemben MA, Thompson RC, Goldman D (2007) Gene expression analysis of zebrafish retinal ganglion cells during optic nerve regeneration identifies KLF6a and KLF7a as important regulators of axon regeneration. Dev Biol 312:596–612

Vihtelic TS, Hyde DR (2000) Light-induced rod and cone cell death and regeneration in the adult *albino* zebrafish (*Danio rerio*) retina. J Neurobiol 44:289–307

Waxman SG, Anderson MJ (1980) Regeneration of spinal electrocyte fibers in *Sternarchus albifrons*: development of axon-Schwann cell relationships and nodes of Ranvier. Cell Tissue Res 208:343–352

Waxman SG, Anderson MJ (1986) Regeneration of central nervous system structures: *Apteronotus* spinal cord as a model system. In: Bullock TH, Heiligenberg W (eds) Electroreception. Wiley, New York, pp 183–208

Waxman SG, Pappas GD, Bennett MVL (1972) Morphological correlates of functional differentiation of nodes of Ranvier along single fibers in the neurogenic electric organ of the knife fish *Sternarchus*. J Cell Biol 53:210–224

Weatherley AH, Gill HS (1985) Dynamics of increase in muscle fibres in fishes in relation to size and growth. Experientia 41:353–354

Weber JT (2004) Calcium homeostasis following traumatic neuronal injury. Curr Neurovasc Res 1:151–171

Weber JT (2012) Altered calcium signaling following traumatic brain injury. Front Pharmacol 3:60

Weiss S, Dunne C, Hewson J, Wohl C, Wheatly M, Peterson AC, Reynolds BA (1996) Multipotent CNS stem cells are present in the adult mammalian spinal cord and ventricular neuroaxis. J Neurosci 16:7599–7609

West-Eberhard MJ (2003) Developmental Plasticity and Evolution. Oxford University Press, Oxford

West-Eberhard MJ (2005) Developmental plasticity and the origin of species differences. Proc Natl Acad Sci USA 102:6543–6549

Williams RW (2000) Mapping genes that modulate brain development: a quantitative genetic approach. In: Goffinet AF, Rakic P (eds) Mouse brain development. Springer, New York, pp 21–49

Wu DM, Schneiderman T, Burgett J, Gokhale P, Barthel L, Raymond PA (2001) Cones regenerate from retinal stem cells sequestered in the inner nuclear layer of adult goldfish retina. Invest Ophthalmol Vis Sci 42:2115–2124

Wullimann MF, Puelles L (1999) Postembryonic neural proliferation in the zebrafish forebrain and its relationship to prosomeric domains. Anat Embryol (Berl) 199:329–348

Yiu G, He Z (2006) Glial inhibition of CNS axon regeneration. Nat Rev Neurosci 7:617–627

Young W (1992) Role of calcium in central nervous system injuries. J Neurotrauma 9:S9–S25

Yu YM, Cristofanilli M, Valiveti A, Ma L, Yoo M, Morellini F, Schachner M (2011a) The extracellular matrix glycoprotein tenascin-C promotes locomotor recovery after spinal cord injury in adult zebrafish. Neuroscience 183:238–250

Yu YM, Gibbs KM, Davila J, Campbell N, Sung S, Todorova TI, Otsuka S, Sabaawy HE, Hart RP, Schachner M (2011b) MicroRNA miR-133b is essential for functional recovery after spinal cord injury in adult zebrafish. Eur J Neurosci 33:1587–1597

Yurco P, Cameron DA (2005) Responses of Müller glia to retinal injury in adult zebrafish. Vision Res 45:991–1002

Zakon HH (1984) Postembryonic changes in the peripheral electrosensory system of a weakly electric fish: addition of receptor organs with age. J Comp Neurol 228:557–570

Zhang Z, Krebs CJ, Guth L (1997) Experimental analysis of progressive necrosis after spinal cord trauma in the rat: etiological role of the inflammatory response. Exp Neurol 143:141–152

Zikopoulos B, Kentouri M, Dermon CR (2000) Proliferation zones in the adult brain of a sequential hermaphrodite teleost species *(Sparus aurata)*. Brain Behav Evol 56:310–322

Zimmerman AM, Lowery MS (1999) Hyperplastic development and hypertrophic growth of muscle fibers in the white seabass (*Atractoscion nobilis*). J Exp Zool 284:299–308

Zottoli SJ, Freemer MM (2003) Recovery of C-starts, equilibrium and targeted feeding after whole spinal cord crush in the adult goldfish *Carassius auratus*. J Exp Biol 206:3015–3029

Zupanc GKH (1999a) Neurogenesis, cell death and regeneration in the adult gymnotiform brain. J Exp Biol 202:1435–1446

Zupanc GKH (1999b) Up-regulation of somatostatin after lesions in the cerebellum of the teleost fish *Apteronotus leptorhynchus*. Neurosci Lett 268:135–138

Zupanc GKH (2001) Adult neurogenesis and neuronal regeneration in the central nervous system of teleost fish. Brain Behav Evol 58:250–275

Zupanc GKH (2006a) Neurogenesis and neuronal regeneration in the adult fish brain. J Comp Physiol A 192:649–670

Zupanc GKH (2006b) Adult neurogenesis and neuronal regeneration in the teleost fish brain: implications for the evolution of a primitive vertebrate trait. In: Bullock TH, Rubenstein LR

(eds) The evolution of nervous systems in non-mammalian vertebrates. Academic Press, Oxford, pp 485–520
Zupanc GKH (2008a) Adult neurogenesis and neuronal regeneration in the brain of teleost fish. J Physiol Paris 102:357–373
Zupanc GKH (2008b) Adult neurogenesis in teleost fish. In: Gage FH, Kempermann G, Song H (eds) Adult neurogenesis. Cold Spring Harbor Laboratory Press, New York, pp 571–592
Zupanc GKH (2009) Towards brain repair: insights from teleost fish. Semin Cell Dev Biol 20:683–690
Zupanc GKH (2011) Adult neurogenesis in teleost fish. In: Seki T, Sawamoto K, Parent JM, Alvarez-Buylla A (eds) Neurogenesis in the adult brain, vol 1, Springer, Tokyo, pp 137–168
Zupanc GKH, Clint SC (2003) Potential role of radial glia in adult neurogenesis of teleost fish. Glia 43:77–86
Zupanc GKH, Clint SC, Takimoto N, Hughes ATL, Wellbrock UM, Meissner D (2003) Spatio-temporal distribution of microglia/macrophages during regeneration in the cerebellum of adult teleost fish, *Apteronotus leptorhynchus*: a quantitative analysis. Brain Behav Evol 62:31–42
Zupanc GKH, Hinsch K, Gage FH (2005) Proliferation, migration, neuronal differentiation, and long-term survival of new cells in the adult zebrafish brain. J Comp Neurol 488:290–319
Zupanc GKH, Horschke I (1995) Proliferation zones in the brain of adult gymnotiform fish: a quantitative mapping study. J Comp Neurol 353:213–233
Zupanc GKH, Horschke I, Ott R, Rascher GB (1996) Postembryonic development of the cerebellum in gymnotiform fish. J Comp Neurol 370:443–464
Zupanc GKH, Kompass KS, Horschke I, Ott R, Schwarz H (1998) Apoptosis after injuries in the cerebellum of adult teleost fish. Exp Neurol 152:221–230
Zupanc GKH, Ott R (1999) Cell proliferation after lesions in the cerebellum of adult teleost fish: time course, origin, and type of new cells produced. Exp Neurol 160:78–87
Zupanc GKH, Sîrbulescu RF (2011) Adult neurogenesis and neuronal regeneration in the central nervous system of teleost fish. Eur J Neurosci 34:917–929
Zupanc GKH, Sîrbulescu RF, Ilieş I (2012) Radial glia in the cerebellum of adult teleost fish: implications for the guidance of migrating new neurons. Neuroscience 210:416–430
Zupanc GKH, Zupanc MM (2006a) New neurons for the injured brain: mechanisms of neuronal regeneration in adult teleost fish. Regen Med 1:207–216
Zupanc MM, Wellbrock UM, Zupanc GKH (2006) Proteome analysis identifies novel protein candidates involved in regeneration of the cerebellum of teleost fish. Proteomics 6:677–696
Zupanc MM, Zupanc GKH (2006b) Upregulation of calbindin-D_{28k} expression during regeneration in the adult fish cerebellum. Brain Res 1095:26–34

Part V
Epigenetics and Cell Cycle in Regeneration

Epigenetics and Regeneration

Nobuyasu Maki and Hironobu Kimura

Abstract During newt lens regeneration a unique transdifferentiation event occurs. In this process, dorsal iris pigmented epithelial cells transdifferentiate into lens cells. This system should provide a new insight into cellular plasticity in basic and applied research. Recently, a series of approaches to study epigenetic reprogramming during transdifferentiation have been performed. In this review, we introduce the regulation of dynamic regulation of core-histone modifications and the emergence of an oocyte-type linker histone during transdifferentiation. Finally, we show supporting evidence that there are common strategies of reprogramming between newt somatic cell in transdifferentiation and oocytes after somatic cell nuclear transfer.

Contents

N. Maki (✉) · H. Kimura
Institute of Protein Research,
Osaka University, 3-2 Yamadaoka, Suita-Shi, Osaka 565-0871, Japan
e-mail: makinobu@protein.osaka-u.ac.jp

N. Maki
PRESTO, Japan Science and Technology Agency, 4-1-8 Honcho, Kawaguchi,
Saitama 332-0012, Japan

Current Topics in Microbiology and Immunology (2013) 367: 237–252
DOI: 10.1007/82_2012_293

© Springer-Verlag Berlin Heidelberg 2012
Published Online: 30 November 2012

1 Introduction

The developmental program is controlled by genetic and epigenetic regulation. Epigenetic regulation provides the diversity of cell differentiation in development. After fertilization, the zygote differentiates into diverse cells depending on interactions between their cell lineage and differentiation signals. Although differentiated cells have identical DNA sequences, they exhibit different profiles of gene expression. Epigenetics involves heritable alterations of gene expression without changes in DNA sequence, and contributes to the diversity of gene expression and memory of cell lineage. One major mechanism of epigenetics is the chemical modifications to the nucleosome, including DNA methylation and histone modification.

It is clear that epigenetics plays a major role in development, the field of epigenetics research during regeneration has just started. There are pioneering studies in DNA methylation during *Xenopus* limb regeneration (Yakushiji et al. 2007) and during zebrafish pancreatic β-cell and liver regeneration (Anderson et al. 2009; Sadler et al. 2007). In this review, we focus on two mechanisms of epigenetic changes, core- and linker-histone regulation, during newt lens transdifferentiation.

2 Newt Lens Transdifferentiation

Urodele amphibians have a strong regenerative ability. In particular, the newt can regenerate almost all tissues in its body including lens, retina, limbs, jaw, tail, small intestine, heart, and brain. Understanding the mechanism of amphibian regeneration will provide crucial information for both basic and applied biology. Especially, the understanding of unique events in regenerative animals will be important. In this chapter, we introduce a unique phenomenon of transdifferentiation identified in newt lens regeneration.

2.1 A Key Biological Event

In newt lens regeneration, dorsal iris pigmented epithelial cells (PECs) transdifferentiate into lens cells (Fig 1a). Newt lens regeneration can be divided into three major steps. The initial step after lentectomy (from day 0 to day 3) involves

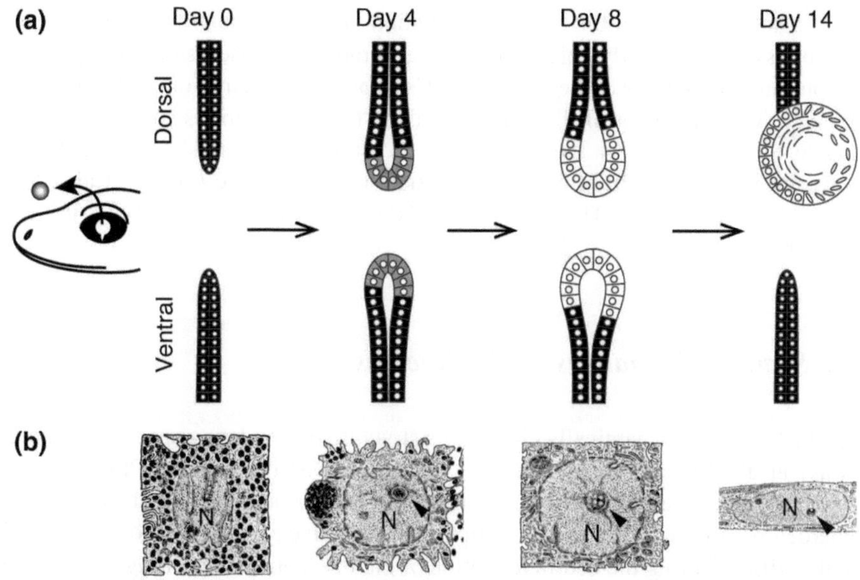

Fig. 1 Newt lens regeneration. **a** Dorsal PECs transdifferentiate into lens cells. About 4 days after lentectomy, PECs begin to re-enter the cell cycle and shed their pigments. PECs change to transparent cells by around day 8. After day 14, lens differentiation occurs from dorsal iris. Although the ventral PECs show depigmentation and cell cycle re-entry, they never regenerate lens. **b** Structural change in the nucleus during lens transdifferentiation. Original PECs have a small and shrunken nucleus. During dedifferentiation, the PEC nucleus swells and its nucleoli become huge. N, nucleus; arrow head, nucleolus. Illustration of nucleus is reproduced from (Eguchi 1980) with permission

molecular and cellular events that precede PECs re-entering the cell cycle. The next step (days 4–12) is the point at which PECs start cell cycle re-entry 4–5 days after lentectomy. At this time, PECs start shedding their pigment granules. PECs continue depigmentation and proliferation and finally change to transparent cells by day 8. This process, where PECs lose their original tissue characteristics, is called dedifferentiation. On days 10–12 depigmented PECs form a vesicle but do not express lens-specific markers.

In the last step, lens differentiation begins. After day 14, posterior cells of the dorsal vesicle elongate and start to express lens markers. The vesicle grows and differentiates into lens, which is of a considerable size and normal morphology by day 20. Embryologically, lens cells and PEC are derived from surface and neural ectoderm origin, respectively (Coulombre 1965), suggesting that lens transdifferentiation is accomplished by a different mechanism from that seen in embryogenesis. The transdifferentiation of PECs has been demonstrated in clonal culture experiments. A single PEC dissociated from dorsal iris transdifferentiates into a lentoid body, which expresses lens-specific markers in culture (Abe and Eguchi 1977).

It is note worthy that the transdifferentiation of PEC is one of the best and most suitable systems to study epigenetics in regeneration because its cell lineage is so simple and so obvious. The ventral PECs show depigmentation and cell cycle re-entry. The number of BrdU-positive cells in the ventral iris is comparable with that in the dorsal iris by day 6 (Maki et al. 2007). In contrast to dorsal PECs, however, the ventral PECs never differentiate to lens in vivo (Grogg et al. 2005; Hayashi et al. 2006). Therefore, there is a dorso-ventral selectivity in lens regeneration.

2.2 Structural Changes in the Nucleus

The PEC nucleus dynamically changes its structure during lens transdifferentiation (Maki et al. 2007, 2010b) (Fig. 1b). The nucleus of the original PEC is small (about 10 μm in a diameter) and shrunken in shape and has highly developed heterochromatin, which is the transcriptionally inactive region of Chromatin (Maki et al. 2010b). During the dedifferentiation of the PEC, nuclear swelling occurs, and finally the nucleus changes its shape to become round with its diameter reaching more than 20 μm by around day 10. In parallel with the nuclear swelling, euchromatic regions, transcriptionally active regions, increase dramatically (Maki et al. 2010b). The structure of the nucleoli also changes during the dedifferentiation (Fig. 1b). Although the nucleoli in the original PEC are small, they become huge during the dedifferentiation. After the onset of lens differentiation, the nuclei of the cells become smaller and elongated, and have small nucleoli. Therefore during transdifferentiation, the nucleus dynamically changes its structure in correlation with the cellular state. It is suggested that the nuclear swelling with the enlargement of euchromatin during the dedifferentiation is due to reprogramming of the cellular state from differentiated to a stem cell-like state.

2.3 Gene Expression

2.3.1 Nucleostemin

Nucleostemin (NS), a member of the nucleolar GTPase family, is highly expressed in stem cells, progenitor cells, and most cancer cells (Baddoo et al. 2003; Nikpour et al. 2009; Ohmura et al. 2008; Tsai and McKay 2002). Both knocking down and over-expression of NS reduces cell proliferation in cultured cells. The major function of NS is as a regulator of proliferation in both p53-dependent (Dai et al. 2008; Tsai and McKay 2002) and p53-independent pathways (Beekman et al. 2006; Romanova et al. 2009).

To understand the cellular state during newt dedifferentiation, expression of NS during the early process of lens regeneration has been analyzed. After lens removal, the expression of NS is activated and NS accumulates in nucleoli of dedifferentiated cells (Maki et al. 2007). Importantly, the increase of NS accumulating cells occurs prior to S-phase re-entry, suggesting that the increase of NS accumulating cells is not due to proliferation of pre-existing stem cells but due to changing of the cellular state of PECs during dedifferentiation.

2.3.2 Stem Cell Pluripotency Factors

Embryonic stem (ES) cells are in a pluripotent state which allows them to differentiate into all types of cells in three germ layers (Evans and Kaufman 1981; Martin 1981). Another significant property of the ES cell is an ability to reprogram somatic cells (Cowan et al. 2005; Tada et al. 2001). By electrofusion with ES cells, somatic cell nuclei can be reprogrammed and express ES cell markers such as Oct4, and the hybrid cells contribute to all three primary layers of chimeric embryos, suggesting the existence of reprogramming factors in ES cells. On the basis of this fact, the reprogramming factors, Oct4, Sox2, Klf4, and c-Myc, have been screened and pluripotent stem cells have been induced from fibroblasts by introducing these four factors (Okita et al. 2007; Takahashi et al. 2007; Takahashi and Yamanaka 2006; Wernig et al. 2007; Yu et al. 2007).

Retinal PECs from chicken embryo dedifferentiate and transdifferentiate into lens cells in culture. It has been shown that dedifferentiated cells from the retinal PECs express c-Myc (Agata et al. 1993). During newt lens regeneration, expression of Sox2 at the lens differentiating stage has been reported (Hayashi et al. 2004). To further investigate newt dedifferentiation, expression of the stem cell factors during dedifferentiation of PECs have been examined (Maki et al. 2009). Although Oct4 and Nanog are not expressed, Sox2, Klf4, and c-Myc are expressed in a stage-dependent manner during dedifferentiation of PEC. Sox2 and Klf4 are upregulated at a very early step (day 2). The expression of c-Myc reaches a peak at a later stage (day 8). In addition to NS, the expression of those stem cell factors suggests that dedifferentiated cells have a stem cell-like state.

3 Epigenetics in Newt Lens Transdifferentiation

3.1 Core Histone Modifications

In the nuclei of eukaryotic cells, genomic DNA is highly organized as chromatin. The nucleosome is a basic unit of chromatin, which consists of a histone octamer, a linker histone, and approximately 147 base pairs of DNA wrapped around the histone octamer. The histone octamer consists of an H3–H4 tetramer and two sets of H2A–H2B dimers (Kornberg 1974). Linker histones are bound to the linker DNA which is found between nucleosomes and are responsible for forming higher-order chromatin structure (Fig. 2a). Core-histones, histone H2A, H2B, H3, and H4, are

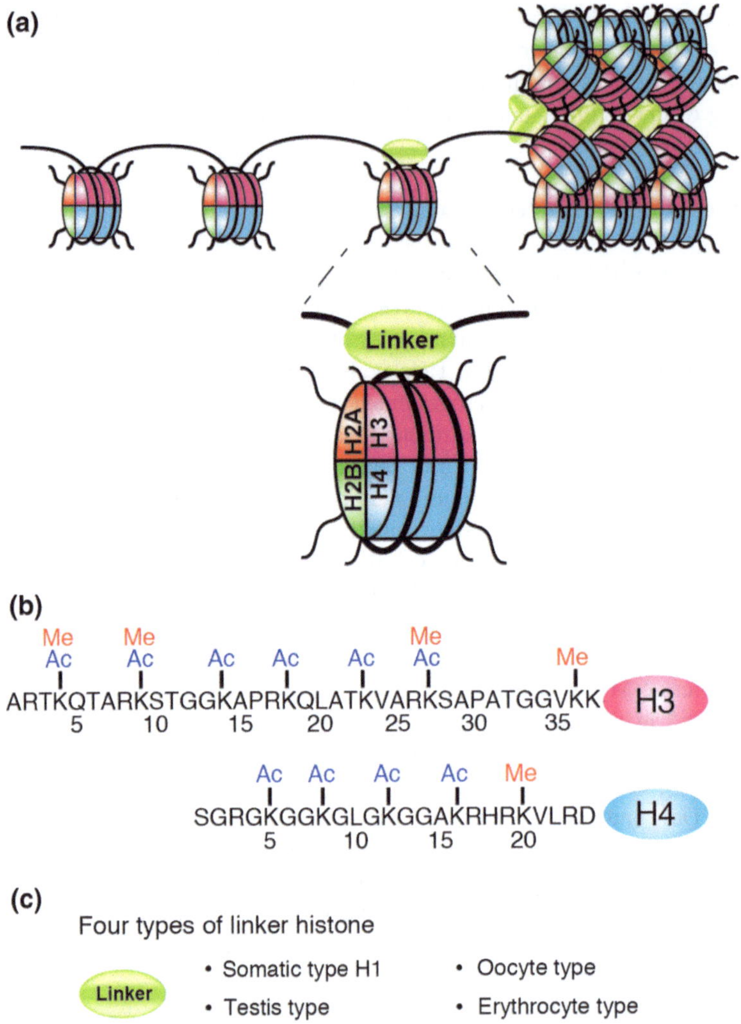

Fig. 2 Chromatin structure and histone modifications. **a** Approximately 147 base pairs of DNA wrap around a histone octamer consisting of an H3-H4 tetramer and two H2A-H2B dimers to form a single nucleosome. The nucleosome is packed with histone H1 to form higher order chromatin structure. **b** Histone tail modifications. Acetylation (Ac) and methylation (Me) of lysine residues at N-terminus of histone H3 and H4 are shown. **c** There are four types of linker-histones identified

small basic proteins consisting of a flexible N-terminus called a "histone tail" and a fold domain that interacts with DNA. The histone tail is subject to several post-translational modifications (Kouzarides 2007; Ruthenburg et al. 2007) (Fig. 2b).

Generally, histone acetylation is a positive mark for gene expression and associated with euchromatin (Shahbazian and Grunstein 2007). Histone acetyltransferases (HATs) transfer an acetyl group to a lysine residue and thus neutralizes the positive charge on lysine, thereby reducing the interaction between DNA and core-histones. As a result, histone acetylation promotes transcriptional activation (Grant et al. 1999; Kuo et al. 1996; Schiltz et al. 1999; Spencer et al. 1997). Contrary to HATs, histone deacetylases (HDACs) remove an acetyl group from lysine and induce transcriptional silencing (Rundlett et al. 1996). The acetylation status of a promoter region, which is accomplished by a balance between HATs and HDACs, regulates gene expression.

Histone methylation is involved in various biological aspects. The comprehensive analysis of methylated histones in the genome reveals that methylated histones are associated with both transcriptionally active and inactive regions (Barski et al. 2007; Bernstein et al. 2006; Guenther et al. 2007; Mendenhall and Bernstein 2008; Mikkelsen et al. 2007; Wang et al. 2008). The lysine residue can be mono-, di-, or tri-methylated and each methylation state is associated with a different effect on gene expression. TriMeH3K4 (tri-methylated histone H3 lysine 4) and TriMeH3K36 are associated with actively transcribed genes (Krogan et al. 2003; Li et al. 2002; Nishioka et al. 2002; Schaft et al. 2003; Wang et al. 2001). In contrast to those active marks, MeH3K9 and MeH3K27 are marks associated with a repressive state (Fischle et al. 2003; Lachner et al. 2003). Mono- and dimethylation on H3K9 are related to facultative heterochromatin, whose formation is developmentally regulated depending on cellular differentiation (Rice et al. 2003). TriMeH3K9 is associated with constitutive heterochromatin such as centromeric heterochromatin (Peters et al. 2003; Rice et al. 2003; Schotta et al. 2004). TriMeH3K27 is associated with facultative heterochromatin. By genome-wide mapping, it has been shown that TriMeH3K27 is associated with more than 1000 silenced genes, including HOX genes, which are repressed for proper embryonic development and cell fate decisions (Bracken et al. 2006).

To understand whether histone modifications are involved in dedifferentiation of PECs and dorsal selectivity of lens differentiation, changes in global histone modification have been analyzed (Maki et al. 2010b). DiMeH3K9 and TriMeH3K9, which are marks for gene repression, are almost constant in both irises during dedifferentiation. However, TriMeH3K27, which is also a mark for gene repression, shows a significant difference between the dorsal and ventral iris during dedifferentiation (Fig. 3). Although not much changes in the dorsal iris, the level of TriMeH3K27 increases in the ventral iris. Because this modification is enriched in the genes which should be repressed for proper development (Bracken et al. 2006), the up-regulation of TriMeH3K27 in the ventral iris suggests its participation in inhibition of lens formation from the ventral iris. TriMeH3K4, AcH3K9, and AcH4 are histone modifications for gene activation. TriMeH3K4 and AcH4 (K5, 8, 12, 16) in the dedifferentiating cell are increased in both irises toward to day 8. In contrast to these modifications, AcH3K9 is decreased during the dedifferentiation in both irises (Fig. 3). Those facts suggest that each histone modification for gene activation is independently regulated during dedifferentiation of

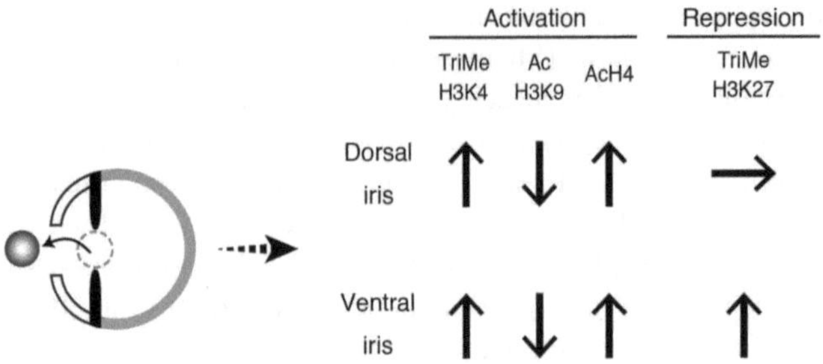

Fig. 3 Summary of changes in global histone modifications of PEC during dedifferentiation in lens regeneration

PEC. The increasing of TriMeK4 and AcH4 could be related to gene activation for cell cycle reentry and reprogramming of cellular fate during dedifferentiation. The decrease of AcH3K9 is an interesting point. It should be noted that Di- and TriMeH3K9 do not change at the same time suggesting that a modification state of H3K9 is not repressive. The combination of such histone modifications, increasing levels of TriMeK4 and AcH4 and decreasing levels of AcH3K9, could be a hallmark of the chromatin state during newt dedifferentiation.

Bivalent histone modification with TriMeH3K27 and TriMeH3K4 is a remarkable feature of the ES cell. The comprehensive analysis of histone modifications shows that a vast majority of genes modified with TriMeH3K27, a repressive mark, are co-modified with TriMeH3K4, an active mark, in ES cells and that the co-modified fraction is enriched in genes that function during development (Azuara et al. 2006; Bernstein et al. 2006; Mikkelsen et al. 2007; Pan et al. 2007; Zhao et al. 2007). The bivalent histone modification is thought to poise genes for later activation, while keep them inactivated (Bernstein et al. 2006). It has been reported that in intact zebrafish developmental regulatory genes are silenced by the bivalent modifications and the silenced genes are activated by loss of TriMeH3K27 modification in the fin regeneration (Stewart et al. 2009). However, during newt lens regeneration, the bivalent modification is not observed. This might be due to a difference in the mode of regeneration between dedifferentiation versus stem cell differentiation. Recently, it has been demonstrated that the zebrafish heart is regenerated by dedifferentiation of cardiomyocytes using a Cre/lox system (Jopling et al. 2010). Thus, histone modifications during dedifferentiation in different regenerative animals could be investigated in the future.

3.2 Oocyte-Type Linker Histone B4

Linker histones are classified into four types, *i.e.*, somatic-, oocyte-, testis-, and erythrocyte-type linker histones, according to their cellular specificity and sequence homology (Fig. 2c). Oocyte-type linker histones have been identified in human (referred as H1oo or H1foo), mouse (H1oo, H1foo), cow (H1foo), newt

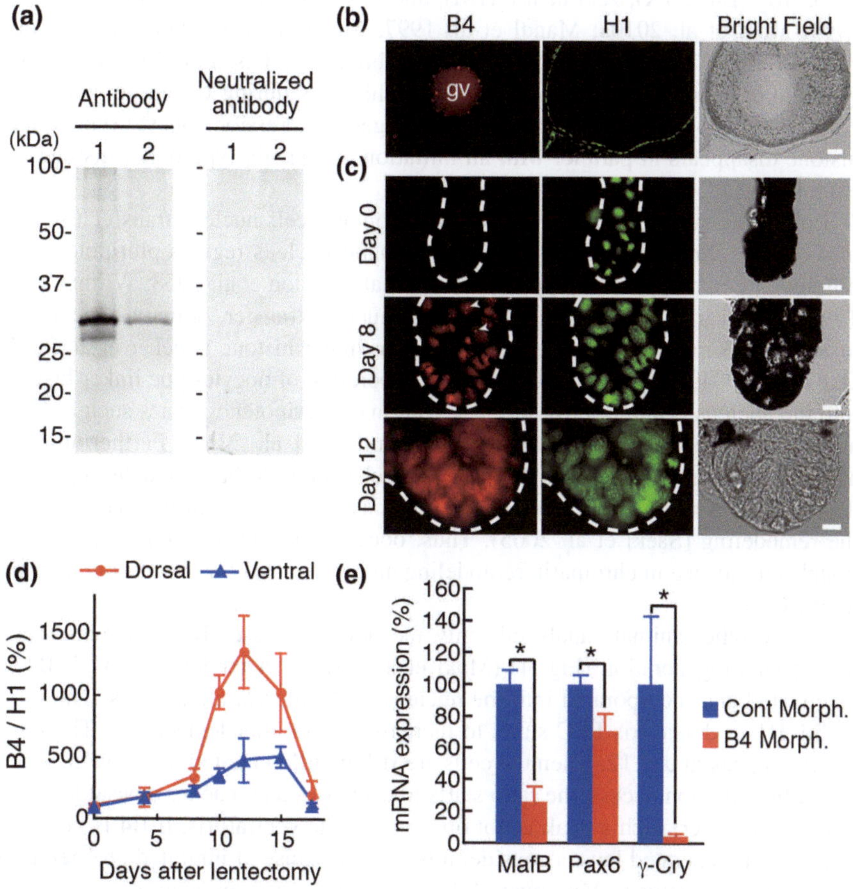

Fig. 4 Oocyte-type linker histone B4 is required for newt lens transdifferentiation. **a** Detection of B4 protein by Western blot analysis using B4 antibody or neutralized antibody with the antigen. Lane 1, ovary; lane 2, dorsal iris 10 days after lentectomy. **b** Immunostaining of ovary using B4 and H1 antibody. Bar, 200 μm. Note that germinal vesicle (GV) was stained by B4 antibody and the nucleus of follicle cells was stained by H1 antibody **c** immunostaining of iris during lens regeneration. Bar, 20 μm. Note that the staining intensities of each panel are not comparative because images were processed to show nuclear distribution of each protein. **d** Changes in the ratio of B4 to histone H1 during lens regeneration. After immunostaining, the intensities of B4 and H1 signals in each nucleus were measured, and the ratio of B4 to histone H1 was calculated. **e** Knocking down of B4 altered gene expression of key genes of lens differentiation. Using a vivo-morpholino technique, the amount of B4 in dorsal iris during lens regeneration decreased by nearly 50 %. In this condition, expression levels of structural and regulatory genes in lens differentiation were analyzed by qPCR. The expression of each gene was normalized with that of ribosomal protein L27. Asterisks indicate a significant difference at $p < 0.0342$, Student's t test

(B4), frog (B4, H1X), zebrafish (H1M), and sea urchin (cs-H1) (Cho and Wolffe 1994; Maki et al. 2010a; Mandl et al. 1997; McGraw et al. 2006; Ohsumi and Katagiri 1991; Tanaka et al. 2001, 2003; Wibrand and Olsen 2002). The oocyte-type linker histones are predominant linker histones during oogenesis and early embryogenesis. After the onset of zygotic gene expression, oocyte-type linker histone disappears in parallel with an initiation of somatic-type linker histone H1 expression.

Epigenetic reprogramming occurs after somatic cell nuclear transfer (SCNT) into oocyte. During reprogramming, the somatic nucleus regains pluripotency to differentiate into all the cell types in the animal (Gurdon et al. 1958; Wilmut et al. 1997; Wakayama et al. 1998). Following nuclear transfer, somatic-type linker histone H1 is rapidly replaced by oocyte-type linker histone (Becker et al. 2005; Gao et al. 2004; Teranishi et al. 2004). Incorporation of oocyte-type linker histone into the nucleus is required for the reactivation of pluripotency genes such as Oct4 and Sox2 in reprogramming after SCNT (Jullien et al. 2010). Furthermore, in assembled chromatin in vitro, B4 allows the chromatin to be remodeled by ATP-dependent chromatin remodeling factor, whereas somatic-type histone H1 prevents the remodeling (Saeki et al. 2005). Thus, oocyte-type linker histone has a functional significance in chromatin remodeling and is required for the reprogramming after SCNT.

Unlike other animals analyzed, only the newt expresses B4 in somatic cells during lens regeneration (Fig. 4) (Maki et al. 2010a). After lens removal, B4 is reactivated and incorporated into the nucleus of dedifferentiating PECs. The ratio of B4–H1 in dorsal iris PEC starts to increase 8 days after lentectomy. The ratio reaches a peak at day 12, when the cells are still undifferentiated. On day 15, when lens differentiation occurs, the ratio starts to decreases and reaches a basal level by day 18. However, such a peak is not observed in the ventral iris. If B4 is knocked down, the regenerated lens is considerably small because of inhibited proliferation and induced apoptosis. Moreover, B4 knockdown represses gene expression of pax6 and MafB, transcriptional factors in lens differentiation, and almost abolishes expression of γ-crystalline, a lens differentiation marker (Fig. 4). Thus, expression of B4 in somatic cells is required in newt lens transdifferentiation and it is suggested that reprogramming in the newt somatic cell during transdifferentiation and in the oocyte after SCNT share common strategies.

4 Discussion

In this review, we have shown a dynamic change of core-histone modifications, emergence of oocyte-type linker histone B4, and expression of stem cell factors during newt lens transdifferentiation. Using those markers, the cellular state during lens transdifferentiation can be dissected in detail. In fact, such changes have modified the previous concepts of dedifferentiation during the process of lens transdifferentiation. In the past, it has been thought that the reprogramming of PEC

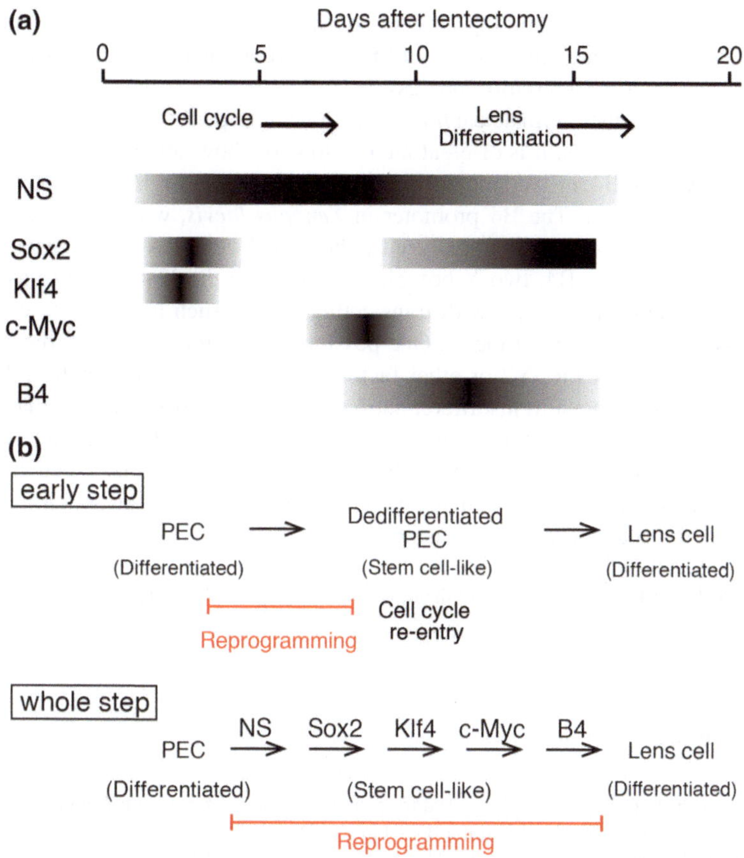

Fig. 5 Reprogramming in newt lens transdifferentiation. **a** Expression prolife of B4 and stem cell factors in dorsal iris PECs during lens transdifferentiation. Note that these genes are activated sequentially through lens transdifferentiation. **b** New model for reprogramming in newt lens transdifferentiation. Thus far, it has been thought that the reprogramming in which differentiated PEC change to stem cell-like cell, occurs only at an early step of lens regeneration (earl step model). However, this model is not based on the gene expression profile in lens transdifferentiation. Based on the expression profile of those genes, we propose a new model in which the somatic nucleus is reprogrammed in a stepwise fashion through the lens transdifferentiation process (whole step model)

is completed by about 8 days after lentectomy and that the reprogrammed cells already have a restored ability for lens differentiation (Fig. 5a), since the cells have lost the morphological characteristics of PEC and have re-entered the cell cycle. In fact, however, those gene markers related to nuclear reprogramming are expressed sequentially throughout lens transdifferentiation and not just limited to the period prior to cell cycle re-entry. Especially, oocyte-type linker histone, which is required for the reprogramming after SCNT, shows a peak of expression on day 12. Based on the expression profile of reprogramming-related genes, we propose a "whole step" reprogramming model during newt lens transdifferentiation (Fig. 5b). In this model, the nucleus of PEC is reprogrammed in a stepwise fashion through the transdifferentiation process.

Even though other animals cannot express oocyte-type linker histone in somatic cells, newts can do this. It is of great interest to know how newts gained the ability of to re-express B4 in somatic cells. Analysis of the newt B4 promoter might answer this question. The B4 promoter in *Xenopus laevis*, which cannot express B4 in somatic cells, has been analyzed (Cho and Wolffe 1994). Consistent with oocyte expression of B4, two Y-box elements exist in the *Xenopus* B4 promoter. The Y-box element interacts with trans-acting factors such as FRGY2, abundant oocyte-specific trans-acting factor. One possible reason for newt B4 expression in somatic cells is that FRGY2 or other factor(s), which can interact with Y-box, is expressed during newt transdifferentiation. The other possibility is that some element(s) for somatic expression is inserted in the newt B4 promoter. Understanding the mechanism of somatic B4 expression will not only shed light on evolutional differences between regenerative and non-regenerative animals, but could also be adapted to future regenerative medicine.

Acknowledgments We would like to thank Kiyoe Ura for critical reading and suggestions, and Rinako Maki for making illustrations.

References

Abe S, Eguchi G (1977) An analysis of differentiative capacity of pigmented epithelial cells of adult newt iris in clonal cell culture. Dev Growth Differ 19:309–317

Agata K, Kobayashi H, Itoh Y, Mochii M, Sawada K, Eguchi G (1993) Genetic characterization of the multipotent dedifferentiated state of pigmented epithelial cells in vitro. Development 118:1025–1030

Anderson RM, Bosch JA, Goll MG, Hesselson D, Dong PD, Shin D, Chi NC, Shin CH, Schlegel A, Halpern M, Stainier DY (2009) Loss of Dnmt1 catalytic activity reveals multiple roles for DNA methylation during pancreas development and regeneration. Dev Biol 334:213–223

Azuara V, Perry P, Sauer S, Spivakov M, Jorgensen HF, John RM, Gouti M, Casanova M, Warnes G, Merkenschlager M, Fisher AG (2006) Chromatin signatures of pluripotent cell lines. Nat Cell Biol 8:532–538

Baddoo M, Hill K, Wilkinson R, Gaupp D, Hughes C, Kopen GC, Phinney DG (2003) Characterization of mesenchymal stem cells isolated from murine bone marrow by negative selection. J Cell Biochem 89:1235–1249

Barski A, Cuddapah S, Cui K, Roh TY, Schones DE, Wang Z, Wei G, Chepelev I, Zhao K (2007) High-resolution profiling of histone methylations in the human genome. Cell 129:823–837

Becker M, Becker A, Miyara F, Han Z, Kihara M, Brown DT, Hager GL, Latham K, Adashi EY, Misteli T (2005) Differential in vivo binding dynamics of somatic and oocyte-specific linker histones in oocytes and during ES cell nuclear transfer. Mol Biol Cell 16:3887–3895

Beekman C, Nichane M, De Clercq S, Maetens M, Floss T, Wurst W, Bellefroid E, Marine JC (2006) Evolutionarily conserved role of nucleostemin: controlling proliferation of stem/progenitor cells during early vertebrate development. Mol Cell Biol 26:9291–9301

Bernstein BE, Mikkelsen TS, Xie X, Kamal M, Huebert DJ, Cuff J, Fry B, Meissner A, Wernig M, Plath K, Jaenisch R, Wagschal A, Feil R, Schreiber SL, Lander ES (2006) A bivalent chromatin structure marks key developmental genes in embryonic stem cells. Cell 125:315–326

Bracken AP, Dietrich N, Pasini D, Hansen KH, Helin K (2006) Genome-wide mapping of Polycomb target genes unravels their roles in cell fate transitions. Genes Dev 20:1123–1136

Cho H, Wolffe AP (1994) Xenopus laevis B4, an intron-containing oocyte-specific linker histone-encoding gene. Gene 143:233–238

Coulombre AJ (1965) The eye. In: DeHaan RL, Ursprung H (ed) *Organogenesis*. New Yolk: Holt, Rinehart and Winston, pp 219–251

Cowan CA, Atienza J, Melton DA, Eggan K (2005) Nuclear reprogramming of somatic cells after fusion with human embryonic stem cells. Science 309:1369–1373

Dai MS, Sun XX, Lu H (2008) Aberrant expression of nucleostemin activates p53 and induces cell cycle arrest via inhibition of MDM2. Mol Cell Biol 28:4365–4376

Eguchi G (1980) Lens regeneration: transdifferentiation of tissue cells. Iwanami shoten, Tokyo

Evans MJ, Kaufman MH (1981) Establishment in culture of pluripotential cells from mouse embryos. Nature 292:154–156

Fischle W, Wang Y, Allis CD (2003) Histone and chromatin cross-talk. Curr Opin Cell Biol 15:172–183

Gao S, Chung YG, Parseghian MH, King GJ, Adashi EY, Latham KE (2004) Rapid H1 linker histone transitions following fertilization or somatic cell nuclear transfer: evidence for a uniform developmental program in mice. Dev Biol 266:62–75

Grant PA, Eberharter A, John S, Cook RG, Turner BM, Workman JL (1999) Expanded lysine acetylation specificity of Gcn5 in native complexes. J Biol Chem 274:5895–5900

Grogg MW, Call MK, Okamoto M, Vergara MN, Del Rio-Tsonis K, Tsonis PA (2005) BMP inhibition-driven regulation of six-3 underlies induction of newt lens regeneration. Nature 438:858–862

Guenther MG, Levine SS, Boyer LA, Jaenisch R, Young RA (2007) A chromatin landmark and transcription initiation at most promoters in human cells. Cell 130:77–88

Gurdon JB, Elsdale TR, Fischberg M (1958) Sexually mature individuals of Xenopus laevis from the transplantation of single somatic nuclei. Nature 182:64–65

Hayashi T, Mizuno N, Takada R, Takada S, Kondoh H (2006) Determinative role of Wnt signals in dorsal iris-derived lens regeneration in newt eye. Mech Dev 123:793–800

Hayashi T, Mizuno N, Ueda Y, Okamoto M, Kondoh H (2004) FGF2 triggers iris-derived lens regeneration in newt eye. Mech Dev 121:519–526

Jopling C, Sleep E, Raya M, Marti M, Raya A, Izpisua Belmonte JC (2010) Zebrafish heart regeneration occurs by cardiomyocyte dedifferentiation and proliferation. Nature 464:606–609

Jullien J, Astrand C, Halley-Stott RP, Garrett N, Gurdon JB (2010) Characterization of somatic cell nuclear reprogramming by oocytes in which a linker histone is required for pluripotency gene reactivation. Proc Natl Acad Sci U S A 107:5483–5488

Kornberg RD (1974) Chromatin structure: a repeating unit of histones and DNA. Science 184:868–871

Kouzarides T (2007) Chromatin modifications and their function. Cell 128:693–705

Krogan NJ, Kim M, Tong A, Golshani A, Cagney G, Canadien V, Richards DP, Beattie BK, Emili A, Boone C, Shilatifard A, Buratowski S, Greenblatt J (2003) Methylation of histone H3 by Set2 in Saccharomyces cerevisiae is linked to transcriptional elongation by RNA polymerase II. Mol Cell Biol 23:4207–4218

Kuo MH, Brownell JE, Sobel RE, Ranalli TA, Cook RG, Edmondson DG, Roth SY, Allis CD (1996) Transcription-linked acetylation by Gcn5p of histones H3 and H4 at specific lysines. Nature 383:269–272

Lachner M, O'Sullivan RJ, Jenuwein T (2003) An epigenetic road map for histone lysine methylation. J Cell Sci 116:2117–2124

Li J, Moazed D, Gygi SP (2002) Association of the histone methyltransferase Set2 with RNA polymerase II plays a role in transcription elongation. J Biol Chem 277:49383–49388

Maki N, Suetsugu-Maki R, Sano S, Nakamura K, Nishimura O, Tarui H, Del Rio-Tsonis K, Ohsumi K, Agata K, Tsonis PA (2010a) Oocyte-type linker histone B4 is required for transdifferentiation of somatic cells in vivo. FASEB J 24:3462–3467

Maki N, Suetsugu-Maki R, Tarui H, Agata K, Del Rio-Tsonis K, Tsonis PA (2009) Expression of stem cell pluripotency factors during regeneration in newts. Dev Dyn 238:1613–1616

Maki N, Takechi K, Sano S, Tarui H, Sasai Y, Agata K (2007) Rapid accumulation of nucleostemin in nucleolus during newt regeneration. Dev Dyn 236:941–950

Maki N, Tsonis PA, Agata K (2010b) Changes in global histone modifications during dedifferentiation in newt lens regeneration. Mol Vis 16:1893–1897

Mandl B, Brandt WF, Superti-Furga G, Graninger PG, Birnstiel ML, Busslinger M (1997) The five cleavage-stage (CS) histones of the sea urchin are encoded by a maternally expressed family of replacement histone genes: functional equivalence of the CS H1 and frog H1 M (B4) proteins. Mol Cell Biol 17:1189–1200

Martin GR (1981) Isolation of a pluripotent cell line from early mouse embryos cultured in medium conditioned by teratocarcinoma stem cells. In: Proceedings of the national academy of sciences of the United States of America, vol 78. pp 7634–7638

McGraw S, Vigneault C, Tremblay K, Sirard MA (2006) Characterization of linker histone H1FOO during bovine in vitro embryo development. Mol Reprod Dev 73:692–699

Mendenhall EM, Bernstein BE (2008) Chromatin state maps: new technologies, new insights. Curr Opin Genet Dev 18:109–115

Mikkelsen TS, Ku M, Jaffe DB, Issac B, Lieberman E, Giannoukos G, Alvarez P, Brockman W, Kim TK, Koche RP, Lee W, Mendenhall E, O'Donovan A, Presser A, Russ C, Xie X, Meissner A, Wernig M, Jaenisch R, Nusbaum C, Lander ES, Bernstein BE (2007) Genome-wide maps of chromatin state in pluripotent and lineage-committed cells. Nature 448:553–560

Nikpour P, Mowla SJ, Jafarnejad SM, Fischer U, Schulz WA (2009) Differential effects of Nucleostemin suppression on cell cycle arrest and apoptosis in the bladder cancer cell lines 5637 and SW1710. Cell Prolif 42:762–769

Nishioka K, Chuikov S, Sarma K, Erdjument-Bromage H, Allis CD, Tempst P, Reinberg D (2002) Set9, a novel histone H3 methyltransferase that facilitates transcription by precluding histone tail modifications required for heterochromatin formation. Genes Dev 16:479–489

Ohmura M, Naka K, Hoshii T, Muraguchi T, Shugo H, Tamase A, Uema N, Ooshio T, Arai F, Takubo K, Nagamatsu G, Hamaguchi I, Takagi M, Ishihara M, Sakurada K, Miyaji H, Suda T, Hirao A (2008) Identification of stem cells during prepubertal spermatogenesis via monitoring of nucleostemin promoter activity. Stem cells 26:3237–3246

Ohsumi K, Katagiri C (1991) Occurrence of H1 subtypes specific to pronuclei and cleavage-stage cell nuclei of anuran amphibians. Dev Biol 147:110–120

Okita K, Ichisaka T, Yamanaka S (2007) Generation of germline-competent induced pluripotent stem cells. Nature 448:313–317

Pan G, Tian S, Nie J, Yang C, Ruotti V, Wei H, Jonsdottir GA, Stewart R, Thomson JA (2007) Whole-genome analysis of histone H3 lysine 4 and lysine 27 methylation in human embryonic stem cells. Cell Stem Cell 1:299–312

Peters AH, Kubicek S, Mechtler K, O'Sullivan RJ, Derijck AA, Perez-Burgos L, Kohlmaier A, Opravil S, Tachibana M, Shinkai Y, Martens JH, Jenuwein T (2003) Partitioning and plasticity of repressive histone methylation states in mammalian chromatin. Mol Cell 12:1577–1589

Rice JC, Briggs SD, Ueberheide B, Barber CM, Shabanowitz J, Hunt DF, Shinkai Y, Allis CD (2003) Histone methyltransferases direct different degrees of methylation to define distinct chromatin domains. Mol Cell 12:1591–1598

Romanova L, Grand A, Zhang L, Rayner S, Katoku-Kikyo N, Kellner S, Kikyo N (2009) Critical role of nucleostemin in pre-rRNA processing. J Biol Chem 284:4968–4977

Rundlett SE, Carmen AA, Kobayashi R, Bavykin S, Turner BM, Grunstein M (1996) HDA1 and RPD3 are members of distinct yeast histone deacetylase complexes that regulate silencing and transcription. Proc Natl Acad Sci USA 93:14503–14508

Ruthenburg AJ, Li H, Patel DJ, Allis CD (2007) Multivalent engagement of chromatin modifications by linked binding modules. Nat Rev Mol Cell Biol 8:983–994

Sadler KC, Krahn KN, Gaur NA, Ukomadu C (2007) Liver growth in the embryo and during liver regeneration in zebrafish requires the cell cycle regulator, uhrf1. In: Proceedings of the national academy of sciences of the United States of America, vol 104. pp 1570–1575

Saeki H, Ohsumi K, Aihara H, Ito T, Hirose S, Ura K, Kaneda Y (2005) Linker histone variants control chromatin dynamics during early embryogenesis. Proc Natl Acad Sci USA 102: 5697–5702

Schaft D, Roguev A, Kotovic KM, Shevchenko A, Sarov M, Neugebauer KM, Stewart AF (2003) The histone 3 lysine 36 methyltransferase, SET2, is involved in transcriptional elongation. Nucleic Acids Res 31:2475–2482

Schiltz RL, Mizzen CA, Vassilev A, Cook RG, Allis CD, Nakatani Y (1999) Overlapping but distinct patterns of histone acetylation by the human coactivators p300 and PCAF within nucleosomal substrates. J Biol Chem 274:1189–1192

Schotta G, Lachner M, Sarma K, Ebert A, Sengupta R, Reuter G, Reinberg D, Jenuwein T (2004) A silencing pathway to induce H3–K9 and H4–K20 trimethylation at constitutive heterochromatin. Genes Dev 18:1251–1262

Shahbazian MD, Grunstein M (2007) Functions of site-specific histone acetylation and deacetylation. Annu Rev Biochem 76:75–100

Spencer TE, Jenster G, Burcin MM, Allis CD, Zhou J, Mizzen CA, McKenna NJ, Onate SA, Tsai SY, Tsai MJ, O'Malley BW (1997) Steroid receptor coactivator-1 is a histone acetyltransferase. Nature 389:194–198

Stewart S, Tsun ZY, Belmote JC (2009) A histone demethylase is necessary for regeneration in zebrafish. Proc Natl Acad Sci U S A 106:19889–19894

Tada M, Takahama Y, Abe K, Nakatsuji N, Tada T (2001) Nuclear reprogramming of somatic cells by in vitro hybridization with ES cells. Curr Biol: CB 11:1553–1558

Takahashi K, Tanabe K, Ohnuki M, Narita M, Ichisaka T, Tomoda K, Yamanaka S (2007) Induction of pluripotent stem cells from adult human fibroblasts by defined factors. Cell 131:861–872

Takahashi K, Yamanaka S (2006) Induction of pluripotent stem cells from mouse embryonic and adult fibroblast cultures by defined factors. Cell 126:663–676

Tanaka M, Hennebold JD, Macfarlane J, Adashi EY (2001) A mammalian oocyte-specific linker histone gene H1oo: homology with the genes for the oocyte-specific cleavage stage histone (cs-H1) of sea urchin and the B4/H1 M histone of the frog. Development 128:655–664

Tanaka Y, Kato S, Tanaka M, Kuji N, Yoshimura Y (2003) Structure and expression of the human oocyte-specific histone H1 gene elucidated by direct RT-nested PCR of a single oocyte. Biochem Biophys Res Commun 304:351–357

Teranishi T, Tanaka M, Kimoto S, Ono Y, Miyakoshi K, Kono T, Yoshimura Y (2004) Rapid replacement of somatic linker histones with the oocyte-specific linker histone H1foo in nuclear transfer. Dev Biol 266:76–86

Tsai RY, McKay RD (2002) A nucleolar mechanism controlling cell proliferation in stem cells and cancer cells. Genes Dev 16:2991–3003

Wakayama T, Perry AC, Zuccotti M, Johnson KR, Yanagimachi R (1998) Full-term development of mice from enucleated oocytes injected with cumulus cell nuclei. Nature 394:369–374

Wang H, Cao R, Xia L, Erdjument-Bromage H, Borchers C, Tempst P, Zhang Y (2001) Purification and functional characterization of a histone H3-lysine 4-specific methyltransfer-ase. Mol Cell 8:1207–1217

Wang Z, Zang C, Rosenfeld JA, Schones DE, Barski A, Cuddapah S, Cui K, Roh TY, Peng W, Zhang MQ, Zhao K (2008) Combinatorial patterns of histone acetylations and methylations in the human genome. Nat Genet 40:897–903

Wernig M, Meissner A, Foreman R, Brambrink T, Ku M, Hochedlinger K, Bernstein BE, Jaenisch R (2007) In vitro reprogramming of fibroblasts into a pluripotent ES-cell-like state. Nature 448:318–324

Wibrand K, Olsen LC (2002) Linker histone H1 M transcripts mark the developing germ line in zebrafish. Mech Dev 117:249–252

Wilmut I, Schnieke AE, McWhir J, Kind AJ, Campbell KH (1997) Viable offspring derived from fetal and adult mammalian cells. Nature 385:810–813

Yakushiji N, Suzuki M, Satoh A, Sagai T, Shiroishi T, Kobayashi H, Sasaki H, Ide H, Tamura K (2007) Correlation between Shh expression and DNA methylation status of the limb-specific Shh enhancer region during limb regeneration in amphibians. Dev Biol 312:171–182

Yu J, Vodyanik MA, Smuga-Otto K, Antosiewicz-Bourget J, Frane JL, Tian S, Nie J, Jonsdottir GA, Ruotti V, Stewart R, Slukvin II, Thomson JA (2007) Induced pluripotent stem cell lines derived from human somatic cells. Science 318:1917–1920

Zhao XD, Han X, Chew JL, Liu J, Chiu KP, Choo A, Orlov YL, Sung WK, Shahab A, Kuznetsov VA, Bourque G, Oh S, Ruan Y, Ng HH, Wei CL (2007) Whole-genome mapping of histone H3 Lys4 and 27 trimethylations reveals distinct genomic compartments in human embryonic stem cells. Cell Stem Cell 1:286–298

Cell Cycle Regulation and Regeneration

Ellen Heber-Katz, Yong Zhang, Khamila Bedelbaeva, Fengyu Song, Xiaoping Chen and David L. Stocum

Abstract Regeneration of ear punch holes in the MRL mouse and amputated limbs of the axolotl show a number of similarities. A large proportion of the fibroblasts of the uninjured MRL mouse ear are arrested in G_2 of the cell cycle, and enter nerve-dependent mitosis after injury to form a ring-shaped blastema that regenerates the ear tissue. Multiple cell types contribute to the establishment of the regeneration blastema of the urodele limb by dedifferentiation, and there is substantial reason to believe that the cells of this early blastema are also arrested in G_2, and enter mitosis under the influence of nerve-dependent factors supplied by the apical epidermal cap. Molecular analysis reveals other parallels, such as; (1) the upregulation of Evi5, a centrosomal protein that prevents mitosis by stabilizing Emi1, a protein that inhibits the degradation of cyclins by the anaphase promoting complex and (2) the expression of sodium channels by the epidermis. A central feature in the entry into the cell cycle by MRL ear fibroblasts is a natural downregulation of p21, and knockout of p21 in wild-type mice confers regenerative capacity on non-regenerating ear tissue. Whether the same is true for entry into the cell cycle in regenerating urodele limbs is presently unknown.

E. Heber-Katz (✉) · Y. Zhang · K. Bedelbaeva
Wistar Institute, 3601 Spruce St., Philadelphia, PA 19104, USA
e-mail: heberkatz@wistar.org

F. Song
Department of Oral Biology, Indiana University School of Dentistry,
Indiana University-Purdue University, Indianapolis, IN 46202, USA

X. Chen · D. L. Stocum
Department of Biology, School of Science, Indiana University-Purdue University,
723 W. Michigan St., Indianapolis, IN 46202, USA

Current Topics in Microbiology and Immunology (2013) 367: 253–276
DOI: 10.1007/82_2012_294
© Springer-Verlag Berlin Heidelberg 2012
Published Online: 22 December 2012

Abbreviations:

AEC	Apical epidermal cap
AGP	Anterior gradient protein
APC	Anaphase promoting complex
CDK	Cyclin-dependent kinase
EGF	Epidermal growth factor
FGF	Fibroblast growth factor
G	Gap
GGF	Glial growth factor
HGF	Hepatocyte growth factor
IGF	Insulin-like growth factor
M	Mitosis
PDGF	Platelet-derived growth factor
RP	Restriction point
S	DNA synthesis
TGF	Transforming growth factor

Contents

1 Introduction: The Cell Cycle

Cell proliferation is an absolute requirement for tissue regeneration, and regulation of cell cycle entry and progression are obviously important for this process. Most studies on cell proliferation in regeneration involve either compensatory hyperplasia (as in liver regeneration) or stem cell/progenitor proliferation of epithelia.

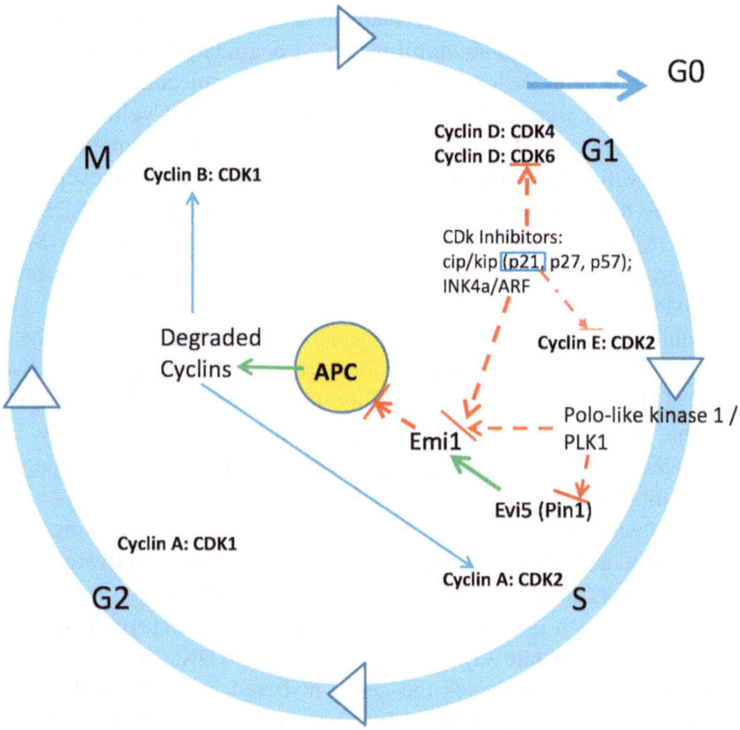

Fig. 1 The phases of the cell cycle and their regulation by cyclin:CDK complexes. *G_1* gap 1, *S* DNA synthesis, *G_2* gap 2, *M* mitosis, *CDK* cyclin-dependent kinase, *Emi1* early mitotic inhibitor, *Evi5* ecotropic viral integration site 5, *PLK1* Polo-like kinase 1, *APC* anaphase promoting complex. The *blue arrows* indicate degraded cyclins. The *green arrow* indicates stabilization of Emi1 by Evi5. *Dashed red arrows/bars* indicate inhibitory interactions

More recently, research has been initiated on cell cycle regulation as part of the regenerative mechanism of complex multi-tissue structures (Arthur et al. 2010). The MRL mouse ear and the urodele limb are two complex structures that share some common features of regeneration after tissue loss. This chapter presents data and ideas that can further our understanding of how cell cycle regulation might explain some aspects of the regenerative ability displayed in these two systems.

The cell cycle has four phases (Fig. 1) that take place sequentially when the cell is activated from its quiescent state, termed G_o. These phases are G_1 during which the cell is rendered competent to synthesize DNA; S-phase, where the DNA is replicated; G_2 during which the cell is certified ready to divide; and M, during which the chromosomes condense and the successive steps of mitosis take place (prophase, metaphase, anaphase, and telophase) to produce two daughter cells (Lodish et al. 2008).

Checkpoints controlled by regulatory proteins inhibit entry into S and M until molecular sensing systems determine that all requirements for DNA synthesis and mitosis are satisfied. Two major mechanisms are used to regulate the cell cycle: protein phosphorylation and ubiquitin-mediated degradation (Lodish et al. 2008). The regulatory proteins belong to three classes of molecules: the cyclins, the cyclin-dependent kinases (CDKs), and the CDK inhibitors. Cyclins are regulatory units and CDKs are catalytic units that must interact with one another for each to function. The cyclin:CDK complexes that drive each phase of the cycle are cyclin D: CDKs 4 and 6 in mid G_1, cyclin E:CDK2 in late G_1, cyclin A:CDK2 in S-phase, Cyclin A:CDK1 in G_2, and cyclin B:CDK1 in M-phase. To exit S or M, cyclins must be degraded and this occurs through the action of the anaphase promoting complex (APC).

CDK inhibitors act primarily in G_1 and belong to two families, the *cip/kip* (*CDK interacting protein/Kinase inhibitory protein*) family, and the INK4a/ARF (Inhibitor of Kinase4/Alternative Reading Frame) family. The *cip/kip* family includes the genes for p21, p27, and p57. The INK4a/ARF family includes $p16^{INK4a}$, which binds to CDK4, and p19ARF, which prevents p53 degradation and thus maintains p21.

The competence of quiescent G_0 cells to enter G_1 requires signaling by growth factors (Morgan and Pledger 1992). These growth factors activate preexisting transcription factors that induce the transcription of early response genes such as c-Fos and c-Jun, which in turn induce the transcription of delayed response genes encoding the mid and late G_1 and S-phase cyclin-CDKs, as well as the E2F transcription factor that activates the genes involved in DNA synthesis (Lodish et al. 2008). The CDKs are inhibited for most of G_1 by the CDK inhibitors p21, p27, and p57, and E2F acts as a transcriptional repressor because it is bound to the hypo-phosphorylated retinoblastoma (Rb) protein. At a point in late G_1 called the restriction point (RP), the concentration of cyclin:CDKs reaches a level where the CDK inhibitors are saturated. As the concentration of cyclin:CDK rises further, it inactivates Rb by hyper-phosphorylation. This frees E2F to now transcriptionally activate the expression of over 800 genes involved in DNA synthesis. E2F also induces the transcription of cyclin E, which interacts with CDK2 to bring about the G_1-S transition, and induces the transcription of S-phase cyclin. Before the cells get to S-phase, DNA damage, oxidative stress, and other DNA damaging events can activate the p53 protein. One major target gene of p53 is $p21^{Cip1/Waf1}$ (el-Deiry et al. 1993; Xiong et al. 1993) and the $p21^{Cip1/Waf1}$ protein will bind to the cyclin/CDK2 complex to block phosphorylation of Rb and thus prevent S-phase transit until DNA repair occurs.

During DNA synthesis, S-phase cyclin A:CDK complexes phosphorylate and activate proteins of pre-replication complexes assembled during G_1, and various enzymes collaborate to replicate DNA strands. Upon completion of DNA synthesis, the cell is temporarily arrested in G_2, which serves as a second checkpoint to detect any DNA damage that needs to be repaired before transiting to M, or if repair cannot be accomplished, to activate apoptotic pathways. The Emi1 (early mitotic inhibitor 1) and Evi5 (ecotropic viral integration factor 5) proteins prevent a premature entry into M. Expression of these proteins is stimulated by E2F at the G_1/S transition, after which they accumulate and localize to the centrosome. Emi1 is an

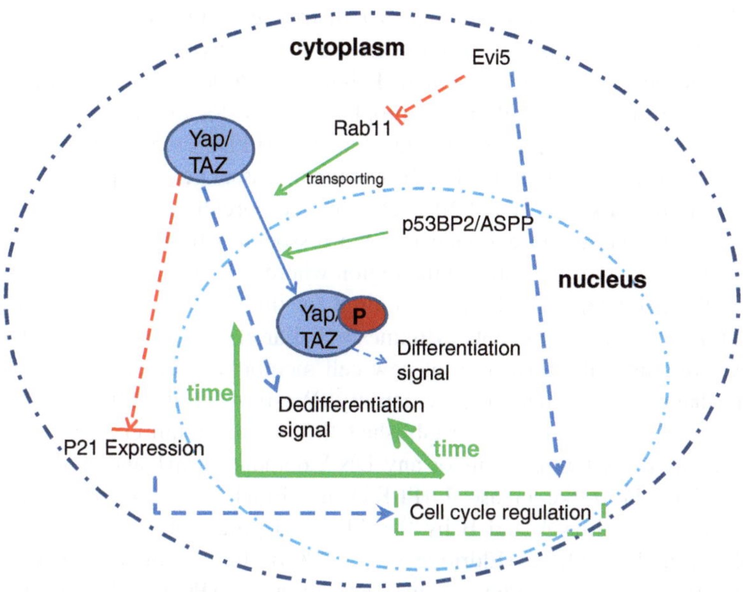

Fig. 2 Potential mechanism of regulation by Evi5 of the Hippo pathway, which plays a role in the cell cycle, cell differentiation and cell dedifferentiation. *p53BP2* tumor suppressor p53-binding protein 2, also known as Apoptosis-stimulator of p53 protein 2 (ASPP2), *TAZ* transcriptional coactivator with PDZ-binding motif, and *Yap* Yes-associated protein. *Red dashed lines* indicate negative regulation; *green solid lines* indicate positive regulation; *blue dashed lines* indicate effects of p21 and Evi5 on cell cycle regulation

F-box protein that interacts with Cdc20 to inhibit the E3 ligase APC and prevent the ubiquitination and degradation of the APC substrates, cyclins A and B, thus maintaining the cell in G_2 (Eldridge et al. 2006). Cells lacking Emi1 suffer DNA damage induced by replication stress and undergo apoptosis (Verschuren et al. 2007). Interphase Evi5 is a 110 kDa protein that stabilizes Emi1 and cyclin A. A second protein, Pin1, has also been shown to stabilize Emi1 (Bernis et al. 2007).

Evi5 also serves as a GAP (GTPase Activating Protein) to inactivate the small GTPase vesicle trafficking protein Rab11 (Dabbeekeh et al. 2007; Westlake et al. 2007). Vesicle trafficking regulates the Hippo pathway by controlling translocation of the Yes-Associated Protein (YAP) from the cytosol to the nucleus (Fig. 2), where YAP can act as a co-activator of Smad-7 to activate genes involved in the cell cycle such as *E2F* and *survivin* (Zhao et al. 2011). In addition, dephosphorylated YAP has been shown to locate in the cytosol and to be upregulated in pluripotent stem cells, whereas inactivated (phosphorylated) YAP is detected in the nucleus and is associated with cell differentiation (Lian et al. 2010). Thus Evi5 might help restrain cells from entering mitosis by inhibiting Rab 11, which would prevent phosphorylated YAP from localizing to the nucleus and acting as a mitotic and differentiation signal, thereby promoting and maintaining a dedifferentiated state.

Evi5 survives during mitosis and is redistributed from the centrosome as a 90 kDa protein and a 20 kDa protein that are formed by cleavage of the 20 kDa form off the far end of the C-terminal (Faitar et al. 2006). The 90 kDa protein is distributed throughout the cell during prophase, but co-localizes at metaphase with α-tubulin of the mitotic spindle. During anaphase 90 kDa Evi5 localizes at the mid zone, where it associates with the chromosomal passenger complex (CPC) consisting of aurora B kinase, INCENP, survivin, and borealin (Ruchaud et al. 2007). During late telophase and cytokinesis, Evi5 dissociates from the CPC and forms two parallel disk-like structures in the region where new membrane formation will separate the daughter cells. Evi5 is required for this separation, as shown by the fact that its knockdown prevents cytokinesis (Faitar et al. 2006). There is evidence that Evi5 regulates the formation of new cell membrane formation at cytokinesis by a mechanism other than GTPase activity (Westlake et al. 2007).

Once sensing systems have established that all the requirements for mitosis have been met, including repair of any DNA damage, Emi1 and Evi5 are phosphorylated by Polo-like kinase 1 (PLK1) and Emi1 is degraded, allowing the degradation of cyclins A and B by the APC and progression through prophase, metaphase, and anaphase (Eldridge et al. 2006). Interestingly, a recent study indicated that p21 downregulates Emi1, activating the APC in cells arrested in G_2 by DNA damage. However, this resulted in stable G_2 arrest, rather than progression into mitosis, indicating that Emi1 plays a role in a novel p21-dependent mechanism to maintain G_2 arrest after DNA damage (Lee et al. 2009).

2 Regeneration of Mouse Ear Tissue

2.1 Events of Regeneration

In 1998, the MRL/lpr and MRL/Mpj mice were shown to have the unique characteristic of closing ear hole punches (Clark et al. 1998) (Fig. 3). Further analysis showed that this occurred without the formation of a scar, with normal tissue architecture, and with the replacement of elastic cartilage. Many of the features described for limb regeneration in amphibians also appeared in this system. Since then many laboratories have confirmed and extended the finding that healing in the MRL mouse is unique and regenerative (e.g. Balu et al. 2009; Buhimschi et al. 2010; Rai et al. 2012; Thuret et al. 2009).

One of the early events in amphibian regeneration is the epidermal response. In the amphibian, rapid coverage of the wound occurs within the first 12–24 h. The epidermis migrates rapidly to re-cover the edges of punch wounds In the MRL mouse and thickens within 3–5 days to form a cap of epidermis resembling the apical epidermal cap (AEC) that forms on the amputated amphibian limb (see ahead). This cap forms a ring around the edge of the punch hole. By contrast, 5–10 days are required to re-cover the wound in the non-regenerating B6 mouse

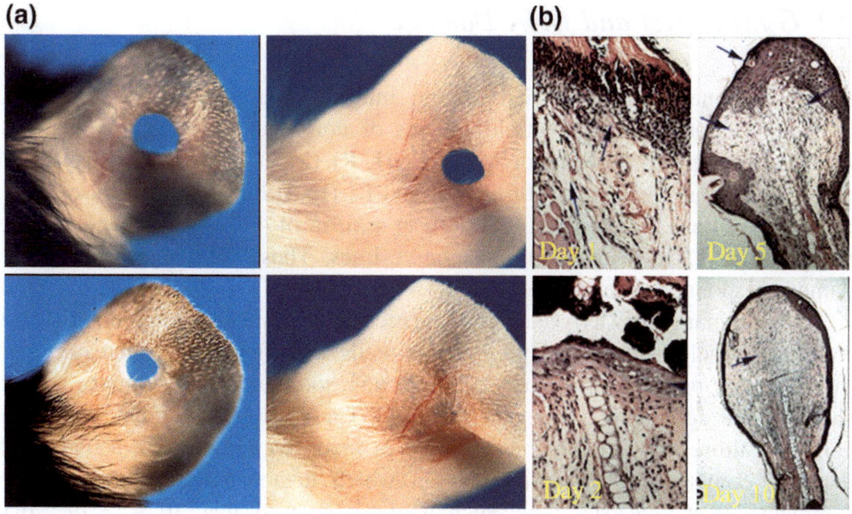

Fig. 3 Punch injuries in the ear pinna lead to different outcomes in MRL versus C57BL/6 ears. **a** Macroscopic visualization of punch holes in C57BL/6 (*left panels*) and MRL (*right panels*) at 1 day post-injury (*upper panels*) and 1 month post-injury (*lower panels*). The hole has been completely closed by regeneration in the MRL by 30 days. **b** Histology of regeneration in MRL ear tissue. At day 1, inflammatory cells can be seen lining up at the edge of the ear punch (*arrow*). By day 2, re-epithelialization is complete. By day 5 hair follicles are forming in the new epidermis (*upper arrow*) and cells are accumulating beneath the epidermis in a blastema-like structure (*lower arrow*). By day 10, the blastema (*arrow*) has expanded significantly

and a structure like the AEC does not form. A mesenchymal ring blastema forms under the thickened MRL wound epidermis and then undergoes a rapid growth phase via cell proliferation. In the amphibian, no basement membrane re-forms to prevent communication between epithelial and mesenchymal cell populations. In the MRL mouse a basement membrane is seen by day 3 and is eliminated by day 5, but is maintained in non-regenerative mice. Cells then accumulate under the "AEC" and proliferate centripetally until the hole is closed (Gourevitch et al. 2003). In vitro, cells derived from the ears of MRL mice were found to be highly proliferative compared to cells from the ears of B6 mice. These cells are mesenchymal-like fibroblasts that potentially contribute to the blastema population.

To determine the genes involved in the MRL regenerative response, extensive mapping studies of genes expressed early after wounding have been performed. The MRL was bred to different non-regenerating mice. MRL is the result of crossbreeding 4 inbred strains, LG/J (75 %), C3H (12 %), AKR-(12 %), and C57BL/6 (0.3 %). The LG/J mouse was shown to contribute the healing phenotype and genetic studies were carried out using this strain as well to compare it to the non-regenerator strain SM/J. In these studies, recombinant RI strains and advanced intercross lines were used for fine mapping and intervals were significantly (20–40-fold) narrowed (Cheverud et al. 2012).

2.2 G_2/M Arrest and DNA Damage

Cell cycle analysis was carried out on multiple sources of regenerator cell populations compared to multiple sources of non-regenerator cell populations. Thus, we compared regenerator MRL, LG/J, RI (LxS) line 6, and a B6 healer congenic to non-regenerative B6, SM/J, and RI (LxS) line 33. In all cases using regenerative cells, G_2/M arrest was seen in a mean of 45 % of the population as compared to 12 % in the non-regenerative lines. These data suggested a DNA damage response in the regenerators.

An early marker of DNA damage is the recruitment and the formation of H2AX foci in the nuclei, which in the regenerator cells was significantly increased with more nuclear foci and more foci/cell. Also p53, the major DNA damage sensor, showed clear upregulation in regenerator cells. These results were found not just in cells in culture but in regenerating tissue in vivo. To demonstrate actual DNA damage, comet assays were carried out which demonstrated shortened DNA fragments in an electric field. Such analysis showed that 80 + % of MRL cells were comet-positive compared with only 6 % of non-regenerator cells (Bedelbaeva et al. 2010).

A major target of p53 activation is $p21^{Cip1/waf1}$ (Xiong et al. 1993; el-Deiry et al. 1993), which is responsible for G_1 arrest, and a first step to senescence. P21 is activated upon DNA damage, but examination of the regenerator cell lines showed that p21 was, contrary to expectation, reduced or not expressed as determined by Western analysis. This finding supported the idea that lack of p21 led not to G_1 arrest but rather G_2 arrest. Further support for this idea came from the finding that the $p21^{-/-}$ B6 mouse is a complete regenerator similar to the MRL mouse. Thus elimination of p21 conferred the ability for ear hole regeneration on a normally non-regenerating mouse (Bedelbaeva et al. 2010). The p21 protein binds to the cyclin E:CDK2 complex as well as other cyclin:CDK complexes, preventing the phosphorylation of Rb, thus blocking E2F family members from activating genes involved in DNA synthesis. Without p21, Rb is phosphorylated, E2F is released and activates these genes.

Why p53 is not inducing the expression of p21 in the MRL mouse remains to be determined, but brings into question the role of p53 in regeneration. Previous studies suggested that p53 was required for hair follicle regenerative growth (Ruzankina et al. 2009). In addition, limb regeneration is blocked in the axolotl by administration of a p53 inhibitor, which inhibited MDM2 (a negative regulator of p53) and GADD45 (a stress response gene upregulated by p53) (Villiard et al. 2007; see ahead). To determine the effect of a p53 deletion in mice, p53 knockout mice were bred to either B6 or B6/129 non-regenerator mice and to MRL mice. In the first case, p53 deletion in a non-regenerator strain in a heterozygous or homozygous state had no effect; i.e., the tissue did not regenerate after ear punch hole wounding. When backcrossed to an MRL, the p53 deletion had no effect on regeneration in MRL males, but did have a positive effect on regeneration in MRL females. Thus, p53 does not appear to be necessary for MRL regeneration and its deletion may have a positive effect, at least in females (Arthur et al. 2010).

To determine the expression of other genes in the $p21^{Cip1/waf1}$ pathway, we examined other knockout mice and found that a TGFβ deletion, on a $RAG^{-/-}$ background to reduce the level of inflammation and promote neonatal survival, showed partial ear hole closure. TGFβ is a potent stimulator of p21 and its elimination should reduce p21 activity and lead to regenerative healing. A recent study identifying an ENU mutant that could close its ear holes proved to be a mutation in TGFβR1 (Liu et al. 2011). Whether this mouse can activate p21 through SMAD activation is unclear. A SMAD3 deletion mutant that could not close its ear holes (Ashcroft et al. 1999) but had inflammatory problems similar to those of the TGFβ mutant, could potentially mask a regenerative response. A study using a SMAD7-overexpressing mouse that inhibited SMAD3, enhanced liver regeneration, suggesting that elimination of TGFβ signaling and thus p21 expression enhances regeneration (Zhong et al. 2010).

Preliminary evidence (Bedelbaeva, MS in preparation) indicates that Evi-5 is upregulated in the MRL as assessed by immunohistochemistry in both normal ear tissue (found in hair follicle stem cells, endothelial cells, and muscle cells) and in injured ear tissue (in blastema cells directly under the AEC and potentially in epidermis as well) (Fig. 4).

2.3 G_2/M Arrest and Cytokinesis

Two genes are candidates for implication in G_2/M arrest. Exonuclease 1 (*exo1*) on chr 1 was found to be mutated in the MRL mouse and this mutation was found to be carried in the congenic regenerator. This molecule interacts with MSH2, removes DNA mismatches during homologous recombination and thus is involved with mismatch repair. In the MRL mouse, mismatch repair leads to metaphase arrest and thus may be involved in the regenerative process (Namiki et al. 2003). A second candidate gene identified on chr 9, Kif23 or MKLP1, a microtubule motor protein, is recruited to the spindle mid-body by INCENP, is phosphorylated by Aurora B kinase and is necessary for cytokinesis (Zhu et al. 2005; Guse et al. 2005). This protein may play an important role in the regenerative response. KIF23 expression is upregulated in the MRL and LG mouse and has several coding sequence changes that may lead to a gain or loss of function and to enhanced or reduced cytokinesis (Cheverud et al. 2012). Such a result could lead to either a rapid 4–2 N shift and increase in cell number during regeneration, or it could be responsible for the G_2/M arrest that is seen. Hepatocytes of regenerating adult liver show a high percentage (approximately 70 %) of either mononuclear or binuclear polyploid cells (Michalopoulos and DeFrances 1997; Guidotti et al. 2003).

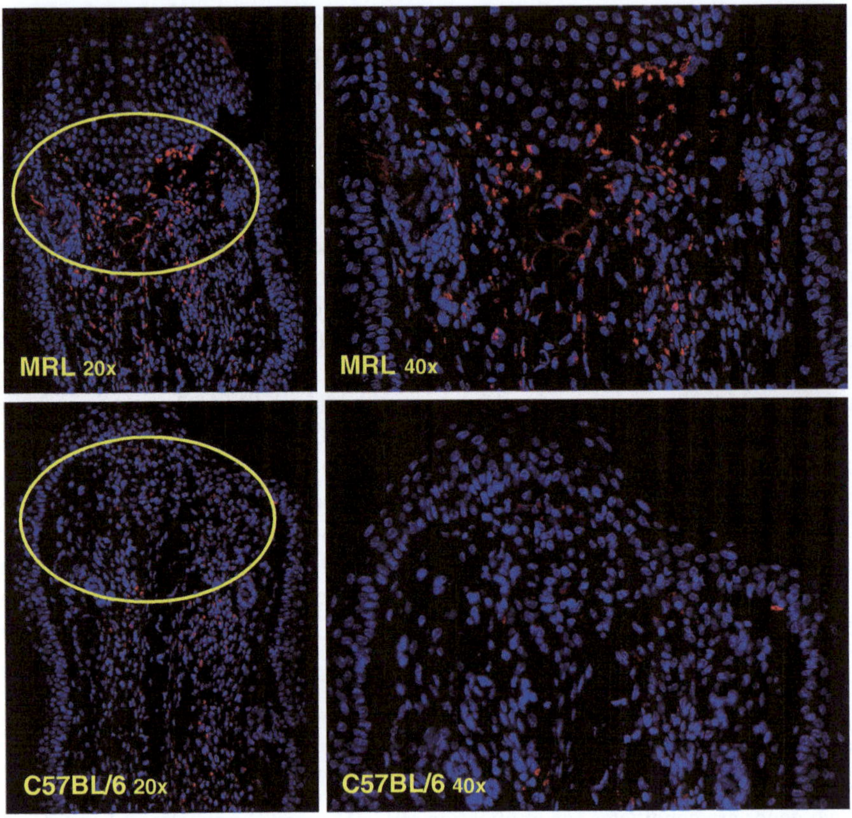

Fig. 4 Evi5 expression in the mouse ear post-injury. Antibody to Evi5 (*red*) and DAPI (DNA, *blue*) was used to stain ear tissue from MRL and C57BL/6 ear tissue 5 days after injury. Evi5 (circumscribed by *yellow ovals*) was found directly under the epidermal cap in the MRL and may also include cells in the epidermis. There were a few positive cells in the C57BL/6 ear tissue distal to the injury

3 Regeneration of the Amphibian Limb

3.1 Blastema Formation and Growth

The urodele (salamanders and newts) limb regenerates after amputation at any level. Amputation induces the formation of a mound-shaped blastema (Fig. 5) consisting of undifferentiated progenitor cells derived by the histolytic liberation and dedifferentiation of dermal and other fibroblasts, Schwann cells, cartilage, and mononucleate myofiber fragments and satellite cells. Within 3–4 days after amputation, the wound epidermis thickens apically to form an AEC several layers thick that has both protective and signaling functions (Christensen and Tassava 2000). Nerve axons and blood vessels regenerate into the blastema and the AEC as

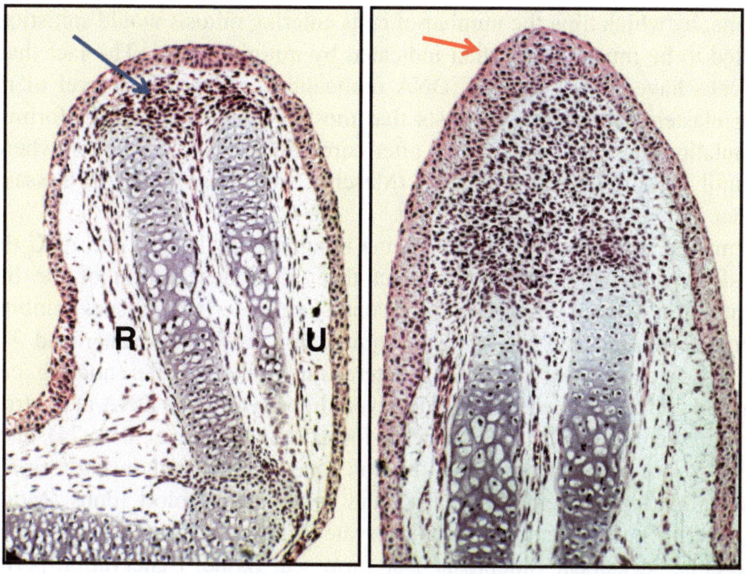

Fig. 5 Blastema formation in the axolotl limb after amputation through the distal radius (*R*) and ulna (*U*). *Left panel* accumulation blastema stage at 4 days post-amputation showing the aggregation of intensely stained dedifferentiated cells (*blue arrow*) under the wound epidermis. *Right panel* three days later, the blastema is a cone-shaped structure as the result of continued dedifferentiation and a significant increase in mitosis. The AEC is indicated by the *red arrow*

they form. The blastema then enters a growth phase and undergoes patterned differentiation that replicates the amputated limb parts. Genetic marking experiments have shown that the blastema cells redifferentiate primarily into their parental phenotype of derivation (i.e., muscle into muscle, cartilage into cartilage, epidermis into epidermis), although blastema cells derived from dermal fibroblasts exhibit transdifferentiation into cartilage and tendons, but never into muscle (Kragl et al. 2009).

During blastema formation, a substantial percentage of blastema cells synthesize DNA. Pulse labeling with ^{3}H-thymidine indicated that ~30 % of blastema cells are labeled by the accumulation blastema stage (Tassava et al. 1974), but continuous labeling experiments indicated that as many as 80 % of blastema cells have entered the cell cycle by that stage (Barger and Tassava 1985; Tassava et al. 1987). The cells do not all cycle synchronously, because there is a transiently quiescent population that appears to feed cells into the actively cycling population. Despite the high labeling index, the cells exhibit a very low mitotic index of ~0.4 %, as assessed by counting mitotic figures in sections. The cell cycle is ~40–45 h in length, with mitosis taking only ~1 h, and varies little throughout regeneration (Tassava and McCullough 1978). The low mitotic index cannot be ascribed just to the long period spanning G_1–G_2 (39–44 h, most of which is taken up by S), or asynchronous cycling, since entry into S begins within 2 days after amputation and it takes 3–5 more days (depending on limb size) to form the blastema, by which time the number of cells entering mitosis would statistically be expected to be much higher than indicated by mitotic counts. The fact that blastema cells have a high level of DNA replication and very low level of mitosis during blastema formation suggests that most of them involved in forming the accumulation blastema arrest in G_2 after completing DNA replication where they stay until the growth phase begins (Mescher and Tassava, 1976; Tassava and Mescher 1975).

Formation and growth of the blastema requires signals from the AEC that are nerve-dependent, as shown by the fact that either denervation of the limb or preventing the wound epidermis from contacting underlying tissues inhibits both blastema formation and the mitosis required for growth (Mescher and Tassava 1976; Tassava and Garling 1979). Interestingly, the dedifferentiating cells of amputated and denervated or AEC-deprived limbs synthesize DNA at control rates and exhibit zero mitosis during blastema formation (Tassava et al. 1974), and there is no difference in the array of proliferation genes upregulated during formation of the accumulation blastema in control and denervated axolotl limbs (Monaghan et al. 2009). The blastema cells of denervated limbs appear to exit the cell cycle and to either disappear (apoptosis?) or form scar tissue. Denervation performed after the blastemal growth phase has begun results in failure of AEC function and both DNA synthesis and mitotic index fall to zero. In this case, however, scar tissue does not form and the blastema cells constitute a mass sufficient to organize into a miniature regenerate (Singer and Craven 1948; Powell 1969; Tassava et al. 1987).

3.2 Regulation of G_o/G_1 and Progression Through G_1

The fact that limb cells enter the cell cycle and synthesize DNA during blastema formation, along with the low mitotic index, suggests that growth factor signaling stimulates entry into and progression through G_1. The growth factors would presumably be the same ones required to exit G_0 by mammalian cells, such as PDGF, FGF-2, EGF, and IGF-1 (Morgan and Pledger 1992). Intraperitoneal injection of IGF-1 has been shown to shorten the time required for blastema formation in newts (Fahmy and Sicard 1998), and insulin was shown to be crucial to ^{3}H-thymidine incorporation by cultured newt blastemas (Vethamany-Globus et al. 1978; Kesik et al. 1986). These signaling molecules may act in combination to stimulate the G_0/G_1 transition. They are probably derived from a combination of sources: platelets, the degradation of ECM, which sequesters growth factors or the inactive forms of growth factors during tissue formation, the wound epidermis and nerves, and the systemic circulation.

Newt and mammalian myotube nuclei differ in their ability to re-enter the cell cycle in response to serum growth factors. In both mammalian and newt myoblasts, pRb is hyperphosphorylated and inactive, allowing them to proliferate in response to serum. As myoblasts fuse and differentiate into myotubes, pRb becomes hypophosphorylated and actively suppresses myonuclear cell cycle activity. Myonuclei of normal mammalian myotubes will not re-enter the cell cycle in response to serum or growth factor stimulation, suggesting that pRb inactivation is not sufficient to reverse cell cycle suppression. In fact, the ARF gene of the mammalian p16$^{\mathrm{Ink4a}}$ locus, which is not expressed in urodeles, must also be inactivated or deleted in mammalian muscle cells to obtain cell cycle re-entry in response to serum (Pajcini et al. 2010).

By contrast, cell cycle re-entry in cultured newt myoblasts and myotubes is promoted by thrombin, which appears to activate a factor present in serum that leads to hyperphosphorylation and inactivation of the Rb protein. Inactivation of Rb is sufficient to allow passage through the RP and to initiate DNA synthesis (Tanaka et al. 1997; Straube and Tanaka 2006). Consistent with the low mitotic index of the forming blastema, the thrombin-activated factor is not sufficient to drive the myonuclei into mitosis, and they arrest in G_2. Cell cycle re-entry is independent of myotube cellularization, since cell cycle-inhibited myotubes implanted into newt limb blastemas cellularize (Velloso et al. 2000). The mechanism of myotube or myofiber fragmentation into single cells is not known, nor is it known whether the thrombin-activated factor is also necessary to drive other cell types as well, or whether this is a feature unique to myofibers. Thus it would appear that muscle cells must be in a mononucleate state to respond to growth factor stimulation, which is consistent with cellularization that takes place during the histolysis of limb regeneration.

The serum factor that promotes phosphorylation of Rb in urodele myonuclei has not been identified, but if limb cells use a mechanism similar to that of mammalian cells to inactivate Rb, one would expect that cyclin: CDKs would increase during G_1 to negate a CDK inhibitor such as p21, resulting in Rb phosphorylation. The Cip/Kip family in most vertebrates consists of p21$^{\mathrm{Cip1/Waf1}}$, p27$^{\mathrm{Kip1}}$, and p57. In Xenopus, three CDK inhibitors of the Cip/Kip family have been identified, p28$^{\mathrm{Xic1}}$, p16$^{\mathrm{Xic2}}$, and p17$^{\mathrm{Xic3}}$ (Daniels et al. 2004). The latter two are homologs of mammalian p21$^{\mathrm{Cip1}}$ and p27$^{\mathrm{Kip1}}$, respectively. The p27$^{\mathrm{Kip1}}$ protein has been identified in the axolotl (Habermann et al. 2004), so it is likely that one or more Cip/Kip proteins is involved in regulating progress through G_1. One of these may be p16, because microinjection of plasmid carrying the human p16$^{\mathrm{Ink4a}}$ gene into nuclei of newt myotubes inhibited uptake of BrdU after serum stimulation (Tanaka et al. 1997). We have detected one CDK inhibitory protein in our proteomic analysis of blastema formation in axolotl limbs (Rao et al. 2009), but it remains unidentified because of poor statistical rank. No protein comparable to p21 has been identified in the axolotl. We have tried to clone axolotl p21 without success using primers from human p21; however, there is a 1713 bp Cip/Kip contig (ID C0207013) on the Sal-Site database (Ambystoma Genetic Stock Center, University of Kentucky; www.ambystoma.org) that is only 44 % identical to human p21, but shares a cyclin:CDK inhibitory site that may offer an opportunity for cloning (Table 1).

Table 1 p21 Homologies, Accession numbers of the protein sequences are as follows: Axolotl (*Ambystoma mexicanum*) ID: *C0207013*); *Xenopus laevis* (NP_001087933); *Xenopus tropicalis* (XP_002935823); *mus* (NP_001104569); humanp21 (NP_000380)

```
                        --------------CDK binding domain-----------------
Am-p21   ----MCSSGGNAPENT-CSKKVCRNLFGELDHEQFKVAAQEMMESCLEEAKQKWNFDFQNGVPLNGDFKWEMVGAVEC
Xl-p21   ----MQSALAIPKQASGNKEKSCRMLFGPVDHEQLRADFDEFMQKSNEEAKAKWNFGFATETPLEGQYDWVKVENN--
Xt-p21   MKEKMQSAIAILKQASGNKEKACRMLFEAVDHEQLKTEFHELMQRSNEEAKAKWNFDFVTETPLEGQYDWEKVEDK--
Mm-p21   ----MSNP-GDVRPVP-HRSKVCRCLFGPVDSEQLRRDCDALMAGCLQEARERWNFDFVTETPLEGNFVWERVRSLGL
Hs-p21   ----MSEPAGDVRQNP-CGSKACRRLFGPVDSEQLSRDCDALMAGCIQEARERWNFDFVTETPLEGDFAWERVRGLGL

                                                                                 ----
Am-p21   HQPLLRSGSPMHCGTPTPQSVESSKTAEKIAHKCSDAEETLKHEGSDAPEAVTHKNSGALQTMKFVGSSAAEALMQKY
Xl-p21   --------------TLNGSSQESQKENQ--------CQDVATER-------CNISP-----SSKAFQNCESSDSGKRK
Xt-p21   --------------TLNCNSQESEKENQ--------CQDVAAEM-------CNINQ-----SHKASQNCESSDSGKRK
Mm-p21   PKVYLSPGSR-SRDDLGGDKRPSTSSAL--------LQGPAPEDHVALSLSCTLVS---ERPEDSPGGPGTSQGRKRR
Hs-p21   PKLYLPTGPRRGRDELGGGRRPGTSPAL--------LQGTAEEDHVDLSLSCTLVPRSGEQAEGSPGGPGDSQGRKRR

         PCNA binding domain--
Am-p21   SGALETLMDEATNAAEALPQKDSGAPETLMPEGNNTAEALTQKDSGALETVTLEGNNAAEAVKWRDGDASRETQA 226
Xl-p21   QKLITDFYPVKRRCSPVPSLHD 138
Xt-p21   QKLITDFYPVKRRCSPKPSPRD 142
Mm-p21   QTSLTDFYHSKRRLVFCKRKP  159
Hs-p21   QTSMTDFYHSKRRLIFSKRKP  164
```

3.3 Regulation of G_2/M

In mammals, the p53 gene is mutated in approximately 50 % of all human cancers. At the G_2 checkpoint, p53 protects against DNA damage by stimulating the expression of p21, and if the damage is irreparable, the p53 binding protein 2 (p53BP2) mediates apoptosis through p53 (Levine et al. 1991). Villard et al. (2007) recently cloned the axolotl p53 gene. They found that endogenous axolotl p53 was activated following DNA damage by UV irradiation or treatment with an alkylating agent. Inhibition of axolotl p53 with pifithrin-α inhibited limb regeneration, as measured by the inhibition of expression of p53 target genes such as *Mdm2* (p53 binding protein 2) and *Gadd45*. However, pifithrin-α may have other targets as well (Komarova et al. 2003). Interestingly, axolotl p53 shows (and tolerates) amino acid changes that mimic those found in p53 variants of human cancer cells. We have detected the p53 binding protein 2 during hind limb blastema formation in Xenopus stage 60 by both immunofluorescence and proteomic analysis (F. Song, X. Chen, unpublished results), but have not examined the expression of the p53 protein.

In our proteomic study of blastema formation during limb regeneration in the axolotl, we found that Evi5 was the most highly upregulated protein at 1, 4, and 7 days post-amputation (Rao et al. 2009). Since the number of cells undergoing mitosis is so low during blastema formation, the Evi5 detected is most likely not the 90 kDa fragment associated with the cytokinetic bridge, but is rather the full-size 110 kDa mammalian protein that accumulates during G_1 and S to stabilize Emi1 and prevent premature entry into M. The high levels of Evi5 during blastema formation again suggest that prior to the growth phase, blastema cells are arrested in G_2. Using a mammalian Evi5 antibody, we found that Evi5 is expressed in both mesenchymal blastema cells and in cells of the wound epidermis/AEC of the regenerating axolotl limb (F. Song, X. Chen, unpublished results). Our current working hypothesis is that Emi1, stabilized by Evi5 (Eldridge et al. 2006) and perhaps Pin1 (Bernis et al. 2007), may restrain cells from entering mitosis until they have dedifferentiated and accumulated sufficiently to constitute a blastema, possibly through interactions with regulatory molecules such as Rab11 (Rao et al. 2009) through the Hippo pathway. Another hypothesis is that Evi5 plays a role in the fragmentation of myofibers into mononucleate cells, comparable to its role in cytokinesis during cell division. In this case, we would expect the 90 kDa form of Evi5 to be expressed during dedifferentiation and blastema cell accumulation.

There is no convincing proof as yet that dedifferentiating cells of the regenerating urodele limb arrest in G_2. In mammals, cell cytometry has been used to distinguish cells in different phases of the cell cycle on the basis of their DNA content (Van Dilla et al. 1969). Cell cytometry has been applied to quantify the DNA content of regenerating control and denervated urodele limbs, but no difference in 2C and 4C populations was found between the two (Mescher and Tassava 1976). This problem could be profitably re-investigated with propidium iodide DNA labeling and flow cytometry analysis to compare cell cycle profiles, as described for the MRL/MpJ mouse ear (Bedelbaeva et al. 2010), and also by the use of G_2 markers such as γH3, cyclins A and B1, Evi5 and Emi1, to determine the proportion of cells in G_2 at the accumulation blastema stage.

3.4 Role of Nerves and Wound Epidermis in Blastema Cell Cycling

As outlined above, dedifferentiating cells of amputated urodele limbs can enter the cell cycle in the absence of either nerves or wound epidermis, but axon-dependent factors expressed by the wound epidermis are necessary for cell cycling during blastema growth. Recently, a protein expressed by epidermal gland cells was identified whose expression by the AEC is axon-dependent, and which can promote complete regeneration in denervated and amputated adult newt limbs after electroporation of its gene into dedifferentiating limb tissue (Kumar et al. 2007). The protein is the anterior gradient protein (AGP), a ligand for the blastema cell surface receptor Prod1. Prod1 is a member of the Ly6 family of three-finger proteins anchored to the cell surface by a glycosylphosphatidylinositol linkage (Morais da Silva et al. 2002; Brockes and Kumar 2008; Garza-Garcia et al. 2009). Neuregulin (GGF-2), along with other growth factors produced by platelets and macrophages (FGFs, PDGF, TGF-β, IL -1, 2, 6) was shown to be mitogenic for Schwann cells in transected mammalian peripheral nerves (Davies 2000). A newt neuregulin gene cloned from spinal cord is expressed in newt dorsal root ganglia, and recombinant human GGF-2 infused into denervated axolotl limb blastemas maintained the DNA labeling index at control levels and was reported to support regeneration to digit stages (Wang et al. 2000). However, little data were given to support the report that GGF-2 supported complete regeneration in these experiments.

AGP is strongly expressed in the distal-most Schwann cells of regenerating newt limbs at 5 and 8 days post-amputation, when histolysis and dedifferentiation are underway (Kumar et al. 2007). By 10 days post-amputation in the newt limb, when dedifferentiated cells are accumulating to form the blastema, AGP expression shifts from Schwann cells to sub-epidermal secretory gland cells of the AEC (Kumar et al. 2007). The wound epithelium of the axolotl does not have sub-epidermal gland cells and here AGP expression is observed in the Leydig cells. Both sets of gland cells appear to discharge secretions by a holocrine mechanism (Kumar et al. 2010). The expression of AGP by Schwann and gland cells is axon-dependent, as shown by the fact that it is abolished in denervated limbs.

Like HGF in liver regeneration (Michalopoulos and De Frances 1997), AGP is a complete mitogen for limb regeneration. Clearly, it binds to Prod1 on the blastema cell surface and is all that is required for DNA synthesis in vitro and in vivo (Kumar et al. 2007). However, FGFs 1, 2 and 8 have also been shown to individually promote proliferation in vitro and in vivo (Chew and Cameron 1983; Boilly et al. 1991; Boi et al. 1997; Han et al. 2001; Christensen et al. 2001, 2002; Yokoyama et al. 2000, 2001; Dungan et al. 2002; Giampaoli et al. 2003). What the functional relationship of AGP is to other mitogens of the AEC in vivo is unclear. AGP might regulate the expression of FGFs by the AEC, or AGP and FGFs might represent redundant mechanisms for proliferation, like the HGF, TGFα/EGF, and TNF-α pathways in liver regeneration, although none of the FGFs have been shown to promote regeneration to digit stages in vivo. Presumably, AGP would function by some intracellular pathway to induce dedifferentiating cells to traverse G_1 to the RP. Answers to these questions will require a great deal more experimentation. The temporal analysis of AGP expression is incomplete, having been carried out for only two time points (stages) of regeneration. Whether or not AGP regulates the expression of FGFs by the AEC might be answered by denervation/AGP therapy experiments. Denervation strongly downregulates FGF-8 expression (Christensen et al. 2001, 2002) and presumably other FGFs of the AEC in regenerating axolotl limbs, but whether AGP expression in denervated limbs restores the expression of these growth factors is unknown.

4 Similarities, Differences, and Unanswered Questions Between the Urodele Limb and MRL/MpJ Mouse Ear Regeneration

The MRL mouse ear tissue and the urodele limb exhibit both similarities and differences with regard to regeneration (Fig. 6). Punch hole injury in the MRL mouse removes a disk of skin, loose connective tissue, and cartilage, whereas amputation of the urodele limb removes whole segments of skin and musculoskeletal tissue. Fibroblasts appear to be a major source of the blastema in the ear, though infiltrating bone marrow-derived cells probably contribute as well

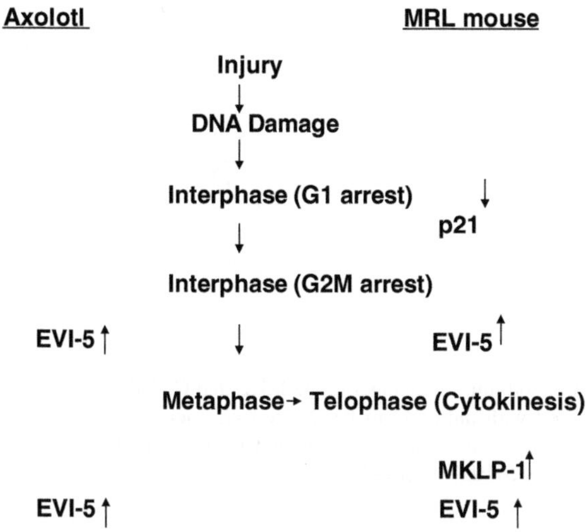

Fig. 6 Cell cycle similarities between MRL mouse and axolotl. The MRL mouse shows a DNA damage response leading to p53 upregulation. However, the G_1 cell cycle checkpoint regulatory gene p21 $^{CIP1/WAF1}$ is not expressed and the cells do not arrest in G_1. Instead, like axolotl cells, MRL cells arrest in G_2 and both species show upregulation of Evi5, a molecule that enforces G_2 arrest by stabilizing the APC inhibitor Emi1. Cytokinesis, when it does occur, requires both Evi5, upregulated in axolotl and MRL, and MKLP1 which is upregulated in the MRL

(unpublished data), and the urodele limb blastema, though the latter gets contributions from muscle (including satellite cells), cartilage, and Schwann cells as well (Stocum 2012, for review). Both form a mesenchymal blastema under a wound epidermis, ring-shaped in the MRL mouse ear, and conical in the amputated urodele limb. While a basement membrane initially forms in the MRL, it degrades after a few days, and is continually degraded in the regenerating limb, allowing communication between the wound epidermis and subjacent mesenchyme. What the molecular nature of this communication in the MRL is, we do not know, but in the limb the wound epidermis secretes AGP and FGFs that act as mitogens for blastema cells. The function of the wound epidermis in the limb is nerve-dependent. Recent experiments indicate that denervation of the MRL/MpJ mouse ear prevents regeneration after a punch hole wound, so there may be a similar nerve-wound epidermis circuit operating in the MRL mouse (Buckley et al. 2012).

Over 40 % of uninjured MRL ear fibroblasts are arrested in G_2, and are thus poised to repair DNA damage and enter M, or initiate apoptosis, immediately after injury. This is not an unusual feature among regeneration-competent tissues. Stem cells in general exhibit a preference for G_2 arrest (Hong et al. 2007; Chuykin et al. 2008; Galvin et al. 2008). Epithelial cells and interstitial gland cells from regenerative hydra are arrested in G_2 (Schmidt and David 1986; Dubel and Schaller 1990; Holstein et al. 1991), and G_2 arrest is a dominant feature in adult mammalian hepatocytes (Michalopoulos and De Frances 1997; Guidotti et al. 2003). Cells of the

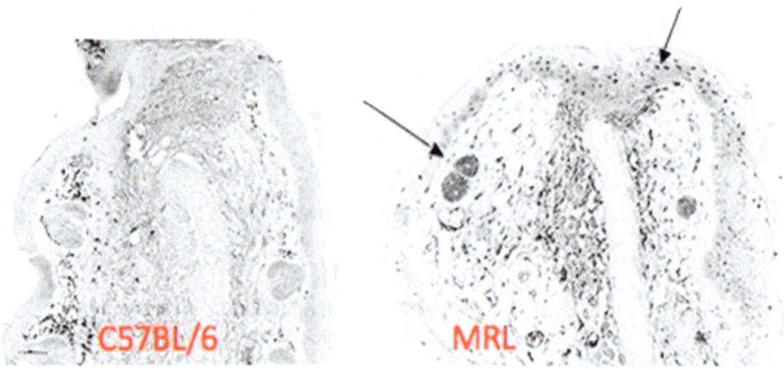

Fig. 7 Expression of sodium ion channel in the MRL and C57BL/6 ear 3 days post-injury. The ear tissue was stained with antibody to the Na ion channel $Nav_V1.2$. Significant immunohistochemical staining was seen in the epidermis and in hair follicles in the MRL (*right panel, arrows*) but not in the C57BL/6 (*left panel*)

unamputated limb are not preferentially in G_2 arrest as far as we know, but collect there as cells liberated by histolysis enter the cell cycle during blastema formation.

The downregulation of $p21^{CIp1/Waf1}$ appears to be sufficient for uninjured MRL ear fibroblasts to phosphorylate Rb and pass the RP checkpoint into S and progress to G_2. The p21 knockout confers regeneration competence on the ear tissue of the non-regenerating wild-type mouse and this could very well be due to the fact that the lack of p21 leads to G_2 arrest instead of arrest in G_1. We would like to know whether p21 or its equivalent is downregulated in regenerating urodele limbs as well, and also whether in MRL ear tissue thrombin activates a factor that leads to downregulation of p21 and phosphorylation of Rb. Deletion of the major regulator of p21, p53, does not confer regenerative ability in B6 mice, but does have a positive effect when deleted in female MRL mice. Inhibition of p53 in the urodele limb, however, may prevent regeneration, suggesting that it is essential for regenerative competence.

The G_1-S checkpoint relies on p53 as a sensor for DNA damage, leading in the worst-case scenario to p21 mediated initiation of apoptosis. In this vein, an interesting finding is that in the tails of Xenopus tadpoles (Tseng et al. 2007) and the South American knifefish (Sirbulescu and Zupanc 2009) there is massive apoptosis during the first 24 h after amputation. This apoptosis is obligatory for Xenopus tail regeneration (it is not known if the same is true for the fish tail) because regeneration is prevented if apoptosis is inhibited. Thus upregulation of p53 and p21 might be expected in the first 24 h after tail amputation in these species. Apoptosis has not been detected in limb regenerate tissue examined at more than 1-day post- amputation (Mescher et al. 2000; Atkinson et al. 2006), but has not been examined within the first 24 h after amputation.

Interestingly, another requisite of tail and limb regeneration is influx of sodium ions and efflux of hydrogen ions (Borgens et al. 1977, 1979; Adams et al. 2007). Inhibition of either sodium or proton channels results in the prevention of regeneration. Xenopus tails prevented from regenerating by shRNA directed at the

$Na_V1.2$ sodium channel mRNA can be rescued by mis-expression of the human $Na_V1.5$ sodium channel gene to transiently create a sodium current (Tseng et al. 2011). The connection of these ionic currents with apoptosis or other early events of regeneration is unknown. Preliminary studies show that the $Na_V1.2$ channel is upregulated in the epidermis in the MRL mouse ear (Fig. 7) and parallels the amphibian tail or limb. Whether this occurs by apoptosis or ionic currents in the first 24 h after punch injury is unknown.

5 Summary

There are multiple parallels presented in this paper showing similarities in the MRL regenerating ear hole and the axolotl regenerating limb. The two major cell cycle pathways involve G_2 arrest and cytokinesis (Fig. 6). First, the axolotl shows clear Evi5 upregulation leading to G_2 arrest stabilization; the MRL downregulates p21, also show Evi5 upregulation, and G_2 arrest. G_2 could very well provide a state in which the accumulating blastema cells are epigenetically modifying their chromatin to attain the state required to respond to growth stimuli.

The next step is mitosis leading to cytokinesis with the upregulation of two molecules during MRL mouse regeneration, specifically Evi5 and MKLP1, and Evi5 in axolotl limb regeneration. Cytokinesis requires three critical complexes associated with the midzone/midbody: the first is associated with microtubules and involves PRC1 and KIF4, the second in Centralspindling and involves CYK-4 and MLKP1; and the third is the CPC and involves INCENP, Aurora B, Borealin, and Survivin where Evi5 is functioning (Glotzer 2005). There is further evidence from another study showing that knocking out the cytokinesis gene mps-1 leads to defects in zebra fish fin regeneration (Poss et al. 2002). However, it is not known whether any modulation of this response such as a delay in cytokinesis contributes to regeneration such as might occur in the liver (Margall-Ducos et al. 2007).

References

Adams DS, Masi A, Levin M (2007) H+ pump-dependent changes in membrane voltage are an early mechanism necessary and sufficient to induce Xenopus tail regeneration. Development 134:1323–1325

Arthur LM, Demarest RM, Clark L, Gourevitch D, Bedelbaeva K, Anderson R, Snyder A, Capobianco AJ, Lieberman P, Feigenbaum L, Heber-Katz E (2010) Epimorphic regeneration in Mice is p53-independent. Cell Cycle 9:3667–3673

Ashcroft GS, Yang X, Glick AB, Weinstein M, Letterio JJ, Mizel DE, Anzano M, Greenwell-Wild T, Wahl SM, Deng C, Roberts AB (1999) Mice lacking Smad3 show accelerated wound healing and an impaired local inflammatory response. Nat Cell Biol 1:260–266

Atkinson SL, Stevenson TJ, Park EJ, Riedy MD, Milash B, Odelberg SJ (2006) Cellular electroporation induces dedifferentiation in intact newt limbs. Dev Biol 299:257–271

Balu DT, Hodes GE, Anderson BT, Lucki I (2009) Enhanced sensitivity of the MRL/MpJ mouse to the neuroplastic and behavioral effects of chronic antidepressant treatments. Neuropsychopharmacology 34:1764–1773

Barger PM, Tassava RA (1985) Kinetics of cell cycle entry in innervated and denervated forelimb stumps of larval *Ambystoma*. J Exp Zool 233:151–154

Bedelbaeva K, Snyder A, Gourevitch D, Clark L, Zhang XM, Leferovich J, Cheverud JM, Lieberman P, Heber-Katz E (2010) Lack of p21 expression links cell cycle control and appendage regeneration in mice. Proc Natl Acad Sci U S A 107:5845–5850

Bernis C, Vigneron S, Burgess A, Labbe J-C, Fesquet D, Castro A, Lorca T (2007) Pin1 stabilizes Emi1 during G2 phase by preventing its association with $SCF^{\beta trcp}$. EMBO Rep 8:91–98

Boilly B, Cavanaugh KP, Thomas D, Hondermarck H, Bryant SV, Bradshaw RA (1991) Acidic fibroblast growth factor is present in regenerating limb blastemas of axolotls and bonds specifically to blastema tissues. Dev Biol 145:302–310

Boi, C, Zenjari C, Boilly B, Hondermarck H, Boilly-Marer Y (1997) Nerve-blastema interactions induce fibroblast growth factor-1 release during limb regeneration in Pleurodeles waltl. Dev Growth Diff 39:15–22

Borgens RB, Vanable JW Jr, Jaffe LF (1977) Bioelectricity and regeneration. Large currents leave the stumps of regenerating newt limbs. Proc Natl Acad Sci U S A 74:4528–4532

Borgens RB, Vanable JW Jr, Jaffe LF (1979) Reduction of sodium dependent stump currents disturbs urodele limb regeneration. J Exp Zool 209:377–386

Brockes JP, Kumar A (2008) Comparative aspects of animal regeneration. Ann Rev Cell Dev Biol 24:525–549

Buckley G, Wong J, Metcalfe AD, Ferguson MW (2012) Denervation affects regenerating responses in MRL/Mpj and repair in C57BL/6 ear wounds. J Anat 220:3–12

Buhimschi CS, Zhao G, Sora N, Madri JA, Buhimschi IA (2010) Myometrial wound healing post-Cesarean delivery in the MRL/MpJ mouse model of uterine scarring. Am J Pathol 177:197–207

Cheverud JM, Lawson HA, Bouckaert K, Kossenkov A, Showe L, Cort L, Blankenhorn EP, Bedelbaeva K, Gourevitch D, Arthur LM, and E. Heber-Katz (2012) Genetics of murine external ear tissue regeneration is due to differences in cell cycle, DNA repair, Cell Adhesion and Migration, and Fibrosis. Heredity

Chew K, Cameron JA (1983) Increase in mitotic activity of regenerating axolotl limbs by growth factor-impregnated implants. J Exp Zool 226:325–329

Christensen RN, Tassava RA (2000) Apical epithelial cap morphology and fibronectin gene expression in regenerating axolotl limbs. Dev Dyn 217:216–224

Christensen RN, Weinstein M, Tassava RA (2001) Fibroblast growth factors in regenerating limbs of *Ambystoma*: cloning and semi-quantitative RT-PCR expression studies. J Exp Zool 290:529–540

Christensen RN, Weinstein M, Tassava RA (2002) Expression of fibroblast growth factors 4,8, and 10 in limbs, flanks, and blastemas of Ambystoma. Dev Dyn 223:193–203

Chuykin I, Lianguzova MS, Pospelov TV, Pospelov VA (2008) Activation of DNA damage response signaling in mouse embryonic stem cells. Cell Cycle 7:2922–2928

Clark LD, Clark RK, Heber-Katz E (1998) A new murine model for mammalian wound repair and regeneration. Clin Immunol Immunopathol 88:35–45

Dabbeekeh JPS, Faitar SL, Dufrense CP, Cowell Jk (2007) The Evi5 TBC domain provides the GTPase-activating protein motif for RAB11. Oncogene 26:2804–2808

Daniels M, Dhokia V, Richard-Parpaillon L, Ohnuma S-I (2004) Identification of Xenopus cuclin-dependent kinase inhibitors, p16Xic2 and p17Xic3. Gene 342:41–47

Davies AM (2000) Neurotrophins: neurotrophic modulation of neurite growth. Curr Biol 10:R198–R200

Dubel S, Schaller HC (1990) Terminal differentiation of ectodermal epithelial stem cells of Hydra can occur in G2 without requiring mitosis or S phase. J Cell Biol 110:939–945

Dungan KM, Wei TY, Nace JD, Poulin ML, Chiu I-M, Lang JC, Tassava RA (2002) Expression and biological effect of urodele fibroblast growth factor 1: relationship to limb regeneration. J Exp Zool 292:540–554

el-Deiry WS, Tokino T, Velculescu VE, Levy DB, Parsons R, Trent JM, Lin D, Mercer WE, Kinzler KW, Vogelstein B (1993) WAF1: a potential mediator of p53 tumor suppression. Cell 75:817–825

Eldridge AG, Loktev AV, Hansen DV, Verschuren EW, Reimann JD, Jackson PK (2006) The evi5 oncogene regulates cyclin accumulationby stabilizing the anaphase-promoting complex inhibitor emi1. Cell 124:367–380

Fahmy GH, Sicard RE (1998) Acceleration of amphibian forelimb regeneration bypolypeptide growth factors. J Minn Acad. Sci 63:58–60

Faitar SL, Sossey-Alaouki K, Ranalli TA, Cowell TA (2006) Evi5 protein associates with the INCENP-aurora B kinase-survivin chromosomal passenger complex and is involved in the completion of cytokinesis. Exp Cell Res 312:2325–2335

Galvin KE, Ye H, Erstad DJ, Feddersen R, Wetmore C (2008) Gli1 induces G2/M arrest and apoptosis in hippocampal but not tumor-derived neural stem cells. Stem Cells 26:1027–1036

Garza-Garcia A, Harris R, Esposito D, Gates PB, Driscoll PC (2009) Solution structure and phylogenetics of Prod1, a member of the three-finger protein superfamily implicated in salamander limb regeneration. PLoS ONE 4(9):e7123

Giampaoli S, Bucci S, Ragghianti M, Mancino G, Zhang F, Feretti P (2003) Expression of FGF2 in the limb blastema of two Salamandridae correlates with their regenerative capability. Proc Biol Sci 270:2197–2205

Glotzer M (2005) The molecular requirements for cytokinesis. Science 307:1735–1739

Goldhamer DJ, Tassava RA (1987) An analysis of proliferative activity in innervated and denervated forelimb regenerates of the newt *Notophthalmus viridescens*. Development 100:619–628

Gourevitch D, Clark L, Chen P, Seitz A, Samulewicz SJ, Heber-Katz E (2003) Matrix metalloproteinase activity correlates with blastema formation in the regenerating MRL mouse ear hole model. Dev Dyn 226:377–387

Guidotti JE, Bregerie O, Robert A, Debey P, Brechot C, Desdouets C (2003) Liver cell polyploidization: a pivotal role for binuclear hepatocytes. J Biol Chem 278:19095–19101

Guse A, Mishima M, Glotzer M (2005) Phosphorylation of ZEN-4/MKLP1 by aurora B regulates completion of cytokinesis. Curr Biol 15:778–786

Habermann B, Bebin A-G, Herklotz S, Volkmer M, Eckelt K, Pehlke K, Epperlein HH, Schackert HK, Wiebe G, Tanaka EM (2004) An *Ambystoma mesicanum* EST sequencing project: analysis of 17,352 expressed sequence tags from embryonic and regenerating blastema cDNA libraries. Genome Biol 5:R67

Han M-J, An J-Y, Kim W-S (2001) Expression patterns of *Fgf-8* during development and limb regeneration of the axolotl. Dev Dyn 220:40–48

Hay ED, Fischman DA (1961) Origin of the blastema in the regenerating newt *Triturus viridescens*. An autoradiographic study using tritiated thymidine to follow cell proliferation and migration. Dev Biol 3:26–59

Holstein TW, Hobmayer E, David CN (1991) Pattern of epithelial cell cycling in the hydra. Dev Biol 148:602–611

Hong Y, Cervanes RB, Tichy E, Tischfield JA, Stambrook PJ (2007) Protecting genomic integrity in somatic cells and embryonic stem cells. Mutat Res 614:48–55

Kelly DJ, Tassava RA (1973) Cell division and ribonucleic acid synthesis during the initiation of limb regeneration in larval axolotls (Ambystoma mexicanum). J Exp Zool 185:45–54

Kesik A, Vethamany-Globus S, Globus M (1986) Effect of insulin on cyclic nucleotide levels and promotion of mitosis by insulin and ionophore A23187 in cultured newt blastemata. In Vitro Cell Dev Biol 22:465–468

Komarova EA, Neznanov N, Komarov PG, Chernov MV, Wang K, Gudkov AV (2003) p53 inhibitor pifithrin alpha can suppress heat shock and glucocorticoid signaling pathways. J Biol Chem 278:15465–15468

Kragl M, Knapp D, Nacu E, Khattak S, Maden M, Epperlein HH, Tanaka E (2009) Cells keep a memory of their tissue origin during axolotl limb regeneration. Nature 460:60–65

Kumar A, Godwin JW, Gates PB, Garza-Garcia AA, Brockes JP (2007) Molecular basis for the nerve dependence of limb regeneration in an adult vertebrate. Science 318:772–777

Kumar A, Nevill G, Brockes JP, Forge A (2010) A comparative study of gland cells implicated in the nerve dependence of salamander limb regeneration. J Anat 217:16–25

Lee J, Kim JA, Barbier V, Fotedar A, Fotedar R (2009) DNA damage triggers p21^{WAF1}-dependent Emi1 down-regulation that maintains G2 arrest. Mol Biol Cell 20:1891–1902

Levine AJ, Momand J et al (1991) The p53 tumor suppressor gene. Nature 351:453–456

Lian I, Kim J, Okazawa H, Zhao J, Yu J, Chinnaiyan A, Israel MA, Goldstein LSB, Abujarour R, Ding S, Guan K-L (2010) The role of YAP transcription coactivator in regulating stem cell self-renewal and differentiation. Genes Dev 24:1106–1118

Liu J, Johnson K, Li J, Piamonte V, Steffy BM, Hsieh MH, Ng N, Zhang J, Walker JR, Ding S, Muneoka K, Wu X, Glynne R, Schultz PG (2011) Regenerative phenotype in mice with a point mutation in transforming growth factor beta type I receptor (TGFBR1). Proc Natl Acad Sci U S A 108(35):14560–14565

Lodish H, Berk A, Kaiser CA, Krieger M, Scott MP, Bretscher A, Ploegh H, Matsudaira P (2008) Molecular cell biology, 6th edn. WE Freeman and Co, New York 1150 pp

Maden M (1978) Neurotrophic control of the cell cycle during amphibian limb regeneration. J Embryol exp Morph 48:169–175

Margall-Ducos G, Celton-Morizur S, Couton D, BrÈgerie O, Desdouets C (2007) Liver tetraploidization is controlled by a new process of incomplete cytokinesis. J Cell Sci 120:3633–3639

Mescher AL, Tassava RA (1976) Denervation effects on DNA replication and mitosis during the initiation of limb regeneration in adult newts. Dev Biol 44:187–197

Mescher AL, White GW, Brokaw JJ (2000) Apoptosis in regenerating and denervated nonregenerating urodele forelimbs. Wound Rep Reg 8:110–116

Michalopoulos GK, DeFrances MC (1997) Liver regeneration. Science 276:60–66

Monaghan JR, Epp LG, Putta S, Page RB, Walker JA, Beachy CK, Zhu W, Pao GM, Verma IM, Hunter T, Bryant SV, Gardiner DM, Harkins TT, Voss SR (2009) Microarray and cDNA sequence analysis of transcription during nerve-dependent limb regeneration. BMC Biol 7:1

Morais da Silva SM, Gates PB, Brockes JP (2002) The newt ortholog of CD59 is implicated in proximodistal identity during amphibian limb regeneration. Dev Cell 3:547–555

Morgan CJ, Pledger WJ (1992) Fibroblast proliferation. In: Cohen IK, Diegelmann RF, Lindblad WJ (eds) Wound healing: biochemical and clinical aspects. WB Saunders Co, Philadelphia, pp 63–66

Namiki Y, Endoh D, Kon Y (2003) Genetic mutation associated with meiotic metaphase-specific apoptosis in MRL/MpJ mice. Mol Reprod Dev 64:179–188

Pajcini KV, Corbel SY, Sage J, Pomerantz JH, Blau HM (2010) Transient inactivation of *Rb* and *ARF* yields regenerative cells from ppostmitotic mammalian muscle. Cell Stem Cell 7:198–213

Poss KD, Nechiporuk A, Hillam AM, Johnson SL, Keating MT (2002) Mps1 defines a proximal blastemal proliferative compartment essential for zebrafish fin regeneration. Development 129:5141–5149

Powell JA (1969) Analysis of histogenesis and regenerative ability of denervated forelimb regenerates of Triturus viridescens. J Exp Zool 170:125–147

Rai MF, Hashimoto S, Johnson EE, Janiszak KL, Fitzgerald J, Heber-Katz E, Cheverud JM, Sandell LJ (2012) Heritability of articular cartilage regeneration and its association with ear-wound healing. Arthritis Rheum 64:(7):2300–2310. doi: 10.1002/art.34396

Rao N, Jhamb D, Milner DJ, Li B, Song F, Wang M, Voss SR, Palakal M, King MW, Saranjami B, Nye HLD, Cameron JA, Stocum DL (2009) Proteomic analysis of blastema formation in regenerating axolotl limbs. BMC Biol 7:83

Ruchaud S, Carmens M, Earnshaw WC (2007) Chromosomal passengers: conducting cell division. Nat Revs 8:798–812

Ruzankina Y, Schoppy DW, Asare A, Clark CE, Vonderheide RH, Brown EJ (2009) Tissue regenerative delays and synthetic lethality in adult mice after combined deletion of Atr and Trp53. Nat Genet 41:1144–1149

Schmidt T, David CN (1986) Gland cells in Hydra: cell cycle kinetics and development. J Cell Sci 85:197–215

Singer M, Craven L (1948) The growth and morphogenesis of the regenerating forelimb of adult *Triturus* following denervation at various stages of development. J Exp Zool 108:279–308

Sirbulescu RF, Zupanc GKH (2009) Dynamics of caspase-3-mediated apoptosis during spinal cord regeneration in the teleost fish, *Apteronotus leptorhynchus*. Brain Res 1304:14–25

Stocum DL (2012) Regenerative biology and medicine, 2nd edn. Elsevier, San Diego 474 pp

Stocum DL, Rao N (2010) Mechanisms of blastema formation in regenerating amphibiasn limbs. In: Atala A, Lanza R, Thomson JA, Nerem R (eds) Principles of regenerative medicine, 2nd edn. Elsevier, San Diego, pp 67–86

Straube WL, Tanaka EM (2006) Reversibility of the differentiated state: regeneration in amphibians. Artif Organs 30:743–755

Tanaka EM, Gann F, Gates PB, Brockes JP (1997) Newt myotubes re-enter the cell cycle by phosphorylation of the retinoblastoma protein. J Cell Biol 136:155–165

Tassava RA, Garling DJ (1979) Regenerative responses in larval axolotl limbs with skin grafts over the amputation surface. J Exp Zool 208:97–110

Tassava RA, McCullough WD (1978) Neural control of cell cycle events in regenerating salamander limbs. Amer Zool 18:843–854

Tassava RA, Mescher AL (1975) The roles of injury, nerves and the wound epidermis during the initiation of amphibian limb regeneration. Differentiation 4:23–24

Tassava RA, Bennett LL, Zitnik GD (1974) DNA synthesis without mitosis in amputated denervated forelimbs of larval axolotls. J Exp Zool 190:111–116

Tassava RA, Goldhamer DJ, Tomlinson BL (1987) Cell cycle controls and the role of nerves and the regenerate epithelium in urodele forelimb regeneration: possible modifications of basic concepts. Biochem Cell Biol 65:739–749

Thuret S, Toni N, Aigner S, Yeo GW, Gage FH (2009) Hippocampus-dependent learning is associated with adult neurogenesis in MRL/MpJ mice. Hippocampus 19:658–669

Tolba RH, Schildberg FA, Decker D, Abdullah Z, Büttner R, Minor T, Von Ruecker A (2010) Mechanisms of improved wound healing in Murphy Roths Large (MRL) mice after skin transplantation. Wound Repair Regen 18:662–670

Tseng A-S, Adams DS, Qiu D, Koustubhan P, Levin M (2007) Apoptosis is required during early stages of tail regeneration in Xenopus laevis. Dev Biol 301:62–69

Tseng A-S, Beane WS, Lemire JM, Masi A, Levin M (2011) Induction of vertebrate regeneration by a transient sodium current. J Neurosci 30:13192–13200

Van Dilla MA, Trujillo TT, Mullaney PF, Coulter JR (1969) Cell microfluorometry: a method for rapid fluorescence measurement. Science 163:1213–1214

Velloso CP, Kumar A, Tanaka EM, Brockes JP (2000) Generation of mononucleate cells from post-mitotic myotubes proceeds in the absence of cell cycle progression. Differentiation 6:239–246

Verschuren EW, Ban KH, Masek MA, Lehman NL, Jackson PK (2007) Loss of Emi1-dependent anaphase-promoting complex/cyclosome inhibition degregulates E2F target expression and elicits DNA damage-induced senescence. Mol Cell Biol 27:7955–7965

Vethamany-Globus S, Globus M, Tomlinson B (1978) Neural and hormonal stimulation of DNA and protein synthesis in cultured regeneration blastema in the newt, *Notophthammus viridescens*. Dev Biol 65:183–192

Villiard E, Brtinkmann H, Moiseeva O, Mallette FA, Ferbeyre G, Roy S (2007) Urodele p53 tolerates amino acid changes found in p53 variants linked to human cancer. BMC Evol Biol 7:180

Wang L, Marchionni MA, Tassava RA (2000) Cloning and neuronal expression of a type III newt neuregulin and rescue of denervated nerve-dependent newt limb blastemas by rhGGF2. J Neurobiol 43:150–158

Watt KI, Judson R, Medlow P, Reid K, Kurth TB, Burniston JG, Ratkevicius A, De Bari C, Wackerhage H (2010) Yap is a novel regulator of C2C12 myogenesis. Biochem Biophys Res Comm 393:619–624

Westlake CJ, Junutula JR, Simon GC, Pilli M, Prekeris, R, Scheller RH, Jackson PK, Eldridge AG (2007) Identification of Rab 11 as a small GTPase binding protein for the Evi5 oncogene. Proc Natl Acad Sci USA 104:1236–1241

Xiong Y, Hannon GJ, Zhang H, Casso D, Kobayashi R, Beach D (1993) p21 is a universal inhibitor of cyclin kinases. Nature 366:701–704

Yokoyama H, Yonei-Tamura S, Endo T, Izpisua-Belmonte JC, Tamura K, Ide H (2000) Mesenchyme with *fgf10* expression is responsible for regenerative capacity in *Xenopus* limb buds. Dev Biol 219:18–29

Yokoyama H, Ide H, Tamura K (2001) FGF-10 stimulates limb regeneration ability in *Xenopus laevis*. Dev Biol 233:72–79

Zhao B, Tumaneng K, Guan KL (2011) The Hippo pathway in organ size control, tissue regeneraton and stem cell self-renewal. Nat Cell Biol 13:877–883

Zhong Z, Tsukada S, Rehman H, Parsons CJ, Theruvath TP, Rippe RA, Brenner DA, Lemasters JJ (2010) Inhibition of transforming growth factor-beta/Smad signaling improves regeneration of small-for-size rat liver grafts. Liver Transpl 16:181–190

Zhu C, Bossy-Wetzel E, Jiang W (2005) Recruitment of MKLP1 to the spindle midzone/midbody by INCENP is essential for midbody formation and completion of cytokinesis in human cells. Biochem J 389:373–381

Index

DOI: 10.1007/978-3-642-35810-4
© Springer-Verlag Berlin Heidelberg 2013

MIX
Papier aus verantwortungsvollen Quellen
Paper from responsible sources
FSC® C105338

If you have any concerns about our products,
you can contact us on
ProductSafety@springernature.com

In case Publisher is established outside the EU,
the EU authorized representative is:
Springer Nature Customer Service Center GmbH
Europaplatz 3, 69115 Heidelberg, Germany

Printed by Libri Plureos GmbH
in Hamburg, Germany